FOUNDATIONS OF COMPUTER TECHNOLOGY

FOUNDATIONS OF
COMPUTER TECHNOLOGY

A. John Anderson PhD C.Eng MIEE

Department of Electrical and Electronic Engineering
University of Paisley, Paisley, UK

CRC Press
Taylor & Francis Group
Boca Raton London New York

CRC Press is an imprint of the
Taylor & Francis Group, an **informa** business

CRC Press
Taylor & Francis Group
6000 Broken Sound Parkway NW, Suite 300
Boca Raton, FL 33487-2742

First issued in hardback 2017

© 1994 A. John Anderson
CRC Press is an imprint of Taylor & Francis Group, an Informa business

No claim to original U.S. Government works

ISBN 13: 978-1-138-41396-2 (hbk)
ISBN 13: 978-0-412-59810-4 (pbk)

Visit the Taylor & Francis Web site at
http://www.taylorandfrancis.com

and the CRC Press Web site at
http://www.crcpress.com

A catalogue record for this book is available from the British Library

Library of Congress Catalog Card Number: 94-71829

Dedicated to

Lesley, Stephen and Laurie

Contents

Part Three   *Software*

Part Four — Operating Systems

Part Five   *Networks*

Preface

DESCRIPTION

The book is an introduction in the easiest possible language to the architecture of computers and various peripherals. Its purpose is to present as clearly and completely as possible the nature and characteristics of modern computer systems and microprocessors through an approach which integrates components, systems, software and design. It also aims to provide a range of information for anyone engaged in the study of computer systems by providing a succinct, systematic and readable guide to the facts on computers. This approach has been taken to try to encourage the reader's understanding of a computer system before moving on to more detailed matters. To provide such an overview the book has been compiled from many different sources.

The book focuses primarily on a description of the hardware elements within a computer system and the impact of software on the architecture. It provides a discussion of practical aspects of computer organization (structure, behaviour and design), gradually developed from the basics, thereby offering the necessary fundamentals for either electrical engineering or computer science readers.

The text not only lists a wide range of terms employed in describing computers but also explains with the aid of appropriate illustrations, the basic operation of the individual components within a computer system. Material on modern technologies is combined with more historical information providing a range of articles on computer hardware, architecture and software, detailing the separate hardware components, programming methodologies and the nature of operating systems. There is also a unified treatment of the entire spectrum of computers from the microcomputers used for personal computing through to articles on supercomputers.

The book is arranged in seven parts:

Part 1   Overview
Chapter 1 introduces the operation of a computer.

Part 2   Hardware
This section presents a hardware description of the elements of a computer system. Chapters 2 to 6 give details on subjects ranging from microprocessors to memory devices and bus systems. Chapter 7 incorporates a description of the hardware development process for computer systems.

Part 3   Software
Information on software and programming languages for computer systems is given in Chapter 8, while a description of the software development process is presented in Chapter 9.

Part 4   Operating Systems
Chapters 10 and 11 present an overview of the functions of an operating system.

Part 5   Networks
Chapter 12 includes information on local and wide area computer networks, detailing both the hardware construction of the networks and the software used in these systems.

Part 6   Advanced Hardware

Chapter 13 gives a presentation of modern technology for advanced computer architectures. The final chapter presents some up to date advanced topics, such as RISC processors and parallel processing, which are dictating how computer systems are being developed.

Part 7   Appendix (Computer History)

The Appendix presents a discussion on the evolution of the computer from mainframes to personal computers.

FEATURES

The text has number of pedagogical features to facilitate understanding and encourage reader enthusiasm.

1. The information is presented in an attractive, intelligible and interesting way. This is achieved by a variety of diagrams, tables and changing layout. The diagrams aid in the description of the computer, while the use of tables allows material to be neatly grouped thereby aiding in comprehension.

2. The book is written using nontechnical terms so that the text focuses on developing a basic understanding of computers.

3. The breaking of the text into the seven sections allows flexibility in the study of particular sections, as appropriate to particular interests or courses. The section on 'Operating systems' for example is intentionally split into two chapters to provide one chapter on an initial introduction to the subject while the second chapter deals with a more detailed exposition. Furthermore, incorporated into the hardware and software sections are chapters dealing with hardware and software design; again, these are extra chapters which can be omitted if considered inappropriate to a particular course.

4. Each chapter has learning objectives and chapter outlines. The learning objectives inform the reader of the major points or concepts to be gleaned from the chapter. Chapter outlines preview chapter topics and organizations in order to allow the reader to see the relationship between the topics covered.

5. Article boxes focus on interesting topics outside the main core material to provide a wider perspective of the subject area.

6. At appropriate points in the text there are small 'Glossary' sections defining certain terms used within the chapter to describe computer systems. The integration of the Glossary into each section of the book means the reader does not have to repeatedly turn to the back of the book to search for a meaning to particular word or expression.

7. Each new key term is put in bold.

8. Each chapter finishes with key terms which are listed alphabetically for reference or review.

9. Review questions check the student's understanding of the main topics in the chapter. They appear at the end of the chapter as a form of self–test. The answers to the majority of the questions can be found through a careful reading of the text. Project questions are also included which require the reader to read outwith the text.

10. A short annotated bibliography is provided at the end of each chapter to direct the reader to additional reading on the various subjects.

11. An index at the end of the book supplies a detailed guide to the text.

UNITS

Most of the units used in the text are SI units, i.e. these are an internationally agreed system of units based on the kilogram, metre, second, Kelvin, ampere, candela and mole. Various units and symbols are presented below:

k	is the prefix for kilo 1000 (e.g. km kilometre).
K	is the symbol used in computer descriptions representing the number 1024.
micron	10^{-6} of a metre, symbol μm.
angstrom	a unit equal to 10^{-10} metre, symbol Å.
mil	1/1000th of an inch.
millisecond	0.001 of a second, symbol ms.
microsecond	10^{-6} of a second, symbol μs.
nanosecond	10^{-9} of a second written as ns.
hertz	unit of frequency equivalent to cycles per second, symbol is Hz.
megahertz	10^6 hertz, symbol MHz.
radian	one radian is the angle subtended at the centre of a circle by an arc equal in length to the radius of the circle 1 radian= 57.296 degree approx. 180 degrees = π radians.
degree	= 1/90 of a right angle.
grade	= 1/100 of a right angle.
bel	measurement of power.
decibel	1/10th of a bel, symbol dB.
inch	an inch = 2.54 cms, symbol in.
foot	a foot = 12 inches, symbol ft.
bits/second	symbol bps.

Units of measure for computer memory:

Memory			
	byte	1	2^0
	Kilobyte (KB)	1024	2^{10}
	Megabyte (MB)	1 048 574	2^{20}
	Gigabyte(GB)	1 073 741 824	2^{30}

ACKNOWLEDGEMENTS

I would like to express my thanks to David Hatter for his enthusiasm in helping to bring this book to the market. I would also express my gratitude to the production staff at Chapman & Hall for their assistance in helping me with the presentation of the final text.

Overview

Part One

Computers: An Introduction

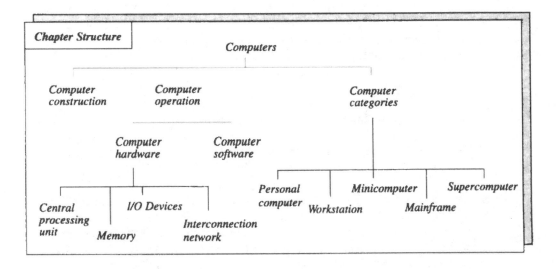

Chapter Structure

Computers — Computer construction, Computer operation, Computer categories

Computer operation — Computer hardware, Computer software

Computer hardware — Central processing unit, Memory, I/O Devices, Interconnection network

Computer categories — Personal computer, Workstation, Minicomputer, Mainframe, Supercomputer

Learning Objectives

In this chapter you will learn about:

1. The constructional details of a personal computer.
2. The basic structure of a computer and its important computing features.
3. The terminology used to describe computers and their components.
4. The operation of instructions on a computer's central processing unit.
5. The major categories of computers.
6. The manufacturers responsible for the diverse computer range.

'This is the era of information and computers are the engines of information. They are to be found in every facet of life, it is therefore important that everyone should study computers.'

H.C. Lucas

Chapter 1

Computers: An Introduction

This chapter presents an overview of computers, their construction and operation. It may be seen as an overture to the rest of the book in that it introduces the different hardware and software themes that will be developed in the later chapters.

In the four decades since the early 1950s the computer has emerged from being a large metal box performing mundane calculations to become a multifaceted machine capable, through various pieces of software, of being used in a myriad of different situations for a large number of jobs. In particular over the past 10 years we have seen an explosion in the use of information technology. **Information technology** may be seen as the organization, manipulation and distribution of information. It is dependent on the ability to process electrical signals at low cost, high speed and with operational reliability. The enabling technology which has brought about the current information explosion includes microelectronics, computation, signal processing and communications. As the activities encountered in information technology are central to almost every computer, the term is used to mean almost the same as computing.

In its most basic form a **computer** is essentially a machine that receives, stores, manipulates and communicates information. It does this by breaking a task down into logical operations that can be carried out on binary numbers (strings of 1s and 0s) and doing millions of such operations per second. A computer is also an electronic system, usually constructed from digital integrated circuits (ICs) that can execute structures of logical operations given to it in a program. A computer system is designed at many levels ranging from its high level software structure down to the transistors in its hardware.

Computers derive their desirable properties from their speed and accuracy in performing complex calculations and in the fact that they possess a memory for retaining programs and data.

1. The speed of a computer is determined by the switching time of the electronic circuits within it and the distance the electric signals have to travel within the unit.
2. The accuracy of the computer results from the sophistication of the arithmetic circuits in the computer and on the repeatability with which it performs its operations. This in turn is due to the reliability of the electronic circuits that make up the system.
3. Computers derive their flexibility from the fact that they can be programmed, where a computer program may be seen as a series of instructions which direct the computer to perform a given task.

1.1 COMPUTER CONSTRUCTION

We will consider in more detail how a computer operates; however, before embarking on this investigation we will note the constructional details of a small desktop or personal computer (PC). The basic units within such a system (shown in Figure 1.1) are:

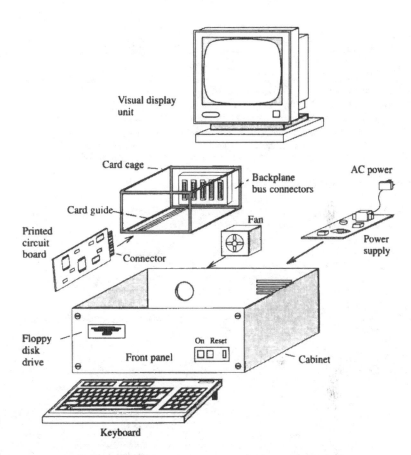

Visual display unit

Card cage

Backplane bus connectors

AC power

Card guide

Printed circuit board

Fan

Connector

Power supply

Figure 1.1

The basic units within a computer.

Floppy disk drive

On Reset

Front panel

Cabinet

Keyboard

1. Visual display unit The computer requires a visual display unit (VDU) to output information and data to the operator. A VDU is similar to a conventional television in appearance. It is connected to the computer by a power cable and a data cable.

2. Keyboard The keyboard is used by the operator to enter commands and data into the computer. It is styled on a typewriter with extra keys and is connected to the computer by a cable.

3. A cabinet There are several styles of cases available for personal computers:

  (a)   Desktop, or table top
  (b)   Compact desktop, or mini desktop
  (c)   Mini–tower
  (d)   Normal size tower
  (e)   Laptops, notebooks, portables.

4. Chassis This is a frame constructed of metal and plastic that houses all the internal components of a computer. The computer also usually has a card cage which holds the printed circuit boards. The card cage has a number of backplane connectors wired together to form a bus.

5. Power supply Most computers use a 5 V direct current (DC) power supply. The power supply has a transformer, a rectifier, a ripple filter and a solid–state voltage regulator. Most modern personal computers come with a 200 W power supply. This is required to power the printed circuit board and all the system peripherals. An indication of the power consumption of various parts of a personal computer system is shown in Table 1.1.

Glossary 1.1 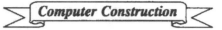 *Computer Construction*

ACTIVE COMPONENT – An active component is one built using semiconductor technology. It is based around the use of semiconductor diodes or transistors.

BATTERY – A battery is a construction of various chemicals and metals which is able to produce a direct voltage between its two terminals. This voltage can be used to drive a current through an electric circuit.

CAPACITOR – A capacitor is a passive component used in electronic systems. Due to its construction it is able to store energy which can drive electric current round a circuit. There are many different forms of capacitors incorporated in electronic systems, depending on the construction material. These include electrolytic, tantalum, polyester and ceramic devices.

CIRCUIT – A circuit is a term used to describe a collection of electronic components designed to perform a particular function.

DIODE – A diode is a semiconductor device which only permits current flow in one direction. It can be used as a switching device to control current flow in an associated circuit.

INTEGRATED CIRCUIT – An integrated circuit (IC) is a small piece of silicon on which a large number of electronic devices are constructed. A common name for this is a chip. The chip is usually enclosed in a plastic or ceramic case and the pads on the chip connected to metal pins on the case.

RESISTOR – A resistor is a passive electronic component used in electronic systems to either control the current through an active device or to divide a voltage down to a lower value.

PASSIVE COMPONENT – Passive components are the various resistors, capacitors and inductors which are necessary for the correct functioning of an electronic system. Any electronic design will invariably have to use passive components as well as the larger integrated circuit active devices.

SURFACE MOUNTING – Surface mounting is a production technique for mounting discrete electronic components directly onto a printed circuit board (PCB). Most electronic assemblies including computers now use surface mounting since it leads to higher density of components, lower assembly costs and improved reliability and speed.

Table 1.1

Power consumption of subsystems in an IBM 486 PC.

	Computer subsystem	Power (Watts)
1.	Main printed circuit board	35
2.	Floppy/hard disk drive controller	18
3.	VDU graphics card	5
4.	Floppy disk drive	5
5.	Hard disk drive	15
6.	Keyboard	2
7.	Input/output boards	4

6. Disk drives The disk drives are the mass storage devices used in the computer to store large amounts of programs and data. They may either use floppy or hard disks. A disk is a circular storage device which uses magnetic material to store information. The disk can either be rigid and fixed into the disk drive, as in a hard disk, or flexible and removable from the disk drive, as in a floppy disk.

Computer construction

7. Circuit boards A personal computer contains a number of printed circuit boards (PCBs). A PCB is a fibreglass board used in electronic circuit construction in which the electrical connections between the components in the circuit are made by copper strips on the board. The copper strips are chemically etched onto the board with the use of a mask. The main printed circuit board within a typical personal computer contains a number of ICs or chips:

(a) A microprocessor IC to perform the basic arithmetic and logic functions and supervise the operation of the entire system. A computer system may have one or more processors either dedicated to the execution of specialized instructions or capable of performing all instructions.

(b) ICs containing the computer's internal memory. The memory is used to hold the instructions and the data to be operated on. There are two main types: random access memory (RAM) which can be written to and read from by the microprocessor and read–only memory (ROM) which can only be read from by the microprocessor.

(c) ICs to control the input/output (I/O) devices (such as a disk drive, display, keyboard, mouse or printer) and to connect the microprocessor and memory with the I/O devices. They either interface with the outside world or are used to access the direct access disk drive on which large amounts of additional data or instructions can be stored and retrieved. The I/O interface ICs are of two forms: the peripheral interface adapter (PIA) for parallel I/O can transfer 8 bits of data into or out of an I/O device at a time; the asynchronous communications interface adapter (ACIA) for serial I/O can only transfer 1 bit of data into or out of an I/O device.

Circuit boards

(d) ICs to control the selection of the memory or I/O ICs, often referred to as glue chips. The layout of the ICs used on the main computer board for the IBM PC personal computer is shown in Figure 1.2.

8. Expansion slots The expansion slots are the internal connectors that allow additional circuit boards to be plugged into a motherboard. Into the expansion slots can go boards with, for example, additional memory, display adapters, disk drive controllers, real time clocks, game adapters, network adapters or multifunction boards. In an IBM PC, for example, the expansion slots are at the back of the computer. The expansion slots can be seen in Figure 1.2.

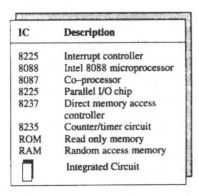

IC	Description
8225	Interrupt controller
8088	Intel 8088 microprocessor
8087	Co–processor
8225	Parallel I/O chip
8237	Direct memory access controller
8235	Counter/timer circuit
ROM	Read only memory
RAM	Random access memory
	Integrated Circuit

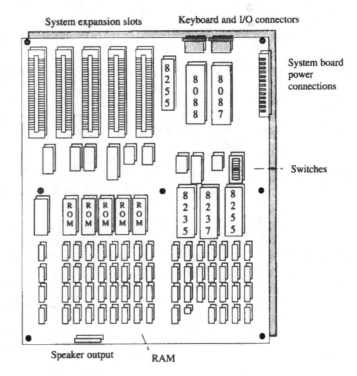

Figure 1.2

Layout of the IBM PC main system board.

The way in which the different components of a computer system are interconnected can be seen from Figure 1.3 which shows a block diagram of a typical personal computer and the interrelationship between the microprocessor and blocks of logic.

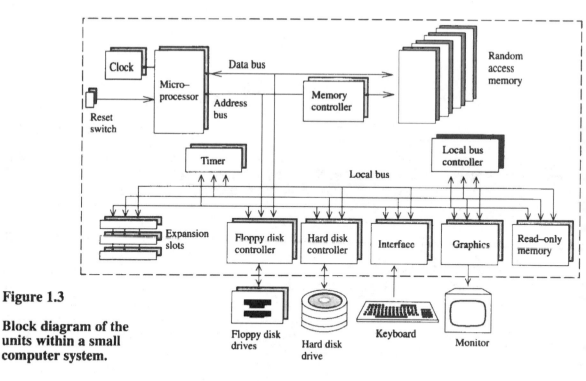

Figure 1.3

**Block diagram of the
units within a small
computer system.**

1.2 COMPUTER OPERATION

A functional block diagram of a generalized computer system is shown in Figure 1.4. The overall functional organization of a computer is divided into four major components. These are the central processing unit, the memory, the I/O devices and the interconnection network. The data and the control paths are also shown to be separate, with the control being centralized in the CPU of the system.

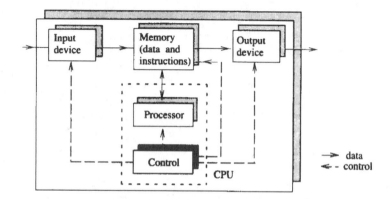

Figure 1.4

**Operation of
a computer system.**

Most modern computer architectures are based on the architectural model shown in Figure 1.4. The model was first proposed by John von Neumann. The principle of the model is that the computer has a main, or primary, memory in which the programs and data are

---

Glossary 1.2 ⟩⟨ **Computer System** ⟩⟨

ARCHITECTURE – There are a number of ways of describing the architecture of a computer:
1. It is the apparent, or virtual, set of resources as seen by the user of a computer system. The word architecture tends to refer to the visible characteristics of the computer as seen either by the person programming the machine or alternatively by the program running on the machine.
2. It is the blueprint which could be used to build a computer.
3. It is the way the computer has been designed, its logical and electrical requirements.
4. It refers to a computer's internal organization, i.e. the way its internal components are organized and interconnected.
5. It describes the functional capabilities of the computer provided by the manufacturer in the design of the device.
    Computer architecture can then be considered to encompass the structure, organization and the implementation of the system. Machines with the same architecture can execute the same programs and can have the same I/O devices. A computer architecture is open when its designers publish its specifications.
    In terms of design, the development of a good computer architecture can be seen as the art of making a specification that will live through several generations of technology.

HARDWARE – The term hardware is usually applied in computer systems to any electrical or mechanical equipment involved in the production, storage, distribution or reception of electronic signals. It refers to the physical part of the computer system implemented by the electronic circuitry and mechanical equipment. The hardware components may be active or passive.

ORGANIZATION – The term organization refers to the dynamic interactions and management of the components within the computer and is usually shown by a block diagram. In considering the organization of a computer the designer will have to consider, for example, how the computer arithmetic, microprogramming, memory and I/O structures are incorporated into the architecture of the machine.

SOFTWARE – The term software is applied in computer systems to any program that can run on the computer.

STRUCTURE – The structure of a computer system is the way in which the resources and components are organized and integrated to form a correctly functioning unit. By studying the internal structure of a computer system it is possible to determine the hardware resources that are available, the purposes they serve, and the relationship between them. The physical structure of a computer is the physical resources used in the installation such as the processor, the input/output (I/O) devices, the printers and the disk drives. In addition to these physical structure elements, a computer will also require an operator console display and possibly a data transmission device.

SYSTEM – The term system refers to an assembly of components united by some form of regulated interaction to form an organized whole. In computer terms the word is used to represent a logical collection of computers, peripherals, software, service routines, terminals and end–users which are organized to achieve some goal.

Computer operation

stored. It usually also has an auxiliary, or secondary, memory which is external to the computer and is stored on magnetic disk, tape or optical disk. The main memory is connected to the CPU such that the CPU takes the instructions from the program stored in the main memory and executes them, i.e. it carries out the instructions. The instructions operate on the data which is also stored in the main memory.

A computer is made up not only from the hardware units necessary to construct a working system but also a correctly written set of instructions, or program. The programs that run on a computer system are often called software.

1.3 COMPUTER HARDWARE

The four main hardware parts are now investigated in more detail.

1.3.1   CENTRAL PROCESSING UNIT (CPU)

The CPU may be regarded as the unit within a computer that interprets instructions from a program and changes data in conformity with these instructions. In almost all computers the CPU is constructed from a single microprocessor IC. It coordinates and controls the activities of all the other units in the computer system together with the interpretation and execution of processing instructions, and also performs the logical and arithmetic processes to be applied to data. The CPU consists of the processor and the control unit. In most conventional CPUs or microprocessors these are incorporated into a single unit and the control section is usually preprogrammed by the manufacturer to execute the selected instruction set.

1.3.2   MEMORY

A computer program is usually far too large to be stored all at once in the CPU, so a computer must have separate memory, the purpose of which is to hold a program's instructions and associated data. A computer has two types of memory, referred to as primary and secondary memory.

1.   Primary Memory – The primary (or main) memory holds program instructions, data, intermediate calculations and final results. Any randomly chosen memory word can be accessed with equal ease; hence, the main memory is also referred to as a random–access memory. In a computer the main memory is made from two types of semiconductor ICs, RAM ICs and ROM ICs:

(a) RAM ICs: These are constructed in such a way that data can be written to them by the microprocessor or read from them by the microprocessor. To access a particular memory cell in a RAM IC the microprocessor specifies the address of the cell, then uses a read control line to read the data from the memory cell or a write control line to write data into the memory cell. RAM ICs are volatile in that once their power is switched off they lose any data or instructions stored in them.

(b) ROM ICs : These can only be read by the microprocessor. The instructions or data stored in the ROM ICs have to be specially programmed into the device and cannot be changed very easily by the microprocessor. ROM ICs are nonvolatile in that if the power is removed from the chips the data or instructions stored in them is not lost.

Both types of IC are organized as blocks, or one–dimensional arrays, of storage cells where each block has a unique name, normally referred to as an address, which the central processing unit can use to locate a particular storage cell. Each of the blocks then corresponds to a location in memory, with the size of the block being a fixed number of bits held in that location. The size of the block decides the number of bits that will be transferred to the CPU each time a memory location is accessed. The number of bits in each location is called the memory width, sometimes called the word size, while the maximum possible number of memory cells is called the memory size or the address space. The address space will always be a power of 2 because the internal representation of the memory address is binary. The difference between the RAM and ROM ICs is that while RAM can be used to read and write data, ROM can only be read by the CPU.

2.   Secondary Memory – Most computers require more storage space than can be provided in main memory. Consequently, they employ secondary memory units, which form part of the computer's I/O equipment. Examples of secondary memory devices are hard or flexible (floppy) disk drives and magnetic tape units. The data on magnetic storage devices are stored as small magnetized areas. To read or write the data from or to the disks requires a disk drive connected to the CPU through various interface units. Magnetic disks and tapes have much higher storage capacities and lower costs per stored bit than main memory. However, secondary memories transfer data at much slower rates than microprocessor to main memory transfers.

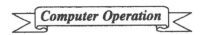

Glossary 1.3

Computer Operation

BIT – The word bit is an abbreviation of binary digit. It is the smallest unit of information in a digital computer and has two possible values, represented by a zero (0) or a one (1). The term bit is also used to classify the storage capacity of memory systems.

BYTE – A byte is a group of bits considered as a single unit of information which is processed or operated on. Most computers use a byte as consisting of 8 bits. It therefore has 256 different possible combinations of the eight 0s and 1s. It is used mainly in referring to parallel data transfer, semiconductor capacity and data storage.

COMMAND – A command is a statement within a computer program which tells a computer what it should do.

CYCLE – A cycle is a time interval in which a characteristic or repeated event occurs.

DATA – Data is a general expression used to describe any group of operands consisting of numbers, alphabetic characters or symbols which denote a condition, value or state. When used in the context of computers, data is digitally represented information, which can be processed or stored by the computer. Information results from the assembly, analysis or summarizing of data into a meaningful form.

INSTRUCTION EXECUTION – The execution of an instruction is the process of the CPU performing the operations indicated by an instruction. The time required to fetch, decode and execute an instruction is referred to as the instruction execution time.

READ – The read process within a computer system is the process used by the central processing unit (CPU) or other control device to copy a data or instruction word from a memory–mapped device into its internal registers. During the read process the contents of the memory location remain unaltered. Read is the opposite of write.

WORD – A computer word is the basic group of bits (either 4, 8, 16 or 32 bits are common) that can be read in, manipulated, stored or written out, by the computer in a single step. Data words contain information to be manipulated while instruction words cause the microcomputer to execute a particular operation.

WRITE – The write process within a computer is the process of sending data from the central processing unit (CPU) to an external storage device such as memory or system peripheral.

1.3.3   INPUT/OUTPUT DEVICES

The third component of a computer system is the input and output devices. These enable the computer to communicate with the outside world, for example to allow an operator to enter instructions or data through a keyboard and to receive information back from the computer through a visual display unit. The input from and output to the I/O mechanisms is handled through interface circuits within the computer.

There are two kinds of I/O device. The first kind can be described as fast and includes magnetic disks and tapes, and high speed communication devices. The second kind are slow devices, including terminals for user interactions, printers, and other interactive peripheral devices. I/O devices present the computational result of the computer system to the outside environment.

The input and output processes in a computer typically consists of several stages, as shown in Table 1.2. It can be seen that once the data has been processed by the computer system and the information extracted, the information is then available to be used within a program. The program will typically manipulate the information from the input before outputing the results again to some I/O device.

1.3.4   INTERCONNECTION NETWORK

A computer system consists of several distinct components: CPU, memories and I/O peripherals. In order for the complete system to function correctly these components must share and exchange instructions and data. This requires a means of transferring information and a mutually accepted set of rules for performing these transfers. The physical medium for the electrical signals and the protocols for the transfer of information in a computer

INPUT	OUTPUT
1. Collection – The first task to be undertaken in the input procedure is to collect the data. Collection involves gathering the data from the various I/O sources and assembling it at one location, usually at the computer or peripheral device. **2. Verify** – After the data has been collected, its accuracy and completeness must be checked. This is important to eliminate the possibility of corrupted data. **3. Code** – Initially the data must be converted into a form that is understandable by the computer so that it can be entered into the data processing system.	**1. Retrieve** – The computer must initially retrieve the information that is to be output from any temporary memory store. **2. Convert** – Conversion is often required to transform the information into a form easily understood by a user, such as a terminal display or printer output. **3. Communicate** – The final task within the output procedure involves sending the information to the right place at the right time.

Table 1.2

Input and output operations.

Inter-connection network

system constitutes a bus. A bus may then be seen as a digital highway or transmission path which allows the CPU, memory, and I/O devices within a typical computer to communicate with one another to exchange data, address and control information.

In a modern personal computer the main bus consists of up to 100 separate wires, lines or tracks, with each line being assigned a particular meaning or function. Although there are many different bus designs for different computer systems, on any bus the tracks can be classified into three functional groups: data, address and control lines. In addition there may be power distribution lines that supply power to the attached modules.

1.4 COMPUTER SOFTWARE

A conventional von Neumann computer works on the basis of processing one instruction at a time in a sequential manner with a centralized control unit to handle program scheduling. This form of single–level serial control with a single machine instruction being processed at a time is common to most computer architectures.

Instructions are given to the CPU by means of a **program**. This is described by a programmer in a high level programming language, i.e. one that the programmer can understand comparatively easily. The program is then converted to a language that the computer machine can use, i.e. machine code, and placed in the computer's memory. When the program is run, the control unit of the CPU locates each instruction and data item in the memory by the use of a unique identifier called an address. The program running on a computer is then essentially a sequence of machine code instructions dictating the direction of computation, where each instruction is a basic command to a computer which will cause it to perform certain operations; machine code instructions are made up of one or more bytes of data.

The machine program for a computer is coded in a **binary** form, since the electronic circuits of a computer have been designed to only be in two different states, i.e. on or off; the two digits 1 and 0 are used to stand for these two states. Binary bits are represented inside the computer by electrical pulses. These pulses are passed around the computer's electronic circuits through switches or logic gates. A gate may be open to pass on a pulse or closed to stop it. Using these gates, pulses can be routed along different paths inside a computer. Electrical pulses can also be combined according to certain logical rules. This is another job done by logic gates. A certain combination of binary pulses into a logic gate gives a certain combination of binary pulses out. The combining of binary inputs to provide certain outputs is called computer logic.

Computer software

Since computers operate on a binary principle all the instructions and data inside a computer have to be represented by binary patterns of 1s and 0s; this is called binary representation. Binary patterns are required for the numbers, letters, the alphanumeric character set in a computer, control codes (for example to clear the screen), instructions and addresses of memory locations which hold data and instructions. All these items have to be represented by their own unique pattern of 1s and 0s.

1.5 COMPUTER CATEGORIES

Since their introduction in the early 1940s computers have evolved and changed considerably. It is possible to identify a number of different classes of computers depending on size, cost and performance. These are shown in Table 1.3 together with examples of the different systems.

Table 1.3

Computer categories and examples.

Computer class	Description of tasks	Examples of systems
Personal computer	Single user system	IBM PS/2
Workstation	Excellent for individual engineers using complex software.	Sun SPARC Station, DEC 5000 Workstation, IBM AS.6000
Minicomputer	Time sharing is a speciality especially for scientific and engineering users. Also good for dedicated on–line transactions. Supports about 30 users.	IBM AS/400, Prime Computers (50 series), Data General Eclipse MV/2000, DEC VAX 6000, IBM system 36 and 38, Hewlett Packard 3000 Series
Mainframe	Good for high volume transactional processing and for maintaining large centralized databases. Supports of the order of 200 users.	IBM 3090, IBM ESA/370, DEC VAX 9000, Bull Information DPS 90/xx series, UNISYS A 15MX
Supercomputer	Used for highly complex computations such as weather forecasting or complex physical simulations. Can support over 500 separate users.	Cray Y–MP/832, NEC SX series, Evans & Sutherlands ES–1Min–Super, Bolt Beranek & Newman TC 2000

Table 1.4 presents a comparison of the architectural features of the different computer categories. We will consider the five main computer classes in more detail.

Table 1.4

Architectural features of different computer classes.

Computer class	Personal computer	Work–station	Mini–computer	Mainframe computer	Super–computer
Example	IBM PS2/50	Sun SPARC	IBM AS400/B60	IBM 3090/600	Cray Y–MP
First installed	1987	1990	1988	1988	1988
Instruction execution rate	2 MIPS	10 MIPS	N/A	102 MIPS	2.6 GFLOPS
Machine cycle time (ns)	100	60	60	15	6
Memory (MB)	1 – 16	16 – 64	32 – 96	128 – 512	256
Price (millions of $)	0.004	0.015	0.3	12.4	20

1.5.1 PERSONAL COMPUTER

The first personal computer was Kenbak, designed and built by John Blankenbaker. An advert for the device appeared in the 1971 September issue of *Scientific American*; however, only 40 were ever sold. It was completely different to what would now be regarded as a PC as it did not use any microprocessors and had a limited input capability through a number of switches, with the output appearing through a series of lights.

The era of true PCs was created in the mid 1970s with the introduction in 1975 of the Altair 8800 PC, developed by MITS Inc. It was the first microcomputer to be available in kit form and had an Intel 8080 microprocessor, 256 bytes of memory, a toggle switch and a light emitting diode (LED) front panel. It cost $498 fully assembled. MITS shipped about 2000 of the machines. In 1976 the Apple 1 personal computer (Article 1.1) was demonstrated by Wonziak and Jobs in California, USA. This computer is widely regarded as being the forerunner to all the other personal computers that have now been developed.

In 1977 a number of completely assembled PCs appeared on the market from manufacturers such as Apple, Radio Shack and Commodore. The TRS–80 (Tandy–Radio Shack) computer was released in 1977 using a Zilog Z80 processor, 4 Kilobytes (KB) of RAM, 4 KB of ROM, a keyboard, display and cassette interface for $600. The device could be programmed in the high level programming language BASIC. The Commodore PET (Personal Electronic Transactor) released in 1977 was an MOS 6502–based machine which cost $595 assembled and included 4 KB of RAM, 14 KB of ROM, keyboard, display, built in cassette tape drive and could be programmed in Microsoft BASIC.

In the United Kingdom the Sinclair ZX80, developed by Clive Sinclair, was released in 1980 and was the first microcomputer to cost less than $200. It contained 1 KB of RAM,

APPLE COMPUTERS

The Apple 1 personal computer was designed and built by Steven Jobs and Stephen Wonziak in 1976 in the USA. It used the MOS 602 microprocessor and ran with a clock rate of 1 MHz. The Apple II computer was launched on April 16th 1977 at the first West Coast Computer Fair, again it used the 6502 microprocessor. By the end of 1977 the Apple II had emerged as one of the most popular microcomputers of its day.

Table 1.5

Initial development of the Apple range of computers.

Date	Model	Cost $	Features
1977	Apple II	1298	16 KB of RAM (expandable to 48 KB) 16 KB of ROM, keyboard, cassette interface, 8–slot motherboard, game paddles and a colour graphics capability.
1980	Apple III	3495	Business version of the Apple II.
1983	Apple IIe	1395	8–bit machine with up to 128 KB of memory.
1983	Apple Lisa	10 000	16–bit 68000 device, featured a graphical user interface and had a mouse.
1984	MacIntosh	2495	See Table 1.9.

Table 1.5 notes the initial development history of the Apple range of computers.

Table 1.6

Features of the Apple MacIntosh computers.

System unit	128K	Plus	SE	SE 30
Date	1984	1985	1987	1989
Microprocessor	68000	68000	68000	68030
Cycle speed (MHz)	8	8	8	16
Processing (bits)	16	16	16	32
Coprocessor	No	No	No	Yes
Standard main memory (KB)	128	1024	1024	2048
Maximum main memory (KB)	512	1024	4096	8192
Standard floppy disk (KB)	400	800	800	1200
Free expansion ports	0	0	1	3

Table 1.6 presents information on the features within the different MacIntosh computers. The MacIntosh was the first personal computer to use a 3.5 in magnetic floppy disk fully enclosed in a rigid plastic case. It was also one of the first to use a graphical user interface incorporating windows, icons, a mouse and pull–down menus. The original computer had an 8 MHz Motorola MC68000 microprocessor, two 3.5 in micro floppy disk drives of 400 KB capacity and a RS449/422A communication port for external communications. It also introduced a mouse and an iconic display in which the user interface was so simple that anyone could use the machine. The initial machines had 128 KB of internal RAM; however, this was expanded in the later machines to 512 KB of RAM. The MacIntosh 2 was introduced in 1987 with a comparable performance to the IBM PS/2 model 80.

4 KB of ROM and had a membrane keyboard. The successor, the ZX–81, released in 1981, had a price of less than $100.

Personal computer

Table 1.7 notes some of the personal computers that have been developed and the microprocessors they used. The most popular range of PCs have been those based on the IBM PC format (Article 1.2). The original IBM PC came with a standard monochrome display adapter (MDA) which has one foreground colour and one background colour, e.g.

Table 1.7

**Personal
computers
and their
microprocessors.**

Personal computers	Microprocessor manufacturer	Chip number	Word size
Apple 1, II & IIe	MOS Technologies	6502	8–bit
Apple IIc	MOS Technologies	65C02	8–bit
IBM Personal computer	Intel	8088	16–bit
Compaq Portable	Intel	80286	16–bit
IBM–AT computer	Intel	80286	16–bit
Apple MacIntosh	Motorola	68000	32–bit
IBM Personal System/2 model 80	Intel	80386	32–bit
Compaq Deskpro	Intel	80386	32–bit

green on black, and a screen resolution of 720 by 350 pixels on the 10 MHz Motorola device and had a 1000 by 800 pixel monochrome graphics display.

The first commercial program specifically written for personal computers, for the Apple II computer, was VisiCalc, a spreadsheet program. It was developed by Dan Bricklin and released on May 11th 1979.

*Personal
computer*

The smallest of the personal computers is the **portable computer**. Most rely on rechargable batteries, allowing them to operate without access to a mains power supply and thereby facilitating their use outwith an office environment. Portable computers also tend to rely on special low power circuitry and contain liquid crystal displays (LCDs).

In 1981 Epson released the HX–20 which was perhaps the first truly portable computer. The machine weighed less than 3 lbs and used a CMOS version of the Motorola 6801, 16 KB of RAM, 20 character by 4 line display and a small cassette unit for tape storage.

Modern portable computers incorporate the latest high performance microprocessors, large internal RAM, have hard disks and floppy disks and use colour displays.

1.5.2   WORKSTATION

A workstation is a term usually given to a high performance computer which a single user will work at. Workstations offer multitasking, high resolution graphics and large memory management facilities. They also have sophisticated expensive software to assist the operator in his or her tasks. Typical application areas that use a workstation are computer aided design (CAD) or computer aided manufacture (CAM). Several examples of workstations and their characteristics are listed in Table 1.8.

Table 1.8

**Examples of
workstations.**

System	Motorola micro– processor	Clock speed (MHz)	Perform– ance (MIPS)	Memory (MB)	VDU resolution (pixels)
Apollo DN3000	68020	16	1.5	2 – 8	1024 x 800
Apollo DN3500	68030	25	4	8 – 32	1024 x 800
Apollo DN4500	68030	33	7	8 – 32	1280 x 1024
Sun–3/50	68020	15	1.5	4	1152 x 900
Sun–3/60	68020	20	3	4 – 24	1152 x 900
NeXT	68030	25	4	8 – 16	1120 x 832

On August 12th 1981 the first IBM PC was introduced to the public. The computer had an Intel 8088 microprocessor, 16 KB of RAM, a keyboard and a connector for a cassette recorder. Including the monitor it cost $1265. A deluxe version with 48 KB of memory, two 160 KB disk drives, a monochrome monitor and a printer sold for $4300. It cost an additional $2000 to purchase the hardware to expand the memory to 256 KB. In 1983 IBM announced the IBM XT which added a 10 MB hard disk drive to the PC frame and three more expansion slots to the standard PC bus. It cost $4995.

In August 1984 IBM introduced the next generation of computer, the AT (advanced technology). It claimed to be 75% faster than the XT model. The AT had an Intel 80286 microprocessor, 256 KB of RAM, a 16–bit data bus, a 20 MB hard disk and a high density floppy disk drive. The internal memory addressing range was limited to 640 KB for compatibility with the PC. Table 1.9 notes the different IBM models.

Characteristic	PC	PC XT	PC AT	Port–able	System 2 model			
					30	50	60	80
Date of introduction	1981	1983	1984	1986	1987	1988	1988	1989
Microprocessor	8088	8088	80286	80C88	8086	80286	80286	80386
Cycle speed (MHz)	4.77	4.77	8	4.77	8	10	10	20
Data transfer (bits)	8	8	16	8	16	16	16	32
Coprocessor	8087	8087	80287	N/A	80287	80287	80287	80287
Standard memory (KB)	256	256	512	–	640	1024	1024	1024
Maximum memory (KB)	640	640	640	–	640	7168	15360	16384
Standard floppy (KB)	360	360	1200	720	720	1440	1440	1440
Free expansion slots	2/3	5/6	5	–	3	3	7	7

Table 1.9

The evolution of the architecture of IBM personal computers.

Comparing the performance of several of these machines we find:
1981 IBM PC – 100 000 instructions/second
1984 IBM AT – 1 million instructions/second
1990 IBM AT with 80486 – 10 million instructions/second

The popularity of the IBM range of computers has spawned a range of compatible machines, these machines are usually much cheaper than comparable IBMs and tend to perform better than the IBM machines. The disadvantage is that certain compatible machines may not in fact be totally compatible with their IBM counterpart.

In 1987 IBM launched its second family of PCs called the System 2 models. Table 1.9 notes the architectural features of the System 2 family. These have has an architecture which is different from the original PCs. The PS/2 family has a number of enhanced features compared to the PCs:

1. Support for the Intel 80386 32–bit microprocessor.
2. The use of very large scale logic circuits for system control.
3. 3.5 in floppy disks.
4. Microchannel bus, this is a more advanced bus compared to the PCs ISA bus.
5. Higher performance graphic display, VGA compared to EGA.
6. A wide range of hard disk sizes available.

Workstation Apollo Computers (now part of the Hewlett Packard company) was the pioneering manufacturer of stand–alone workstations. Apollo's first offering was the Domain DN1000 which contained a Motorola 68000 microprocessor. A couple of years later Sun Microsystems entered the market with the Sun 100, which was also based on the 10 MHz Motorola device and had a 1000 by 800 pixel monochrome graphics display.

Workstation

Modern workstations have a 32–bit microprocessor, a floating point co–processor, at least 16 MB of main memory, a 19 in colour VDU monitor with 1024 by 800 pixel resolution, at least 100 MB of hard disk storage, local area communication facilities and a UNIX operating system. Most workstations have opted for the Motorola range of microprocessors but the SPARC reduced instruction set microprocessor has also proved popular.

1.5.3   MINICOMPUTER

Up until 1965 computer systems were viewed by many as general–purpose data processors. However, in 1965 the computer company DEC introduced what is generally known as the world's first minicomputer, the DEC PDP–8 (programmed data processor). It cost £16 000. Since its introduction the minicomputer has developed into a cheaper version of the mainframe computer aimed at applications not as intensive as those run on mainframes and at companies and universities who require computing power but at a fraction of the cost of a mainframe computer.

The advent of low cost logic using integrated circuits led to the explosive growth of the minicomputer industry in the 1970s, with minicomputers making their impact in the field of engineering and scientific applications. The distinction between the mini and larger machines has now become increasingly blurred due to advances in hardware and software technology. The introduction of DEC's 32–bit 11/780 minicomputer in 1978, which evolved from the PDP–11, marked the emergence of the super minicomputer class that combined the design concepts of mainframes with relatively lower–cost technology. Further examples of modern mini computers are the IBM AS/400 and the VAX 8800.

1.5.4   MAINFRAME

The term mainframe is widely used to refer to the central processing unit of a large general purpose computer. A mainframe usually has a wide range of facilities and a powerful time shared operating system, i.e. one that shares the computer's time between a number of different users, thereby allowing many users to simultaneously access the resources offered by this centralized computing facility. Mainframes are less expensive than supercomputers

Table 1.10

Comparison between mainframes, minicomputers, and workstations.

Computer	Year released	Processor size (bits)	Physical address size (bits)	Data bus width (bits)
Mainframe				
IBM S/360	1964	32	24	128
IBM S/370XA	1983	32	32	128
Minicomputer				
DEC PDP–11/45	1973	16	18	32
DEC PDP–11/70	1976	16	22	32
DEC VAX–11/780	1976	32	32	64
IBM ESA/370	1988	32	32	128
Workstation				
IBM RISC System/6000	1990	32	32	64–128
HP Precision	1986	32	32	32–64

and so tend to dominate the business and scientific applications so that many large companies and universities, for example, have a mainframe machine usually linked to a multitude of time sharing terminals. IBM is widely regarded as having been the main suppliers of this type of computer for a large number of years. For large computer *Mainframe* mainframes it is common practice to house the actual computer system in an air–conditioned *computer* and dust free environment. Examples of mainframe computers are CDC Cyber 9600, IBM System/390 (IBMs most powerful mainframe) and the ES 9000 Model 580.

Table 1.10 presents a comparison between mainframes, minicomputers and workstations in terms of their processor size and their address and data bus widths.

1.5.5 SUPERCOMPUTER

Supercomputers are the largest, the most powerful, most complex and most expensive of the different computer classes. The CDC 6600, released in 1964, is widely regarded as the first supercomputer. The term had already been used to describe the IBM Stretch machine and the UNIVAC LARC, both developed in the late 1950s; however, the CDC computer was the first commercially available device. As to what constitutes a supercomputer is open to debate but it is generally considered that it should have a number of features:

1. It should be able to perform many hundreds of **MFLOPS** (millions of floating point operations per second). A floating point operation is either the addition, subtraction, multiplication or division of two real numbers.
2. It should have a word length in the order of 64–bits, i.e. it should be able to process 64–bit data items.
3. It should have main memory size measured in excess of tens of millions of 64–bit words.

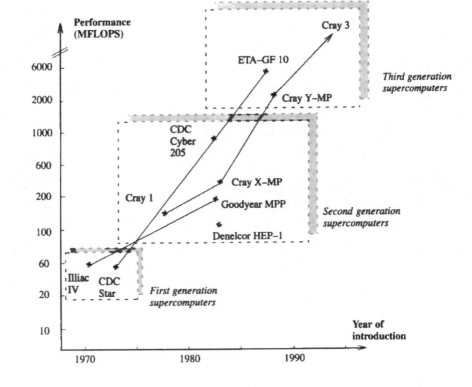

Figure 1.5

Comparative performance of several generations of supercomputers.

Since the development of supercomputers they have been widely used by many research institutes and government sponsored establishments, mainly for scientific and engineering tasks to perform highly complex tasks involving massive amounts of computation. For example the theoretical behaviour of some aspects of nature expressed in mathematics can form the basis for a computer program which simulates nature, such a program typically requires the immense computing power of a supercomputer to handle it. Scientists use supercomputers in the areas of physics and chemistry to study chemical or nuclear reactions in order to validate some theory about the behaviour of matter. Supercomputers are also finding more and more applications in the region of simulation work for aircraft manufacturers.

Super-computer

It is estimated that by 1993 there were between 400 and 500 supercomputers installed world wide. Today's supercomputers have large memories of 1 to 32 GB with a number of independent high performance CPUs which are sometimes called functional units.

A comparative performance of several supercomputers is given in Figure 1.5 which also indicates the development line for three of the supercomputer families. It can be seen, for example, that the Cray 1 (Article 1.3) has been developed through several generations of supercomputers to the Cray–3 while the ETA–GF 10 supercomputer can trace its design back to the earlier CDC Star computer.

The most important feature of a supercomputer is its performance. As such there is intense rivalry between computer manufacturers to claim to have the world's fastest computer. Table 1.11 indicates the top ten in performance for a number of the more modern supercomputers. It will be appreciated that this table is simply a snapshot in time as it will continually change with the introduction of new models. The performance is measured in 1000s of millions of floating point operations (GFLOPS).

Table 1.11

Top 10 in supercomputer performance.

Position	Computer	Speed (GFLOPS)	Description
1.	Hitachi S–3800	32	Has four processors. It also has a 2ns clock cycle and uses specially constructed ECL ICs (gate arrays using 25 000 logic gates).
2.	NEC SX–3	26	Uses four processors. It has a 2.5 ns clock cycle.
3.	Thinking Machines (CM–200)	20	This large scale multiple processor has 65 536 small processing elements.
4.	Cray Y–MP C90	16	16–processor version. It has a 4.2 ns clock cycle and uses ECL gate arrays with 20 000 gates.
5.	ETA Systems 10 (Model G)	10	8–processor configuration. Released in 1988.
6.	Fujitsu VP2600	5	Two processor machine It has a 3.2 ns clock cycle and uses ECL gate arrays with 15 000 gates.
7.	Intel iPSC/860	2.6	This is a 128–node multiple processor using a hypercube structure.
8.	Amdahl 1400E	1.7	
9.	NEC SX2–400	1.2	Has a 6 ns cycle time.
10.	Cray 2	1	Released in 1985 with a 4.1 ns cycle time.

CRAY COMPUTERS

The Cray Corporation of the USA specializes in the production of very powerful supercomputers. The original Cray 1, designed by Seymour Cray and released in 1976, has been perhaps the most widely publicized supercomputer. It uses emitter coupled logic (ECL). ECL is a semiconductor method for producing digital logic gates for a computer. It produces the highest speed available with silicon technology but requires complex cooling arrangements to dissipate the large amount of heat generated through the ECL gates.

The Cray 1 consists of a central processing unit, power and cooling equipment, a maintenance control unit and a mass storage disk subsystem. The physical appearance of the device is of a central 4.5 ft diameter cylindrical column, 6.5 ft high and surrounded by a circular seat bringing the diameter to 9ft. The unit comprises three segments each of which has four wedge–shaped columns holding up to 144 circuit modules. Each circuit module in turn comprises a pair of circuit boards mounted on opposite sides of a copper heat exchanger. The gas freon is circulated through the unit to keep it cooled to 21° C. The total power consumption of the unit is 128 KW.

Internally the Cray consists of a number of processing sections, or functional units, with each unit being designed to carry out a specific task such as multiplication, addition and logical operation. Each of the 13 functional units in the Cray vector unit can function independently meaning the Cray can be carrying on 13 different independent operations in parallel.

The Cray is a vector computer in that it achieves its maximum performance when dealing with vectors of data, i.e. large strings of data items, which can be efficiently dealt with by the specialized vector hardware in the device.

The Cray has regularly achieved measured performance rates of 130 MFLOPS (millions of floating point operations per second) on suitable problems and has made such an impact on the world of computing that it has for many years been regarded as the standard by which all other supercomputers are judged. The Cray 1 has an operating system, text editor, an assembler, a high level programming language (FORTRAN) compiler and a number of application packages.

To form a supercomputer complex requires many more computers and devices as well as the supercomputer itself. In this respect the Cray 1 is not a stand–alone machine in that it also requires a front–end processor to interface the I/O terminals to the CPU. In order for terminals to have access, the front–end processor usually also requires several layer of intermediate computers with an Ethernet network often being used to link the number of smaller workstations through the front–end mainframe computer into the Cray device. A mainframe computer is also used to act as a file server to the system, i.e. to control the hard magnetic disks used to store the programs and data to be run on the Cray. In total 63 Cray 1 machines were sold and although discontinued the Cray 1 is still widely used.

The Cray 1S was introduced in 1979 and is very similar to the original Cray except that it used bipolar semiconductor technology (38 machines were sold). In 1981 a two processor Cray X–MP, which evolved from the Cray 1, was launched while the four–processor version was released in 1983. The Cray X–MP/48 contains four processors with a memory store of 8 million 64–bit words and a clock speed of 8.5 ns. The four–processor Cray 2 computer was introduced in 1985 while an eight–processor version of the Cray Y–MP was released in 1988 with 32 million 64–bit words of memory.

Super-computer   A second selling point for supercomputers is their very large main memory. As well as their performance increasing with time, the size of the their main memories has also increased. This is shown in Figure 1.6.

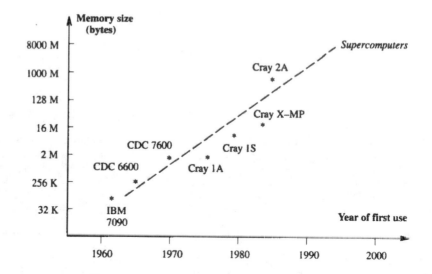

Figure 1.6

**Memory sizes
for several
supercomputers.**

1.6 KEY TERMS

Active component	Instruction execution
Apple computers	Integrated circuit
Architecture	Keyboard
Battery	Mainframe
Binary	Memory
Bit	Million floating point operations
Byte	per second (MFLOPS)
Cabinet	Minicomputer
Capacitor	Organization
Central processing unit (CPU)	Output operation
Chassis	Passive component
Circuit	Personal computer
Circuit board	Portable computer
Command	Power supply
Computer	Primary memory
Computer categories	Program
Computer construction	Random access memory (RAM)
Computer hardware	Read
Computer operation	Read–only memory (ROM)
Cray computers	Resistor
Cycle	Secondary memory
Data	Software
Diode	Structure
Disk drives	Supercomputer
Expansion slots	Surface mounting
Hardware	System.
IBM Personal Computer	Visual display unit
Information technology	Word
Input operation	Workstation
Input/output devices	Write
Interconnection network	

1.7 REVIEW QUESTIONS

1. What is the purpose of an information system?
2. What is the difference between information and data?
3. What is the basic function of a computer?
4. Describe a computer in less than 25 words.
5. Describe briefly any two characteristics of a computer.
6. To what do general purpose computers owe their flexibility?
7. For a personal computer identify the major subsystems in its construction.
8. What does the term 'computer architecture' mean?
9. What do you understand by the term 'system'?
10. Draw a block diagram of the main units within a computer hardware system. Show the flow of data between the various peripheral devices attached to the central processor.
11. Within a computer system what do the following terms refer to:
    (a) CPU, **(b)** ROM, **(c)** RAM, **(d)** PIA.
12. What is the basic unit of information in a digital computer?
13. What is the connection between bit, byte and word?
14. What are the main functions of the central processor of a computing system?
15. What are the main functions of the I/O subsystem within a computer?
16. In computer systems what is a 'bus'?
17. Why do computers manipulate data and execute instructions in the binary number system?
18. In a computer system how is information represented?
19. What are the main categories of computers?
20. What is a supercomputer?
21. Name four uses to which computers are put in science and engineering.
22. What is the meaning of the term 'mainframe computer' and what purpose does it serve in business applications?
23. Why are microcomputers sometimes called personal computers?
24. Describe how desktop computers now offer the resources of mainframe computers.
25. Describe the difference between a personal computer and a mainframe computer.

1.8 PROJECT QUESTIONS

1. What benefits would you expect a company to obtain from using a computer?
2. Describe what you consider constitutes useful or valuable information.
3. Can you name any information systems which do not contain a computer.
4. List three things that computers cannot do.
5. Describe the problem and benefit areas that you think might arise in the future from the continuous and increasingly wider use of computers.
6. Computer data storage banks hold more information about peoples' personal affairs every year. Discuss the possible dangers and benefits of this fact and suggest safeguards to limit the danger.
7. Discuss the future influence of microtechnology upon society, stating why this influence may be for the good of society or otherwise.
8. Do you think the computer aids or hinders peoples involvement in society?

9. Do you agree or disagree with the statement 'Computers are only good for handling repetitive jobs'? Explain your answer.
10. Distinguish between a general–purpose computer and a special–purpose computer.
11. Computers are only capable of responding to input in a predictable fashion. True or false?
12. Do you envisage a time when everyone will be able to direct computers to perform a range of tasks? If so what facilities and features would the computer require?
13. Explain why the study of computers should be part of every college curriculum.
14. Do you agree with the assertion that 'Success in the future will increasingly depend on an individual's computer literacy'?
15. Devise an introductory program of studies for someone who is unfamiliar with the operation of a computer. You need only list the topics you consider important for inclusion in the study program.
16. In what ways could you see the fact that computerization may cause repetitive monotonous working conditions.
17. Information is often regarded as a commodity which can be traded. Give three examples of this trade.
18. Investigate the use of computerised information in the stock market.
19. In what ways will technology management become more important in the future?
20. What skills does the use of microtechnic technology demand?
21. Which of the following has more importance on the development of the computer: social, economic, political or philosophical?
22. Select a computer magazine or computer journal and describe who it is aimed at and the types of articles it contains.
23. What responsibilities should the vendor of a computer system have to a purchaser of a computer?
24. Find an article in a computer magazine describing a hardware or software computer product and produce a 300–word synopsis.
25. Do you agree with the statement 'Technology constitutes a practical application of knowledge'? Explain your answer.

Project questions

1.9 FURTHER READING

This chapter has presented a wide range of terms and ideas on the construction and operation of a computer. Subsequent chapters will develop the themes that have been introduced. There are a number of magazines and books which can be used to provide more information on computer systems, a selection of which are noted below.

Byte: This monthly magazine offers a wide range of detail on every computer subject, from chip details to high level programming. It has an easy to understand reading style.
ACM Computing Survey: Excellent clearly written introductory articles on a variety of subjects.
Communications of the ACM: Advanced journal about theoretical matters and events within computing. Not recommended for the 1st–level student.

IEEE Computer: Clearly written articles on high level topics

IBM–PC Magazine, etc: There are a wide range of machine–specific magazines, most of which are dedicated to the IBM PC. However the Apple, Amiga and Commodore computers all have their own magazine.

There are also other recommended publication such as *Computer World, Datamation, IEEE Micro, PC Computing* and *PC World*.

Introduction to Computer Systems (Books)

1. *Computers and Application Software*, R. A. Szymanski, D. P. Szymanski, N. A. Morris and D.M. Pulschen, Merrill, 1988.
2. *The World of Computing*, R. E. Anderson and D. R. Sullivan, Houghton–Mifflin, 1988.
3. *Elements of Computer Organization*, G Langholz, J. Francioni and A. Kandel, Prentice Hall, 1989.
4. *Computing Essentials*, T. J. O'Leary, B. K. Williams and L. O'Leary, McGraw–Hill, 1989.
5. *Computers and Computing*, S. Hock, Houghton–Mifflin, 1989.
6. *Computer Fundamentals, Concepts*, W. S. Davis, Addison–Wesley, 1991.
7. *Computing Fundamentals: Concepts and Applications Software*, L. Ingalsbe, Macmillan, USA, 1992.
8. *Prentice Hall's Illustrated Dictionary of Computing*, J. C. Nader, 1992.
9. *Computer Fundamentals*, D. A. Adams, Course Technology, 1992.
10. *Desktop Computers in Perspective*, R. A. Henle and B. W. Kuvshinoff, Oxford University Press, 1992.
11. *Computing in the Information Age*, N. Stern and R. A. Stern, Wiley, 1993.
12. *Computers and Information Systems: Tools for the Information Age*, 3rd Edn, H. L. Capron and J. D. Perron, Addison–Wesley, 1993.
13. *Computer Organization and Architecture: Principles of Structure and Function*, 3rd Edn, W. Stallings, Macmillan, 1993.
14. *Introducing computers: Concepts, Systems and Applications*, R. H. Blissmer, Wiley, 1993.

Supercomputer (Articles)

1. Supercomputing, J. P. Riganati and R. B. Schneck, *IEEE Computer*, October 1984, pp. 97–113.
2. Supercomputing–the view from Japan, *IEEE Micro*, February 1993, pp. 67–70.

Workstations (Articles)

1. The current crop, B. Nicholls, *Byte*, February 1989, pp. 235–244.

Personal Computers (Books)

1. *Insight into Personal Computers*, A. Gupta and H. D. Toong, IEEE Press, 1985.

IBM (Articles)

1. The creation of the IBM PC, D. J. Bradley, *Byte*, September 1990, pp. 414–419.

Portable computers (Articles)

1. Destination laptop, D. Gephardt and M. C. Klonower, *Byte*, February 1991, pp. 239–246.

Hardware

Part Two

Parts	
Part One:	*Overview*
Part Two:	*Hardware*
Part Three:	*Software*
Part Four:	*Operating Systems*
Part Five:	*Networks*
Part Six:	*Advanced Hardware*
Part Seven:	*Appendix*

Chapter 2

Digital Integrated Circuits

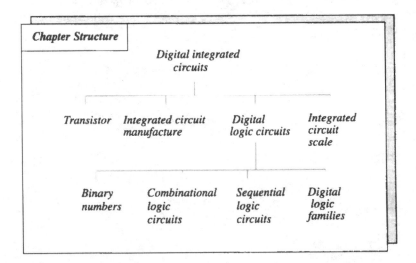

Chapter Structure

Digital integrated circuits

Transistor · Integrated circuit manufacture · Digital logic circuits · Integrated circuit scale

Binary numbers · Combinational logic circuits · Sequential logic circuits · Digital logic families

Learning Objectives

In this chapter you will learn about:

1. The advances in the manufacturing of active electronic components which made microcomputers possible.

2. The manufacturing processes required to develop integrated circuits.

3. The basic logic gates used in digital electronic circuits.

4. The types and operating characteristics of the implementation technologies used for digital ICs.

'It is unlikely that the great development in computer hardware can be paralleled by the same development expansion in software. Programming is a human intellectual task that is not subject to the same kinds of inventiveness that produced transistors and ICs.'

H.C. Lucas

Chapter 2

Digital Integrated Circuits

The decrease in size and cost together with the increase in performance and use of computers at all levels in society has been made possible due to the ever advancing sophistication of the integrated circuits (ICs) which make up the computer. The development of IC technology is shown in Figure 2.1.

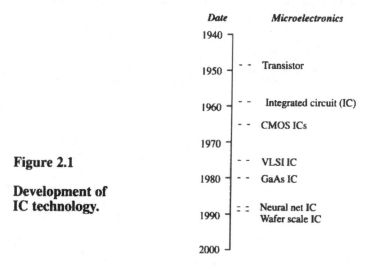

Figure 2.1

Development of IC technology.

The introduction of the integrated circuit in 1959 revolutionized the design of computer systems and resulted in the creation of a new field of science called **microelectronics**. This science is a branch of physics and electronic engineering which specializes in the production and incorporation of new materials and techniques into an integrated circuit environment.

It can be noted that the interaction between integrated circuit technology and computer architecture is complex and bidirectional. The characteristics of various implementation technologies affect the decisions that computer architects make by influencing performance, cost and other system attributes. Developments in computer architecture on the other hand also impact the viability of different IC technologies.

In this chapter we will consider the various properties and operational features of digital integrated circuits. We will begin by looking at the transistor, the building block of ICs.

2.1 TRANSISTOR

On December the 23rd 1947 John Bardeen, Walter Brattain and William Shockley of Bell laboratories in the USA first demonstrated the transistor effect. They observed was that the current through two terminals of a semiconductor germanium device could be controlled by the current into the third terminal. The original transistors built in the early 1950s were made from a material called germanium. However, it was soon found that the element silicon was more abundant and has far more useful properties than germanium for the construction of transistors.

The term transistor is a combination of the word transfer and resistor in that it regulates the flow of electric current through a combination of conductivity and resistance. A transistor is a three–terminal device which can enable the flow of current through two of its terminals to be controlled by the current or voltage at the third electrical connection. Transistors are constructed from silicon into which is implanted other materials. The addition of different types of materials, or dopants, to the silicon determines whether n–type or p–type silicon is produced. Transistors come in two basic types, bipolar transistors and field effect transistors (FETs).

Bipolar transistors are made by sandwiching two layers of n–type material with an intermediate layer of p–type to form an npn transistor or alternatively using two layers of P type material and an intermediate layer of n–type to produce a pnp transistor.

Transistor

FETs are constructed from either a piece of p–type material with n–type regions added, thus forming a p–channel device or alternatively, using a piece of n–type material with p–type regions added, thus forming an n–channel device. In practice the FET acts as a switch connecting two of its terminals (the source and the drain) such that the switch is either on (closed) or off (open) depending on the voltage applied between its third terminal (the gate) and the source. In this way the current through the transistor is controlled by an electric field around the gate of the transistor.

Bipolar transistors are harder to fabricate than FETs but they exhibit a much higher frequency response and can therefore switch faster. A further problem with the bipolar device is the large amount of energy they dissipate as heat during their operation, thereby limiting the number of bipolar transistors that can be employed on the one integrated circuit. Very large scale integrated circuits are only made possible by using FETs.

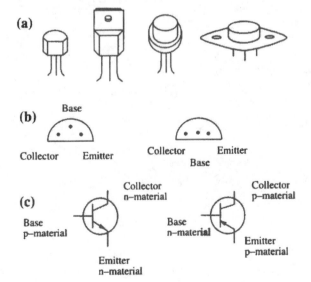

Figure 2.2

Transistors:
(a) Package shapes
(b) Lead assignment
(c) Circuit symbol for a
**   bipolar transistor.**

The most popular package shapes used for discrete transistors is shown in Figure 2.2(a). The lead assignment for two of the bipolar devices is given in Figure 2.2(b) while Figure 2.2(c) presents the circuit symbols for the bipolar npn and pnp devices. For a digital logic device, transistors allow the current through the device to be turned on or off. By wiring transistors in series and parallel or in combinations, both the NAND and NOR gates that are the basis for digital logic and memory can be constructed.

Transistor
Silicon transistors became commercially available in the mid 1950s and were rapidly introduced into computers. The wide acceptance of transistor technology in computers was due to the fact that transistors brought increased speed and reliability and a decrease in physical size and cost, compared to the older valve technology. The second generation computers, 1963 to 1972, were based on transistors and with these components the price of the computers could be brought to a level which permitted many companies to use them.

2.2 INTEGRATED CIRCUIT

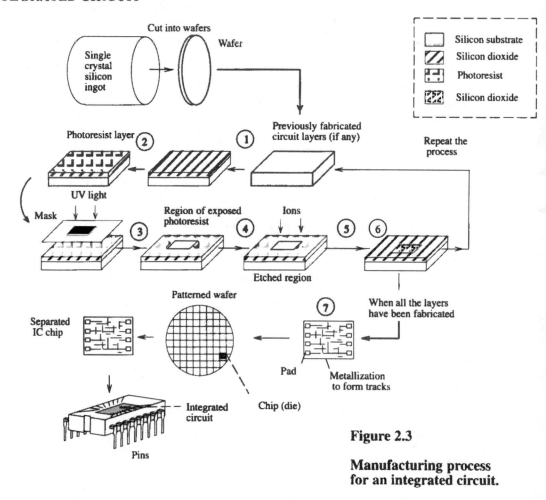

Figure 2.3

Manufacturing process for an integrated circuit.

A milestone in computer technology took place on September 12th 1958 when the first integrated circuit was demonstrated to work by Jack S. Kilby of Texas Instruments. Kilby had interconnected two simple circuits on a single piece of germanium semiconductor. In April 1959 Fairchild Semiconductor perfected a method of depositing aluminium as interconnections between a series of transistors on a silicon wafer.

Glossary 2.1

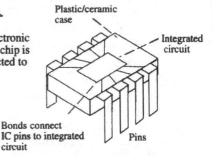

CHIP – A chip is a small piece of silicon on which a large number of electronic devices are constructed. A common name for this is an integrated circuit. The chip is usually enclosed in a ceramic or plastic case and the pads on the chip connected to metal pins on the case. This is shown in Figure 2.4 for a dual in–line case.

Figure 2.4

Integrated circuit chip embedded in a dual in–line case.

DIL – A DIL (dual in–line) package indicates a semiconductor integrated circuit package having two rows of metal pins perpendicular to the edges of the package. These pins are used to enable the device to be connected into a larger electronic circuit. The pins are internally connected onto the pads of the IC. A DIL package is shown in Figure 2.4.

GATE – A gate in a digital circuit is a logic element, usually an electronic switch or circuit where the binary value of the output depends on the values of the inputs and certain logic rules.

SEMICONDUCTOR – A semiconductor is a material, based on a layer (or substrate) of silicon or germanium, which can be arranged as an insulator or conductor by the introduction of various materials, or dopants. It is used in the construction of integrated circuits (ICs).

WAFER – A silicon wafer is a thin disk of silicon material on which many integrated circuit devices are fabricated at the same time. It can be up to 8 inches in diameter. Figure 2.5 shows a silicon wafer made up from a number of small integrated circuit (IC) ships. The ICs are eventually broken from the wafer and mounted in an appropriate IC package.

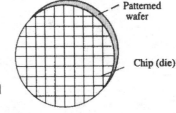

Figure 2.5   **A silicon wafer.**

This planar process resulted in the first practical IC. An IC, or chip, consists of an interconnected array of transistors, diodes, resistors and capacitors where the various circuit components are made on a semiconductor silicon substrate.

2.2.1   INTEGRATED CIRCUIT MANUFACTURE

IC chips are made by a complex series of steps, several of them repeated many times. The steps build transistor parts and other circuit elements into the **silicon substrate** and also create insulating layers and metallized paths atop the silicon substrate. From Figure 2.3 we can see that the silicon chips themselves are actually manufactured on thin slices or wafers of silicon. The crystal of silicon is sliced into round wafers and then by suitable masking and doping features, which selectively alter the electrical behaviour of small regions within the silicon, many transistors can be fabricated on each wafer slice.

The IC development procedure requires a number of different processes; these are shown in Figure 2.3.

1.   Thermal Oxidation – This is the first stage of the process in which a layer of silicon dioxide, about 1 micron thick, is grown on the silicon substrate as hot gas flows over it. The oxide prevents the doping materials, or dopants, penetrating the silicon.

2. Lithography – The silicon dioxide layer is next coated with a photoresistant, which is a material that resists chemical attack after it has been exposed to ultraviolet light. The light is projected onto the photoresist through a mask containing a pattern of open areas. The pattern of open areas controls the formation of the different components on the silicon. The unexposed areas covered by the opaque areas of the mask are soft and are dissolved by a solvent, leaving an aperture in the photoresist; the exposed areas of the photoresist form an etch–resistant layer on the substrate.

3. Etching – A plasma of reactive gas is applied to remove the exposed photoresist and the oxide below it, opening portions of the silicon surface. In this way open areas are established in the silicon dioxide so that the dopants can enter the silicon substrate to form n–type and p–type semiconductor regions for transistor parts.

IC manufacture

4. Ion Implantation – High energy ions of dopant elements, impurities, are then fired at the silicon substrate to penetrate the open areas on the silicon surface. Boron ions are used to form n–type regions while phosphorous ions are used to form the p–type regions. Where the photoresist and oxide remain the ions are absorbed before they enter the silicon.

5. Thermal redistribution – Heating the substrate in a furnace drives the ion–implanted dopants further into the silicon.

6. Insulation – Silicon dioxide is then deposited on the substrate to insulate the underlying structure electrically. Unlike the oxide grown before lithography, the insulating oxide forms at a relatively low temperature to minimize further redistribution of the dopants.

7. Metallization – After steps 1 to 6 have been repeated several times to build up the different component layers, a thin film of metal is finally sputtered on the oxide insulation and etched photolithographically into a pattern of electrical connections among the components connecting them through windows in the oxide layer.

Steps 1 to 7 may be repeated in various combinations until all components are deposited in the silicon substrate and all insulating and connecting layers are built on it.

Typical chip sizes range from about 40 x 40 mills, a mill is 0.001 inch, to about 300 x 300 mils, depending on the complexity of the circuit. Anything from a few to thousands of components can be fabricated on a single chip.

There are two main classes of integrated circuits: those designed to handle analog signals and those designed to process digital signals. There are also a small number of ICs designed to process both kinds of signals, e.g. analog to digital converters and digital to analog converters.

Analog electronics deals with the processing of signals and information which can have continuously variable values. This contrasts with digital information which has a set of discrete values. Analog, or linear, ICs are characterized by the inputs and outputs to and from the IC being primarily analog signals. Examples of analogue ICs are various types of operational amplifiers, analog comparators, voltage regulators and analog multiplexors.

Digital electronics on the other hand is concerned with the understanding, the design and the development of electronic circuits using digital integrated circuits. These circuits use digital data, where this data is the representation of information in a discrete quantized form, i.e. the data is discontinuous in time, as opposed to the analog representation of information as continuous variables.

One of the major advantages that digital systems have over analog systems is the ability to easily store large quantities of information and data for short or long periods of time.

In order to provide some background of the use of digital electronics in a computer we will briefly examine the different forms of digital logic circuits and their implementation using IC technology.

2.3 DIGITAL LOGIC CIRCUITS

Digital information is represented by a fixed number of discrete symbols called digits such that a digital quantity such as number is formed as a finite sequence of digits. For example the ten decimal digits 0, 1, 2,..., 9 can be used to construct any number. In terms of manipulating information in a computer it is preferable to restrict digital signals to being represented by only two distinct values or states. We call two–valued digital signals and the digital systems that process them, binary.

The main reason for restricting digital systems to binary manipulation is that switches are the basic building blocks of digital systems. They are binary in nature in that they can either be off (open) or on (closed). The standard approach is to use the digit symbols 0 and 1 to represent the two possible values of a binary quantity at any point in time. These symbols are referred to as bits, a contraction of the term binary digit. In terms of switches this can be represented by saying that when the switch is off it is in a state 0 and when it is on it is in a state 1 (see Figure 2.6).

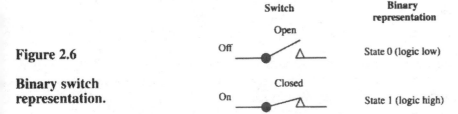

Figure 2.6

Binary switch representation.

The binary manipulation of information in a digital system has three distinct advantages over other forms of representation. Firstly digital information systems can be easily constructed from switches which are binary devices. Secondly the basic decision–making process can be constructed from two states i.e. yes or no, on or off; and finally binary signals are more reliable than those formed by more than two quantization levels.

2.3.1   BINARY NUMBERS

The binary number system is simply another way to count. It is less complicated than the decimal system because it is composed of only two digits. Just as the decimal system with its ten digits is a base10 system then the binary system with its two digits is a base2 system. The two binary digits (bits) are 1 and 0. The position of the 1 and 0 in a binary number indicates its weight or value within the number just as the position of a decimal digit determines the magnitude of the digit, i.e. the weight of each successively higher position (to the left) in a binary number is an increasing power of two.

Any number in the decimal system is made up from the ten digits 0,...,9. The number 1101 for example in the decimal system is represented as:

Base10 $1101_{base10} = (1 \times 1000) + (1 \times 100) + (0 \times 10) + (1 \times 1)$

Each digit to the left represents a higher power of 10. This can be written in another way using the fact that $1000 = 10^3$, $1000 = 10^2$, $10 = 10^1$, $1 = 10^0$.

Base10 $1101_{base10} = (1 \times 10^3) + (1 \times 10^2) + (0 \times 10^1) + (1 \times 10^0)$.

We can perform a similar operation assuming that instead of the decimal system we make use of the binary system which only has the two numbers 0 and 1. The binary system is therefore represented in base2.

Binary numbers

Base2 $1101_{base2} = (1 \times 2^3) + (1 \times 2^2) + (0 \times 2^1) + (1 \times 2^0)$.

Now in the decimal system $2^3 = 8$, $2^2 = 4$, $2^1 = 2$, $2^0 = 1$. This therefore gives us that :

$$1101_{base2} = 13_{base10}$$

A similar conversion method has been used to convert the binary numbers in Table 2.1 into their decimal equivalents.

Table 2.1

Decimal equivalent for a selection of binary numbers.

Decimal number	Binary number	Decimal number	Binary number
0	0000	8	1000
1	0001	9	1001
2	0010	10	1010
3	0011	11	1011
4	0100	12	1100
5	0101	13	1101
6	0110	14	1110
7	0111	15	1111

A final point worth noting is that since digital systems manipulate numbers encoded in a binary form, but the external world is decimal in nature, various conversions between the decimal and binary systems must be performed.

As we have stated digital circuits work on a binary principal, i.e. only handling strings of 1s and 0s. They manipulate these 1s and 0s through a series of digital logic functions. In this context logic may be viewed as a form of reasoning that tells us a certain proposition (declarative statement) is true if certain conditions or premises are true. The concept of a function, on the other hand, is well known to everyone even if not by this name. For example the + of addition is a function that takes two numbers as input and produces a sum as output. Just as there are several basic functions with arithmetic there are several with logic.

The principal of handling logic and arithmetic in terms of 1s and 0s makes use of Boolean algebra. **Boolean algebra** is a mathematical system for formulating logical statements with symbols which makes use of an algebraic notation to express logical relationships in the same way that conventional algebra is used to express mathematical relationships. Boolean algebra is named after the mathematician George S. Boole, 1815 to 1864, who in 1854 published the fundamentals of Boolean algebra.

In general, a Boolean operation is one in which the result of giving each of a set of variables one of two values (0 or 1) is itself one of two values (0 or 1). It may also be seen as a mathematical system for formulating logical statements with symbols so that problems can be written and solved in a manner similar to ordinary algebra.

The hardware devices which manipulate the 1s and 0s to perform Boolean algebra are referred to as **digital logic gates**. These logic gates in turn are constructed from a small network of transistors and resistors. The simplest, or basic, logic gates can subsequently be used to build larger scale digital circuits such as data encoders, digital multiplexors, microprocessors, memory circuits and peripheral controllers. These large scale digital circuits are the basic building blocks for constructing complex digital systems such as a computer.

Digital gates, especially when viewed at the logic level, can be divided into two broad classes: those without memory, which are called combinational circuits, and those with memory, which are called sequential circuits.

2.3.2   COMBINATIONAL LOGIC CIRCUITS

A combinational logic circuit has a behaviour that lists for every combination of its inputs the corresponding output combination, i.e. the output is only a function of the present inputs. A combinational circuit has no memory, implying that no matter when a set of inputs is presented to it an identical output will be generated, i.e. the circuit is time invariant. Such a circuit also maps input data to output data in a single step; in practice, there may be a slight delay before the output signals change in response to new input values.

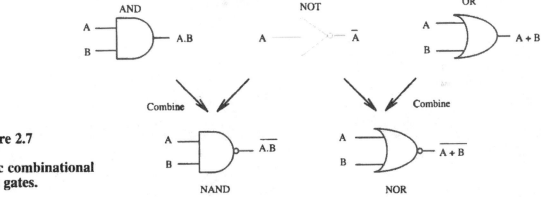

Figure 2.7

Basic combinational logic gates.

Figure 2.7 shows the symbols for the three most basic combinational logic gates: the AND, NOT and OR gate. Each of these gates processes the binary logic 1s and 0s at its inputs in a different way according to the laws of Boolean algebra.

NOT – The simplest of the basic logic elements is the NOT circuit. The primary function of this circuit is to produce a logic 1 level at its output from a logic 0 level at its input and to produce an output logic 0 level for an input logic 1. This circuit is commonly called an invertor. Its Boolean symbol is shown in Figure 2.7.

AND – The second basic logic element is the AND gate. The primary function of this circuit is to produce a logic 1 condition on its output if and only if all of its input conditions are at logic 1. The AND function is expressed in Boolean terms by A.B if A and B are the inputs.

OR – The third basic logic element is the OR gate. The primary function of this circuit is to produce a logic 1 on its output when one or more of its input conditions are at logic 1. A + B is the expression for the OR function operating on the two inputs A and B.

By combining the AND and NOT logic gates we can form a NAND gate, while the combination of an OR and a NOT gate gives a NOR.

*Combin-
ational
logic
circuits*

Manufacturers of digital ICs produce a range of different basic logic gates depending on the number of inputs. Table 2.2 presents a small selection of the possible types indicating the number of inputs and the number of gates per IC. The table also gives the generic code, or designation number, for the ICs. Figure 2.8 shows the internal construction of a number of the ICs. The NAND and the NOR gates are used as the basis for the construction of larger scale digital integrated circuits.

Table 2.2

Examples of basic combinational logic gates.

Type of logic gates	Number of inputs	Number of gates per IC	Designation number
NOT	1–input	6	7404
AND	2–input	4	7408
	3–input	3	7411
	4–input	2	7421
OR	2–input	4	7432
NAND	2–input	4	7400
	3–input	3	7410
	4–input	2	7420
	8–input	1	7430
	13–input	1	74133
NOR	2–input	4	7402
	3–input	3	7427
	5–input	2	74260
EX–OR	2–input	4	7486
EX–NOR	2–input	4	74266

Figure 2.8

Examples of the internal construction of basic combinational logic gates.

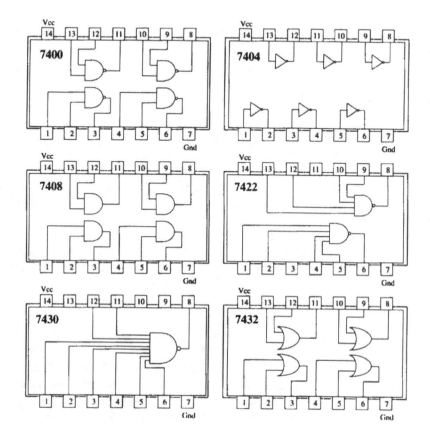

Logic function	Description
Comparator	The comparator compares two input binary numbers and indicates if they are equal or not equal.
Adder	The adder adds two input binary numbers together with an input carry and generates an output sum and an output carry.
Subtractor	A subtractor requires three inputs: the two input binary numbers to be subtracted and an input borrow. It produces an output difference and an output borrow.
Multiplier	A multiplier multiples two input binary numbers to produce an output binary product. Since multiplication is simply a series of additions with shifts in the positions of the partial products, it can be performed using an adder.
Divider	A divider takes two binary inputs and generates a quotient and a remainder. This operation can be performed by a series of subtractions, comparisons, and shifts, and thus it can also be performed using an adder.
Encoder	An encoder converts input information at its input lines into some coded binary form at its output. For example, a certain type of encoder converts each of the decimal digits, 0 through 9, to an output binary code.
Decoder	A decoder converts input binary information into a particular combination of its output lines. For example, a particular type of decoder convert a 4–bit binary code into the appropriate decimal digit.
Multiplexor	A multiplexor allows information to be switched from several input lines onto a single output line in a specified sequence. A simple multiplexor can be represented by a switch operation that can sequentially connect each of the input lines with the output.
Demultiplexor	The inverse of the multiplexing function is called demultiplexing. Here, logic data from a single input line can be sequentially switched onto several output lines.

Table 2.3

Example of combinational logic functions.

Combinational logic circuits

The larger scale ICs can perform a number of more complex logic functions. Table 2.3 notes a number of the more complex logic functions that can be obtained in IC form together with a brief description of the logic function.

The functions obtainable with larger scale combinational ICs noted in Table 2.3 can be divided into four main groups:

1. Arithmetic operations – Arithmetic functions include operations such as addition, subtraction, multiplication, and magnitude comparison. Arithmetic chips can also be used for additions and subtractions in various number representations. These types of chips can also be arranged in an array to perform multiplications and divisions.

2. Data format conversions – Data format conversion refers to the conversion of various data formats, such, as binary to binary coded decimal (BCD), BCD to display codes, etc. Data converters ICs include encoders, decoders, and various kinds of display drivers.

3. Data selections – In a digital circuit, data selection ICs are frequently used to govern and control the flow of information through the circuit. Data selectors include multiplexors and demultiplexors.

4. Data communication – Data communication combinational ICs are available to handle parity generation and checking. These are used for identifying communication errors.

Table 2.4 presents a small selection of the possible types of larger scale combinational circuits indicating the designation numbers of the various types.

Logic category	Logic function		Designation number
Arithmetic units	Adders	1–bit full adder	80, 183
		2–bit full adder	82
		4–bit full adder	83, 283
	Comparator	Magnitude comparator	85
	Arithmetic	4–bit accumulator	281
	Logic Units	4–bit ALU	181
	Multipliers	Multiplier	261
Data format converters	Encoders	BCD priority encoder	147
		Octal priority encoder	148, 348
	Decoders	BCD–decimal decoder	45,145
	Converters	BCD–7 segment	46, 246
		Binary–BCD	185
		BCD–binary	184
Data selectors	Multiplexors (MUX)	16 to 1 line	150
		8 to 1 line	151, 251, 351
		4 to 1 line	153, 253
		2 to 1 line	98
	Demultiplexors (DEMUX)	4 to 16 line	154, 159
		4 to 10 line BCD	42
		3 to 8 line	138
		2 to 4 line	139, 155

Table 2.4

Examples of MSI TTL logic gate designation.

2.3.3   SEQUENTIAL LOGIC CIRCUITS

As we noted, in combinational logic circuits the output at any point in time is only a function of its present inputs at the same point in time, whereas sequential logic circuits are constructed to store or memorize binary digits for an indefinite period of time. In this type of logic circuit the outputs are functions of both present and past inputs. Sequential circuits are built using the basic combinational logic gates together with a number of devices called flip–flops. **Flip–flops** can retain logic levels after the input conditions have been removed and so they can be used to form a basic memory circuit. Actually, the flip–flop is constructed from combinations of basic combinational logic gates, but it may be treated as a distinct logic element because of its ability to retain information.

Logic function	Description
Counter	A counter counts events represented by changing levels or pulses at its inputs to generate a particular sequence of binary values at its outputs. Counters contain storage or memory capability, based on flip–flops, to enable them to remember their present state.
Register	A register is a digital circuit used for the temporary storage and shifting of information. For instance, a number in binary form can be stored in a register and then its position within the register can be changed by shifting it one way or the other. Flip–flops are used in registers.
Memory	Memory circuits can be formed by combining a large number of registers with some suitable decoding circuitry to allow access to each of the registers on receipt of a suitable address.

Table 2.5

Example of sequential logic functions.

Combinational logic gates can be combined with flip–flops and used to construct more complex logic circuits. Several examples of the functions that can be created in this way are noted in Table 2.5.

As a last point it is worth noting that because of manufacturing costs and other logistic problems, there are several restrictions regarding the availability of logic functions on digital ICs.

1. The cost of a particular IC depends directly on the production volume; that is, the higher the production volume, the lower the cost per chip. It is therefore not cost effective for a manufacturer to produce ICs covering every possible combination of basic logic gates combined on an IC. Hence manufacturers only manufacture the most widely used logic functions.

2. In order to be helpful in circuit board packaging and layout, ICs have to conform to standard sizes with standard numbers of pins.

3. Since every input and output terminal of a gate must be accessible outside the chip for connections (the input and output connections are made accessible by providing pins on the chip) the number of gates that can be packed into a chip is related directly to and largely dependent on the number of available pins on the chip.

2.3.4   DIGITAL LOGIC FAMILIES

In fabricating a digital IC chip, several semiconductor technologies are available, each tailored to a specific working environment. In terms of constructing a digital system the major considerations for selecting which logic family should be used are speed, power consumption, cost and availability. There are two major fabrication methods for producing digital integrated circuits, these are to use bipolar or metal oxide semiconductor (MOS) techniques. This is indicated in Figure 2.9 together with a note of their major variants.

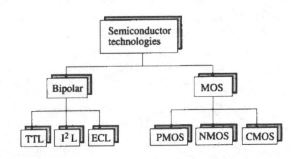

Figure 2.9

Semiconductor technologies for digital ICs.

Symbol	Description
TTL	Transistor transistor logic
I^2L	Integrated injection logic
ECL	Emitter coupled logic
PMOS	p–Channel metal oxide semiconductor
NMOS	n–Channel metal oxide semiconductor
CMOS	Complimentary metal oxide semiconductor

TTL – Transistor transistor logic (TTL) is the most widely used bipolar technology for manufacturing digital integrated circuits. This is mainly due to the fact that it was one of the first to be developed. Because of their widespread use, TTL chips are also the least expensive and most accessible of the logic types. Within the category of TTL gates, there are several different performance classes. These are noted in Table 2.6 together with a brief description of their features. See Article 2.1 for the way TTL chips are labelled.

Table 2.6

Different types of TTL logic.

TTL class	Description
Standard	The standard chips were one of the first types on the market. The typical propagation delay time through a gate is about 10 ns, and the usual power consumption per gate is around 10 mW.
H	The high–speed versions of TTL chips can operate at a higher frequency compared to standard TTL at the expense of twice the power consumption.
L	The low–power TTL version consumes the least amount of power among the TTL family, requiring about 1mW per gate. L chips are comparatively slow.
S	The Schottky (S) chips are among the fastest ones in the TTL family.
LS	The most popular type in the TTL family is the low power Schottky (LS) class. It achieves speeds comparable to the standard chip but consumes almost as little power as the low power version.
AS	The advanced Schottky (AS) chips are faster and consume less power than the S–type chips. This is made possible by newer technology.
ALS	The advanced low–power Schottky (ALS) is another family of fast chips made possible by newer technology.

It can be noted from Table 2.6 that the primary distinctions between the different TTL classes is speed and power. To illustrate this point Table 2.7 presents more detailed information on the power dissipation and speed for four classes of TTL device.

Table 2.7

TTL characteristics.

Property	Low speed 74L	Medium speed 74	High speed 74H	Very high speed 74S
Average power dissipated per gate (mW)	1	10	22	20
Typical gate delay (ns)	31	8	4	1.5

Digital logic families

ECL – Emitter coupled logic (ECL) is a form of bipolar semiconductor technology primarily used in high–frequency applications, with clock rates greater than 80 MHz, because it produces chips with short propagation delays (gate delays as short as 2 ns compared to 10 ns for TTL chips). ECL ICs are, however, expensive to manufacture and dissipate large amounts of heat. ECL ICs are often used in supercomputers.

MOS – Metal oxide semiconductor (MOS) is a semiconductor process that uses field effect transistors (FETs). There are several different MOS technologies, such as PMOS, NMOS, CMOS and HMOS. ICs made using MOS technology can operate with a variety of power supply voltages ranging from less than 5 V to 15 V.

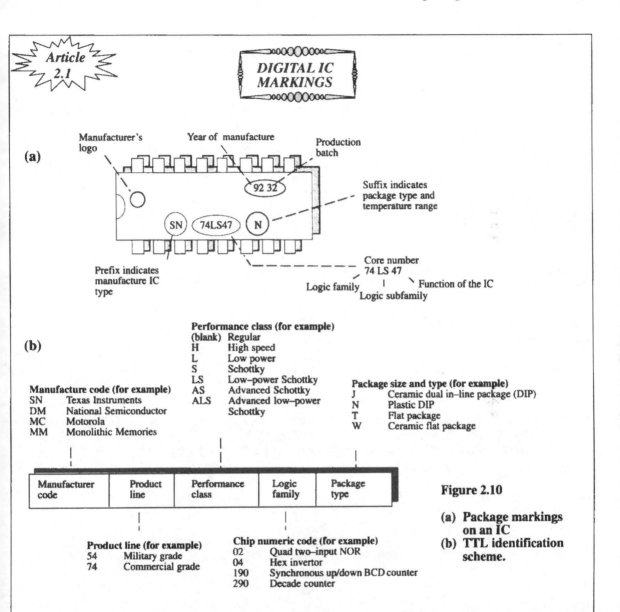

Figure 2.10

(a) Package markings on an IC

(b) TTL identification scheme.

Digital integrated circuits are usually labeled in a particular manner. An example of this is shown in Figure 2.10(a) for a digital IC. In order to identify the different types of chips, a common identification scheme consists of five parts: the manufacturer's code, the product line, the performance class, the logic family, and the package type. This is shown in Figure 2.10(b). This nomenclature is adopted by most chip manufacturers with minor variations and serves as a means of identifying the contents of a chip and provides all the necessary information for the user to find a suitable replacement for a particular chip when the need arises.

For TTL a typical code may appear as follows. The first two or three alphabetic characters of the identification code specify the manufacturer and its product line. The second part of the identification code is a two digit number that specifies the product line of the chip. Some manufacturers also use other alphanumeric codes to indicate various grades such as temperature range. The next part specifies the performance class and is usually related to parameters such as speed and power. The fourth part is a numeric code that is unique for each kind of chip. This code identifies the type and number of gates in a particular chip and is often the same among all IC chip manufacturers to avoid confusion. The last part of the TTL identification code specifies the package size and type.

Glossary 2.2

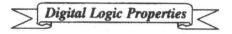

Digital Logic Properties

DELAY (LOGIC) – In general there are three types of delays associated with signals in a digital system: logic, capacitive and transit time.
1. A logic delay (or propagation delay) is the time required for the output of a digital element to switch with respect to the input.
2. A capacitive delay is caused by capacitive loading on the output of the logic element.
3. A transmit time delay is the time necessary for the signal to travel along a wire or printed circuit board trace.

FAN OUT – Fan out indicates the number of digital logic gate inputs a particular logic gate output can drive. In the design of circuits boards using logic integrated circuits (ICs), the designer should adopt a design rule which limits the fan out of IC gates to a small number. By imposing this restriction, the IC power dissipation can be decreased thereby reducing any thermal effects and thus lowering the probability of a hard failure developing. Fan out limitation also increases the effective noise margin at the inputs of subsequent gates and thus decreases the possibility of a transient fault.

LOADING – The currents that logic gates can sink, i.e. receive, when the output is at a logic 0 state, from other gates, and can source, i.e. transfer, when the output is in a logic 1 state, to other gates of both transistor transistor logic (TTL) and complimentary metal oxide semiconductor (CMOS) logic families are shown in Table 2.8. When used to refer to logic gates the drive and source current capability of the logic gate is often referred to as loading .

Family	Type	Sink current	Source current
TTL	74	1.6 mA	40 μA
TTL	74LS	0.39 mA	20 μA
TTL	74S	2.0 mA	50 μA
NMOS		10 μA	10 μA
CMOS		10 μA	10 μA

Table 2.8

Loading for various logic families.

LOGIC LEVELS – The voltage levels interpreted as a logic high or logic low by the logic gates varies between the different logic families. The typical voltage properties of two logic families are given in Table 2.9.

		Description	TTL	CMOS
Voltage	v_{OH}	Minimum logic 1 output voltage	2.4 V	4.6 V
	v_{OL}	Maximum logic 0 output voltage	0.4 V	0.4 V
	v_{IL}	Maximum acceptable logic 0 input voltage	0.8 V	1.5 V
	v_{IH}	Maximum acceptable logic 1 input voltage	2.0 V	2.0 V

Table 2.9

Voltage levels of two logic families.

Digital logic families

MOS chips are slower than their bipolar counterparts with typical gate propagation delays of 20 ns; however, they have an extremely low power consumption. For example, a typical **CMOS** IC consumes only 0.1 mW when active and 0.005 mW when on standby, compared to 10 mW for TTL chips and 20 mW for ECL chips. Due to its low power consumption and its ability to produce high density ICs CMOS and NMOS technologies are used to fabricate today's microprocessors and memory devices.

It is possible to compare the different semiconductor technologies in terms of their level of integration against the speed of the logic gates they produce. This is plotted in Figure 2.11. It can be seen for example that at present the fastest silicon logic is to be made using the ECL process; however, it also has the lowest level of integration, i.e. circuits with the smallest number of gates.

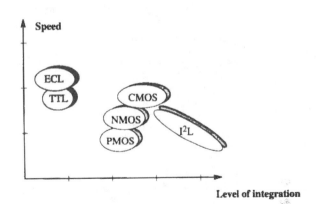

Figure 2.11

Relative levels of integration versus speed for various semiconductor technologies.

2.4 INTEGRATED CIRCUIT SCALE

In addition to differentiating ICs by technology, complexity is another measure that can be used to classify IC chips. Complexity is usually measured in terms of the number of equivalent logic gates involved in making a particular IC. The number of equivalent gates is that number of discrete elementary two–input logic gates required to achieve the same function as a particular IC. Although the number of gates or transistors is not always a good estimate of complexity, it is nevertheless a convenient and readily available measure. Basically, there are four categories.

Small–Scale Integration (SSI) – The least complex digital ICs are placed in the SSI category. These circuits are usually a collection of basic logic gates and flip flops, up to twelve equivalent gate circuits on a single chip, performing elementary operations. The full range of ICs in this category includes elementary logic gates such as the AND, OR, NOT, NAND, and NOR with different numbers of input variables. Gates that implement two–level logic such as AND–OR gates and AND–OR–INVERT gates are also available.

Medium–Scale Integration (MSI) – Medium–scale integration ICs are circuits with complexities ranging from twelve to one hundred equivalent gates on a chip. MSI circuits include the more complex logic functions such as encoders, decoders, counters, registers, multiplexors, arithmetic circuits, small memories and others.

Large–Scale Integration (LSI) – Large–scale integration chips are standard building blocks with rather complex and specific functions or utilities. Circuits with complexities of 100 to 1000 equivalent gates per chip, including small memories and some 8–bit microprocessors generally fall into the LSI category.

Very Large–Scale Integration (VLSI) – Integrated circuits with complexities ranging from 1000 to 100 000 equivalent gates per chip and beyond are generally considered VLSI. Very large memories, microprocessors, and single chip computer are in this category. The first signs of the VLSI era appeared with the introduction of the 64 Kbit dynamic random access memory (DRAM) and the 16–bit single chip microprocessor introduced in the late 1970s.

Table 2.10 notes the different integration scales together with some examples of the types of logic chips which can be associated with the different integrated circuit scales.

Table 2.10

Classification of the number of gates in an IC.

Date of introduction	Integration scale	Maximum gates per chip	Examples
Early 1960s	Small–scale integration (SSI)	12	4 two–input NAND gate packages. 3 three–input NAND gate packages. 6 invertor packages. 4 two–input XOR packages. 2 master slave JK flip–flop packages.
Late 1960s	Medium–scale integration (MSI)	100	4–bit ALU with up to 32 functions. 8–input multiplexor/selector. Decimal decoder. 4–bit synchronous binary counter. 8–bit serial–in parallel–out shift register.
Early 1970s	Large–scale integration (LSI)	1000	16 Kbit (2048 x 8) read only memory. 120–input, 12 output programmable logic array. 300 gate array. 4–bit ALU/register bit slice device. 4–bit microprogram sequencer. 8–bit microprocessor.
Late 1970s	Very large–scale integration (VLSI)	100 000	16–bit and 32–bit microprocessor. Programmable systolic chip.
Mid 1980s	Ultra large–scale integration (ULSI)	>100 000	Large random access memories.

2.5 KEY TERMS

Analog electronics
AND gate
Binary number
Bipolar transistors
Boolean algebra
Chip
Combinational logic circuits
Complimentary metal oxide semiconductor (CMOS)
Delay
Digital electronics
Digital IC markings
Digital logic circuits
Digital logic gates
Digital logic families
Dual in line (DIL)
Emitter coupled logic (ECL)
Fan–out
Field effect transistor (FET)
Flip–flop
Gate

Integrated circuit
Integrated circuit manufacture
Integrated circuit scale
Large–scale integration (LSI)
Loading
Logic levels
Medium–scale integration (MSI)
Metal oxide semiconductor (MOS)
Microelectronics
NOT gate
OR gate
Semiconductor
Sequential logic circuits
Silicon substrate
Small–scale integration (SSI)
Transistor
Transistor transistor logic (TTL)
Very large–scale integration (VLSI)
Wafer

2.6 REVIEW QUESTIONS

1. Define the terms diode, transistor, integrated circuit, gate and wafer.
2. What are ICs? What major development did they lead to?
3. What material is the most common for the production of semiconductor components?
4. The manufacture of an integrated circuit is a complex process. Explain.
5. Why are digital logic circuits used in a computer as opposed to analog circuits?
6. Contrast the characteristics of combinational and sequential logic circuits.
7. What are the three most basic digital logic functions?
8. Note four complex digital logic functions that can be created using MSI logic circuits.
9. What is the distinguishing feature of a flip–flop which can be used to create a memory cell?
10. Why is it not possible to buy a 6–input exclusive–OR TTL logic circuit?
11. What is the main difference in logic voltage levels between TTL and MOS logic gates?
12. Why are there so many varieties of TTL logic types?
13. Why are CMOS devices so popular for VLSI memory devices?
14. Why are ECL chips popular for constructing supercomputers?
15. What does the term 'level of integration' mean ?

2.7 PROJECT QUESTIONS

1. Investigate in more detail than provided in the text the different constructional details between a bipolar and MOS transistor.
2. By obtaining data sheets for different transistors explain why transistors come in different sizes.
3. The size of the transistors in an IC is determined in part by the type of radiation used at the lithography stage. Investigate the different types of radiation that can be used.
4. Obtain an IC and compare its markings with those shown in the text (see Article 2.1).
5. Explain how the miniaturization of electronic circuits has aided in the evolution of the computer.

2.8 FURTHER READING

There are a large range of books on the market describing microelectronic subjects. Most, however, are aimed at students studying microelectronics and tend to describe the IC development process in great detail, noting in depth the construction techniques required for IC production. Similarly the journals in microelectronics do not provide an overview to the subject and are more targeted at those already doing work in the field. If you wish to read more on the subject it is suggested that you select an introductory book on digital electronics. Most provide an introduction to digital ICs, their characteristics and uses.

Introduction to Digital Logic (Books)

1. *Digital, Logic Techniques' Principles and Practice*, 2nd Edn, T. J. Stonham, Chapman & Hall, 1987.
2. *Digital Systems Logic, and Applications*, T. A. Adamson, Delmar, 1989.
3. *Digital Electronics*, 2nd Edn, J. W. Bignell and R. Donovan, Delmar, 1989.

Microprocessors and Custom Logic

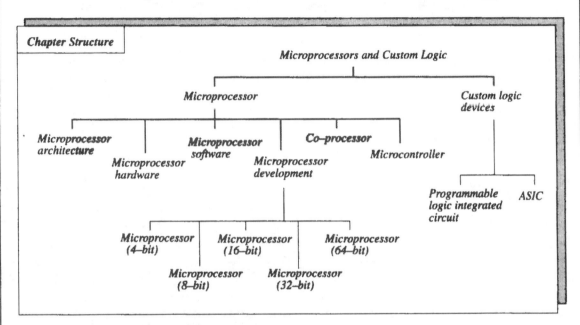

Chapter Structure

Learning Objectives

In this chapter you will learn about:

1. The operation of a microprocessor through its internal architecture and external buses.
2. The evolution of the microprocessor in terms of its hardware complexity.
3. The features of a number of the more important 8–, 16–, 32– and 64–bit microprocessors.
4. The internal architecture of a micro–controller IC.
5. The different types of programmable integrated circuits which can be used in microprocessor systems to handle system logic.

'First think about what the ultimate would be. Then take a step away from that and another; until you get to something you can build.'

Matt Reedyl

Chapter 3

Microprocessors and Custom Logic

In Chapter 2 we noted the development of integrated circuit (IC) technology to the point where very large scale integrated (VLSI) chips can be readily produced. The most complex of the ICs so far developed are the microprocessors. This chapter considers the operation and development of microprocessors and the recently produced programmable logic IC devices. Both offer programmability in a computer system, the microprocessor through software, the programmable logic ICs through hardware.

A **microprocessor** is the single electronic integrated circuit capable of receiving and executing code instructions which forms the central processing unit (CPU) of most computer systems. A large number of different microprocessors have been developed by many different companies. Some of the features of a number of microprocessors are shown in Table 3.1. The performance of the microprocessors is given in terms of millions of instructions per second (MIPS) that the device can process.

Company	Micro– processor	Address bus (bits)	Data bus (bits)	Max RAM (bytes)	Clock rate (MHz)	Max speed (MIPS)
Zilog	Z80	16	8	64 K	4	N/A
Intel	8088	20	8	1 M	8	0.9
	8086	20	16	1 M	12	1.2
	80286	24	16	16 M	16	3
	80386 SX	32	16	4 G	25	5
	80386 DX	32	32	4 G	40	8.5
	80486	32	32	4 G	50	18
Motorola	68000	24	16	16 M	20	1.3
	68020	32	32	4 G	33	7
	68030	32	32	4 G	50	12
	68040	32	32	4 G	50	17

Table 3.1

Characteristics of several types of microprocessors.

Microprocessors and their close relatives the single chip microcomputers have become the standard building blocks for digital systems and are incorporated into almost every conceivable application in domestic and industrial environments. The availability of these complete processors as low cost building blocks has brought increased attention to digital structures where dedicated microprocessors can be fully programmable and execute their own programs to perform some specified task.

3.1 MICROPROCESSOR ARCHITECTURE

From Chapter 1, we can recall that a microprocessor essentially reads (or fetches) each instruction in a program, one at a time, from memory and performs the data manipulation specified by the instruction. It also reads data from input devices, and writes (or sends) data to output devices. To perform these operations, the microprocessor requires a group of internal logic circuits and a set of external data, address and control signals to transfer the information.

Though the architecture of different microprocessors may appear to be vastly different, there are features that are common to all microprocessors. In this section we will expand on the brief details presented in Chapter 1 to consider how a microprocessor, in our case the **Zilog Z80**, coordinates its operations within a complete microprocessor system, i.e. one containing memory and input/output (I/O) integrated circuits as well as the Z80 itself. In particular we will detail the internal construction of the device and note the function and purpose of its external connections (pins).

The microprocessor company Zilog was founded in November 1974 by Federico Faggin and Ralph Ungermann, both of whom were ex–engineers from Intel. Faggin developed the Z80 microprocessor starting in December 1974 and finally released the device in 1976. The 8–bit Z80 IC contains 40 pins and is an enhanced version of the earlier Intel 8080. A block layout of the Z80 microprocessor is given in Figure 3.1.

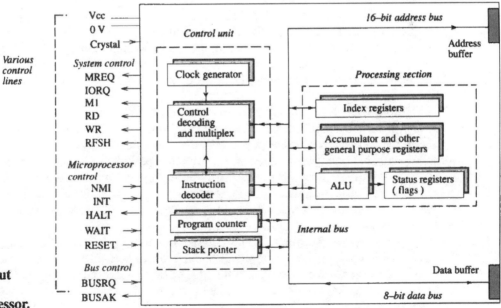

Figure 3.1

Block layout of the Z80 microprocessor.

The figure shows the internal registers. A register is a small very high speed memory store which is usually only large enough to contain a word or two of information, and functional units that are incorporated into the microprocessor. It can also be seen that the device has 16 address lines, 8 data lines, 13 control lines, 1 clock line and 2 power lines, a +5 Vcc and ground. The Z80A runs at 4 MHz while the Z80B runs at 6 MHz.

3.2 MICROPROCESSOR HARDWARE

The microprocessor is the most important part of any computer in that it is responsible for processing information and controlling every part of the computer system. Internally, the device is usually divided into two parts, the arithmetic logic unit (ALU) and the control unit (CU). The ALU is responsible for handling the computational and data transfer aspects of the microprocessor, while the CU is responsible for decoding and controlling the execution of the program instructions by sending out control signals to activate the appropriate parts of the computer at the proper time.

3.2.1　　CONTROL UNIT

All microprocessors have a control unit that controls the activity in the microprocessor dictating the correct sequencing of the operations of the device. The control unit does not process or store data. Rather, it fetches instructions from memory, interprets them and subsequently produces signals that act both internally on the other parts of the microprocessor and externally to the other circuits in the system. The control unit also communicates with input devices to initiate the transfer of instructions and data from memory; it also communicates with output devices to initiate the transfer of results to memory.

To execute a program the control unit must read in and obey each instruction from the program in the correct sequence, make any decisions required by the instruction, and then issue commands or control signals to the rest of the hardware within the computer to cause it to execute certain functions. The internal sequence of the control unit alternates between a read cycle, when the instruction is read in from the memory store, and an obey cycle when the order is decoded and obeyed. A microprocessor operating in this way is generally referred to as a two–beat machine. The control unit is typically made up from the following basic logic subsystems.

1. **Timing Circuits** – Microprocessors work on a synchronous mode of operation. This means there is a central clock which dictates when the different operations within the microprocessor can take place. A crystal–controlled oscillator plus a pulse squarer circuit, delay units and counters are used to generate the periodic timing reference pulses to all elements within the microprocessor. This allows the various activities of the system to be synchronized and coordinated. The regular timing signals from the central clock also govern transitions in the other ICs in the microprocessor system, essentially determining the speed of the system.

2. **Microprogram Memory** – The control unit usually has a special control memory (called a microprogrammed memory) which is used to help interpret the instructions in the program that the microprocessor is running. The microprogrammed memory is a read–only memory which is programmed by the manufacturer of the microprocessor who specifies the types of instructions that the microprocessor will understand. The program, called a microprogram, interprets each instruction read, and converts it into a series of operations to control the internal logic units in the microprocessor.

3. **Program Counter** – The program counter is the address register within the control unit that points to the next instruction that is to be executed. In this way it specifies where the next instruction is to be found in memory, i.e. its address. Under normal conditions the program counter is incremented after an instruction has been decoded, by the control unit. Instructions are then executed sequentially until the sequence is explicitly changed.

Any abrupt change in program direction requires a new address to be loaded into the program counter.

4. <u>Stack Pointer</u> – A stack pointer is a register that contains the address of a random access memory stack. The memory stack is a known area of free memory which the microprocessor can use for temporary storage of its internal registers. The memory stack is organized as a last in, first out (LIFO) data structure such that as items are put on top of the stack, the stack pointer is decremented accordingly. In this way that the stack pointer is always pointing to the last available item. As the most recently stacked information is retrieved, the stack pointer is also changed to point to the next available item on the stack.

Control unit

3.2.2 PROCESSING SECTION

The second main part of a microprocessor is the processing section, responsible for performing arithmetic and logical computations. It contains a number of general purpose and special purpose registers, an accumulator and an arithmetic logic unit.

1. <u>Registers</u> – Small blocks of registers are located inside the microprocessor to facilitate the execution of instructions, acting as temporary holding areas for both instructions and data. A register may have a specific function, or it can be a general purpose register which may be used by the arithmetic logic unit to store intermediate values during the calculation of a particular operation. The information in a register is rapidly accessible by the microprocessor since there are no external memory accesses required to fetch it.

The Z80 is a register–orientated processor with eighteen 8–bit and four 16–bit registers. Its general purpose 8–bit registers are grouped in pairs namely B and C, D and E, and H and L. Many of the instructions have been designed to use the contents of a register pair to point to the address of data in memory. An unusual feature of the Z80 is its alternative register set. Although only one set of registers can be active at any time, the register set can be easily swapped, this can be useful in facilitating multiple tasking, i.e. running two tasks and periodically swapping between them.

2. <u>Index Registers</u> – Index registers are general purpose registers. In the Z80 they are 16 bits wide. They can be useful for providing offsets for address calculations and in program counting. The main characteristic of index registers is that they can be incremented or decremented explicitly by an instruction, or sometimes implicitly whenever they are being referenced by an instruction.

3. <u>Arithmetic Logic Unit</u> – The ALU is the section of the processing unit logic that executes the standard set of arithmetic operations available with the particular microprocessor, for example addition, subtraction, multiplication and division using binary and binary coded decimal (BCD) numbers. Note that a particular processor may not have all the arithmetic functions listed above built into its hardware. The ALU is also designed to perform logical operations such as AND, OR, compare, shifts and rotate operations. The results of the operations from the ALU are usually placed in a register called the accumulator.

4. <u>Accumulator</u> – The accumulator, or A register, is the microprocessor's main register and takes part in all logical and arithmetic operations. It is used whenever numbers are manipulated or have to be temporarily stored, and for transfer to or from memory or I/O devices. It is normally used in conjunction with the ALU of the microprocessor. In the Z80 microprocessor it is eight bits wide.

5. <u>Flag Register</u> – The flag register contains a number of flags. These are 1–bit memory cells, which indicate specific conditions for 8– and 16–bit operations, e.g. to indicate whether or not the result of an arithmetic operation is equal to zero. Other flags are an overflow flag, carry flag and sign flag. For example, after a subtraction operation it is necessary to indicate whether the result is positive, negative or zero; this is done through the appropriate flags. The contents of the flags can be used by certain instructions to determine the proper direction of program flow.

3.2.3   INTERNAL BUSES

From Figure 3.1 it can be seen that there are internal buses with a microprocessor to connect the various hardware elements in the control unit and processing section together. Microprocessors also have a number of external connections to allow them to be connected to memory and interface ICs.

3.2.4   EXTERNAL BUSES

Microprocessors have three sets of communication lines called buses. The first group is called the data bus and is used to transfer data, the second is the address bus which is used to identify particular memory locations and the third group, the control bus, is used to produce timing and control signals.

1.   Data Bus – The data bus provides a path for moving data between the different units in a microprocessor system. It is a bidirectional bus in that data can move from the microprocessor to the memory and I/O devices, or can move from the memory and I/O devices to the microprocessor. The direction of data flow in the data bus is governed by additional signal lines included as part of the control bus. The width of the data bus may be 8, 16, 32 or 64 bits, depending on the type of microprocessor. Since each track in the data bus can carry only 1 bit at a time, the number of tracks determines how many bits can be transferred at once. For example, if the data bus is 8–bits, then the microprocessor can transfer 8 bits of data to and from the memory on each clock cycle. The data bus pins in the Z80 are labelled D0 to D7.

2.   Address Bus – The address bus is unidirectional, coming from the microprocessor to the memory and I/O subsystems. In this context an **address** is a digital number that may be seen as an identification code within a microprocessor system to allow one memory location or input/output port to be distinguished from another. It can then be used by the microprocessor to select a specific location in memory to read or write a piece of data or access a program instruction. For example, if the microprocessor wishes to read a word of data from memory, it puts the address of the desired word on the address bus and activates the read control signal. The correct location in memory is then accessed and the contents of the memory location are copied onto the data bus for transfer back to the microprocessor.

The number of bits (or address lines) used for addressing by the microprocessor determines the maximum number of memory resisters it can identify. Most popular 8–bit microprocessors have a 16–bit address bus that can address up to $2^{16} = 65\ 536$ memory locations. Other recent microprocessors have an address bus that is 20, 24, or even 32 bits wide. The relationship between the number of address lines and the maximum number of memory locations that can be accessed is shown in Table 3.2. The 16–bit address bus in the Z80 microprocessor (the pins are labelled A0–A15) is used for three purposes.

(a) It is used for addressing memory ICs, using the MREQ control signal. When used to address the memory the address bus contains a 16–bit address which allows any of 64 K different memory locations to be read from or written to.

Address bus width (bits)	Maximum address (words)
8	256
10	1 K = 1024
12	4 K
14	16 K
16	64 K
18	256 K
20	1 M = 1024 K
22	4 M
24	16 M
26	64 M
28	256 M
30	1 G = 1024 M
32	4 G

Table 3.2

Relationship between the address bus width and the maximum number of possible memory locations.

(b) It is used for addressing an I/O device (using the IORQ control signal). In the I/O mode, the address bus contains an 8–bit I/O address, the port number, that is used to address one of 256 different I/O devices. The I/O port number appears on the least significant eight address bits (A0 to A7).

Address bus

(c) It is used for addressing memory to refresh any dynamic random access memory (DRAM) in the system, using the RFSH control signal. During a refresh operation, the address bus contains a seven–bit refresh address on address pins A0 to A6.

3. <u>Control Bus</u> – The external control bus in a microprocessor is used to control the access to, and the use of, the data and address buses. This is required since the external data and address buses are shared by all components within a microprocessor system, hence there must be a means of controlling their use in order to provide an orderly transfer of information between the microprocessor and the memory or I/O peripherals. Data transfers may be synchronous with respect to a system clock, or asynchronous, relying solely on other control signals for timing reference. Control signals also transmit timing information between system modules to indicate the validity of data and address information.

The control signals available with the Z80 microprocessor are shown in Figure 3.1. It can be seen that there are a number of external pins, some of which are configured as outputs and some as inputs. Output control signals from the microprocessor are used to inform peripherals about the processor's impending actions while input control signals report the status of peripherals and pending data transfers. Table 3.3 notes the different control signals together with a brief description of their functions.

3.3 MICROPROCESSOR SOFTWARE

Microprocessor systems are constructed to process a sequence of instructions. The sequence is intended to implement an algorithm or procedure, a well defined set of behavioural rules, which has been devised by the system's programmers to handle a particular task. The sequence is stored in the microprocessor system memory in the form of

Control Line	Description
Power supply	The Z80 requires a supply current of approximately 200 mA with a voltage that must be plus or minus 5% for proper operation. If the supply voltage falls outside of this range, the microprocessor will not function properly. If the voltage exceeds 7 V the circuit will become damaged.
Clock pin	An external clock is fed into this pin. The clock is generated from a crystal–controlled oscillator with typical frequencies being 4 MHz or 6 MHz.
MREQ	The MREQ signal is used to access a memory IC. It indicates that the address bus contains a valid memory address for a memory read or write operation.
IORQ	The IORQ signal is used to access an I/O device. It indicates that the address bus contains a valid I/O port number for an IN or an OUT instruction.
M1	The M1 pin becomes a logic zero whenever the Z80 fetches an opcode from the memory. M1 also goes low during an interrupt acknowledge along with the IORQ pin.
RD & WR	The RD (read) and WR (write) control signals are used in conjunction with the MREQ and IORQ signals to enable the memory or I/O for read and write operations.
RFSH	The RFSH (refresh) pin is used to indicate that the lower seven bits of the address bus contain a refresh address from the internal R register.
NMI	The NMI pin signals an interrupt has occurred. An interrupt is a hardware signal that interrupts the normal execution of the program running by calling a special program (called an interrupt service subroutine). The NMI pin is non maskable, which means that it is always active and cannot be switched off.
INT	The INT pin is also used to signal an interrupt has occurred. The INT input is maskable, that is, it can be turned on and off by the EI (enable interrupt) and DI (disable interrupt) instructions.
HALT	The HALT pin becomes a logic 0 whenever the HALT instruction is executed. HALT remains at a logic 0 until the microprocessor is either interrupted or reset where it continues to execute instructions.
WAIT	The WAIT input is used to indicate that external memory or I/O is ready to transfer data. By controlling this pin, the Z80 can be made to wait for slow memory or I/O.
RESET	The RESET input is used to initialize the Z80 after power is applied and also to reinitialize it if a HALT occurs or some other catastrophic event.
BUSRQ	The BUSRQ pin is used by an external device to gain access to the Z80 systems address, data, and control buses.
BUSAK	The BUSAK signal is an indicator to the external device that the Z80 has switched off its control of its external buses (open circuited its buses).

Table 3.3

Control signals on the Z80 microprocessor.

Micro–processor software

a **program**, a list of organized instructions that control the behaviour of the system. The advantages of using a microprocessor system that is programmable, as opposed to a hardwired digital logic system, is that programs can readily be changed (either during manufacture or in the field) so that the same basic design can be made to perform many different functions. The programs that run on a computer system are collectively called **software**.

The number of unique instructions required to direct a microprocessor to process a program is referred to as the microprocessor's **instruction set**. It is the repertoire of general purpose instructions available with a particular microprocessor to which the microprocessor will produce a known response during the instruction fetch cycle. These instructions control the basic logical and arithmetic procedures of the microprocessor such as addition, subtraction and comparison, as well as allowing the microprocessor to control the different parts of its complete system. In general, different microprocessors have different instruction sets.

The instruction set for any microprocessor can be broken down into different instruction groups as shown in Table 3.4. The instructions within each group cause data to be moved, indicate which arithmetic and logic functions are to be performed, dictate the control of certain I/O devices, or make decisions as to which instruction is to be executed next. Table 3.4 also notes a number of example instruction types in each group.

Table 3.4

Instruction groups.

Data manipulation	Data movement
Add	Load/input
Subtract	Store/output
Logical	Indexing
Shift/rotate	Stack operation
Bit arithmetic	
Program manipulation	**Program status manipulation**
Call subroutine	Carry and overflow control
Branch operations	Zero and negative status
Program jumps	Test and branch
Interruption	Stack overflow control
	Interrupt enable

Depending on whether a microprocessor is very complex or relatively simple in design will determine whether there are many or few instructions in the instruction set. It can be noted for example that the first operational stored program computer, the Manchester Mark 1 (1948), had a very small instruction set, with only six instructions: jump, load accumulator negative, subtract, store accumulator, test for zero, and stop. The other first–generation devices, e.g. UNIVAC 1 and the IBM 701, also had small and simple instruction sets. Modern complex instruction set microprocessors (CISMs) on the other hand may have instruction sets of over 200 instructions.

*Micro–
processor
software*

The Z80 has an instruction set which is compatible with all 78 of the 8080A instructions. There are also block transfer instructions for moving an entire block of data and instructions to and from memory. In total there is provision for 694 different instructions, variations of 158 basic types.

Each instruction has two parts: the **Operation code** (opcode) which is that part of a microprocessor's binary instruction word that tells the control unit what operation or function is to be performed, and the **Operand** which is the part of a microprocessor's instruction which tells the control unit the data or location of the data to be operated on. It is sometimes also used to refer to the quantity that normally results from the execution of a microprocessor instruction.

In dealing with each instruction the microprocessor goes through a cycle of events: fetching, decoding and executing the instruction. This is known as the **instruction cycle**.

1.   Instruction Fetch – To fetch an instruction, the microprocessor places a memory address on the address bus and reads binary information using the data bus. Therefore, it needs a register that can hold memory addresses and increment these addresses after the fetching, is completed, this is the program counter.

2.   Instruction Decode – Once an instruction byte has been fetched, the control unit must decide whether the instruction complete. If not, how many more bytes need to be fetched, what type of operation is required and on what data? To perform these functions, the control unit needs an instruction decoder that can interpret the fetched binary information.

Micro-processor software

3.   Instruction Execute – The execution of the instruction is performed in the arithmetic logic unit.

This sequence of events will normally require several clock or machine cycles to execute it, depending on the type of instruction. The instruction cycle is shown in Figure 3.2, where the instruction fetch part has been expanded to show the individual operations within that part of the cycle. The instruction fetch is the portion of a computer's instruction cycle concerned with addressing the memory and then reading into the processor the information stored at that location. Each of the other parts of the cycle has a number of different internal microprocessor operations associated with it.

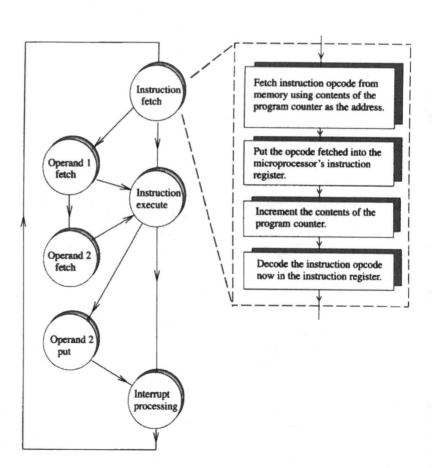

Figure 3.2

The instruction cycle.

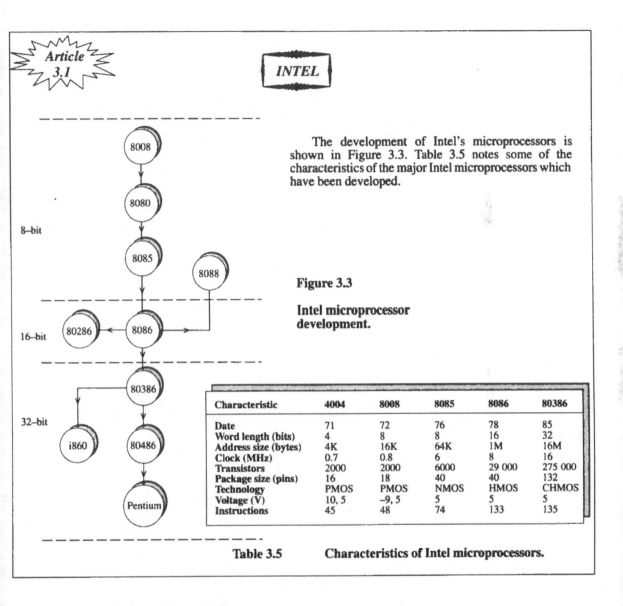

The development of Intel's microprocessors is shown in Figure 3.3. Table 3.5 notes some of the characteristics of the major Intel microprocessors which have been developed.

Figure 3.3

Intel microprocessor development.

Characteristic	4004	8008	8085	8086	80386
Date	71	72	76	78	85
Word length (bits)	4	8	8	16	32
Address size (bytes)	4K	16K	64K	1M	16M
Clock (MHz)	0.7	0.8	6	8	16
Transistors	2000	2000	6000	29 000	275 000
Package size (pins)	16	18	40	40	132
Technology	PMOS	PMOS	NMOS	HMOS	CHMOS
Voltage (V)	10, 5	−9, 5	5	5	5
Instructions	45	48	74	133	135

Table 3.5   Characteristics of Intel microprocessors.

3.4 MICROPROCESSOR DEVELOPMENT

The microprocessor development curve shown in Figure 3.5 details the growth of transistor density over time, it also indicates the introduction date of some of the more popular microprocessors

3.4.1   MICROPROCESSOR (4–bit)

In the early 1970s the combination of IC technology and ever increasing need for more processing power led to the development of the 'computer on a chip' or microprocessor. While considering the design of a calculator chip set, Ted Hoff of Intel proposed that instead of using a dedicated set of ICs to perform a given function it would be better to design a general purpose programmable IC which would perform the calculator functions as well as

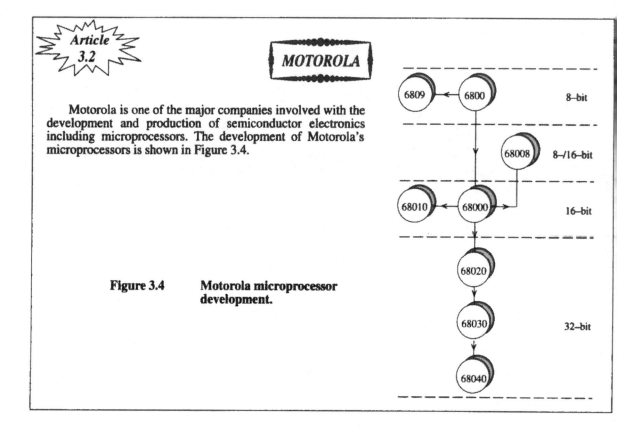

Article 3.2

MOTOROLA

Motorola is one of the major companies involved with the development and production of semiconductor electronics including microprocessors. The development of Motorola's microprocessors is shown in Figure 3.4.

Figure 3.4   Motorola microprocessor development.

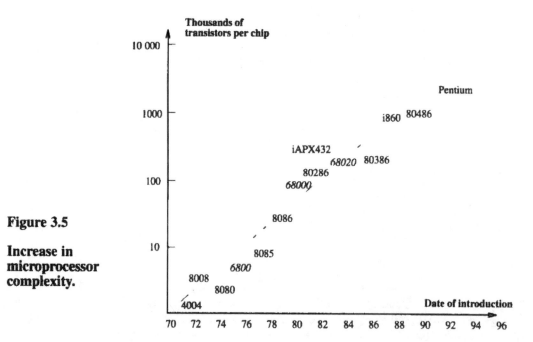

Figure 3.5

Increase in microprocessor complexity.

many other logical functions. In 1971 the **Intel 4004** single chip microprocessor was produced. The 4004 which was part of a four–chip set, broke new ground in the era of large scale integrated (LSI) ICs and was the first real attempt to use the technology of ICs to reduce the requirements of a computer to a single IC (Article 3.1).

The group of four LSI devices that made up the simple complete computer were known as the MCS–4 microcomputer set.

1. The 4004 central processing unit IC, contained the arithmetic logic unit and control unit on a single chip. It processed 4–bit data quantities.

2. The 4001 was a read–only memory which contained 256 8–bit words of memory and a 4–bit input/output port. The port provided the opening or gateway for the transfer of data between the microcomputer and the external world and so allowed the MCS–4 microcomputer set to communicate with circuits and devices that were not a part of the microcomputer set.

3. The 4002 IC was a random access memory.

*Micro–
processor
(4–bit)*

4. The fourth chip in the set was the 4003–SR shift register IC. This was a 10–bit serial input and either parallel output or serial output shift register which also provided additional I/O for the system.

The 4004 microprocessor executed 60 000 instructions per second and, although it never appeared in a commercial desktop machine, its descendants appeared in the first microcomputers. It is worth noting that the term microprocessor usually refers to the main CPU chip; however, in order to construct a microcomputer it is essential to have other IC chips available, as well as the attendant interconnection hardware and power supplies.

In 1972 Rockwell announced the PPS–4, similar to the MCS–4 but packed in 42–pin packages, while by 1972 Gary Boone and Michael Cochran of Texas Instruments had produced the TMS 1000 which, like the 4004, was a 4–bit microprocessor. The difference was that the TMS 1000 incorporated support functions directly on the chip, earning its title the first 'computer on a chip'. Shortly after the development of the 4004, Intel introduced the Intel 8008 an 8–bit microprocessor device.

3.4.2   MICROPROCESSOR (8–bit)

The second generation of microprocessors appeared in 1972 with the introduction of the Intel 8008 8–bit microprocessor. Acceptance of the microprocessor really came with the incorporation of 8–bit devices into general purpose computing systems. Compared to the 4–bit devices the second–generation microprocessors had enhanced features, shown in Table 3.6.

Table 3.6

Features of 8–bit microprocessors compared to 4–bit devices.

1. The 8–bit microprocessors were such that their basic electronic components were of smaller size densities of 4 K–5 K transistors/chip, were faster, with gate delays of about 15 ns, and dissipated less power than the initial 4–bit devices.

2. They had the ability to address larger memory spaces (64 KB).

3. They had more I/O pins, meaning that they could have separate buses for the data and the addresses. Therefore, multiplexing between data and address was not required and most of the support circuits were eliminated, thus reducing the number of IC packages required for a complete system.

4. They had more powerful instruction sets.

5. They were able to reduce the power supply requirements to just a single +5V supply.

6. They also placed the clock generator on the same microprocessor chip.

MICROPROCESSOR DEVICES (8–BIT)

MOTOROLA 6800 – Motorola's 6800 was one of the early devices that was widely adopted in control systems. It was launched in 1974, six months after the Intel 8080 device. The 6800's architecture has been used as the basis for whole families of powerful 8–bit microcomputer ICs.

MOTOROLA 6809 – The 6800 was later superseded by the 6809 (released in 1978) which contained additional registers, an expanded 6800 instruction set (59 basic instructions) that incorporated a number of 16–bit instructions and more addressing modes while maintaining compatibility with the 6800 device. The 6809 however failed to make a large impact in the 8–bit market due to its late introduction.

MOS 6502 – The 6502 was developed by MOS Technologies as an improved 6800 device. In 1975 MOS technologies announced the 6501 and 6502 microprocessors which cost only $20 and $25, respectively. The device was brought to tremendous popularity by its use in several generations of desktop computers, notably the Commodore PET and the Apple machines. The processor uses the same bus structure and is broadly compatible with the same range of peripheral devices as the 6800.

Table 3.7 notes a number of the 8–bit processors together with some of their features. It can be seen that all the processors used 40–pin ICs with 5 V power supply; in addition, the 8080 required a –5 V and 12 V power supply. Article 3.3 gives more information on 8–bit microprocessors.

Table 3.7

Features of a number of 8–bit microprocessors.

Property	MOS Tech 6502	Motorola 6800	Intel 8080	Intel 8085	Zilog Z80
DIL package (pins)	40	40	40	40	40
Max power (watts)	0.8	0.5	0.8	0.8	0.85
Voltage (volts)	5	5	+5, –5, 12	5	5
Address bus and data bus multiplexed	No	No	No	Yes	No
Max clock rate (MHz)	3	2	3	5	4

Micro–processor (8–bit)

Intel released the 8080 in 1974. It proved to be extremely popular and was second sourced by many manufacturers including Texas Instruments and NEC. The instruction set and programming architecture of the original 8080 has been perpetuated in later products resulting in a huge wealth of knowledge about programming this type of device.

For those applications requiring low power consumption, CMOS devices were also developed, requiring power of the order of several hundred micro watts per chip compared to several hundred milliwatts per chip for NMOS devices.

3.4.3   MICROPROCESSOR (16–bit)

The third generation of microprocessors can be regarded as being initiated in 1976 when Texas Instruments announced the TMS9000, the first 16–bit microprocessor. Other initial devices were the National Semiconductor PACE, GIM CP1600, and the Texas Instruments TMS 9900. The 16–bit microprocessors not only represented a significantly more complex breed of device than the previous generation of 8–bit devices but they also achieved a much higher performance than their 8–bit predecessors through extra features and higher clock frequencies (from 4 to 10 MHz).

The 16–bit devices attempted to rectify other criticisms of the 8–bit processors, these extra features are noted in Table 3.8.

Table 3.8

Features of 16–bit microprocessors compared to 8–bit devices.

1.	An increased number of registers to provide the programmer with more internal memory stacks. For example, four registers in a general purpose CPU will give a much better performance than if only two registers are included, provided the compiler can take advantage of the additional registers.
2.	Multiply and divide arithmetic in hardware.
3.	Microprogrammed control sections.
4.	An increased memory accessing capability to allow them to handle very large physical memory spaces, from 1 to 16 MB. They also had a variety of flexible and powerful addressing modes.
5.	More powerful interrupt handling capabilities, able to deal with both external multi–level vectored interrupts and internal interrupts, called software traps.
6.	Privileged instructions with both user and supervisor modes of operation.
7.	Instruction look ahead to increase the system throughput.
8.	Support for multiple processor configurations.

It had also been noted by the microprocessor manufacturers that in the construction of a microprocessor system, since the partitioning of the design onto multiple chips had to adhere to limits on package pins and the signalling speed between the chips, it was preferable to consolidate an entire instruction processor onto a single chip together with as much of the lowest levels of the storage such as registers, instruction caches and data caches as possible. The effect of microprocessor chips being designed with localized communication between the different internal functional parts means that the number of signals which have to pass out through the IC package pins can be reduced. On–chip storage, for example, is effective in reducing the frequency of relatively slow off–chip storage references. It can also be noted that as more and more on–chip features have been provided in the microprocessor the distinction between microprocessors and single chip microcomputers has become blurred. Table 3.9 indicates the features of three 16–bit microprocessors.

Table 3.9

Characteristics of three 16–bit microprocessors.

Characteristics	Intel 8086	Zilog Z8000	Motorola 68000
Internal data bus (bits)	16	16	32
External data bus (bits)	16	16	16
Memory space (MB)	1	8	16
I/O space separated from memory space (KB)	64	64	none
Instruction queue	yes	yes	no
Number of instruction types	111	110	56
8–bit version of the microprocessor	8088	—	68008

In June 1978, Intel released the 16–bit 8086 microprocessor which was a 40–pin (5 MHz) device containing of the order of 29 000 transistors. It was manufactured in three–micron technology. Figure 3.6 shows the internal registers of the device. It has a 16–bit internal data bus, an 8–bit external data path and can address up to 1 MB of external memory, with the additional feature that the bytes can be paired to form 16–bit words.

The instruction set for the device includes most of the earlier 8–bit 8080 instruction set but with extra facilities to increase the power of the chip. It operates from a single 5 V power supply. The internal logic of the 8086 is shown in Figure 3.7.

Intel also launched the 8088 microprocessor in 1978. It is a low–end version of the 8086 which only has an 8–bit internal data bus although it still has a 20–bit address bus and an instruction set which consists of about 100 instructions. The original IBM personal computer, the IBM PC (released in 1981) used an 8088 running at 4.77 MHz; however, it is possible to get 8088s with up to 8 MHz clock speeds and 8086s with clock speeds of 16 MHz.

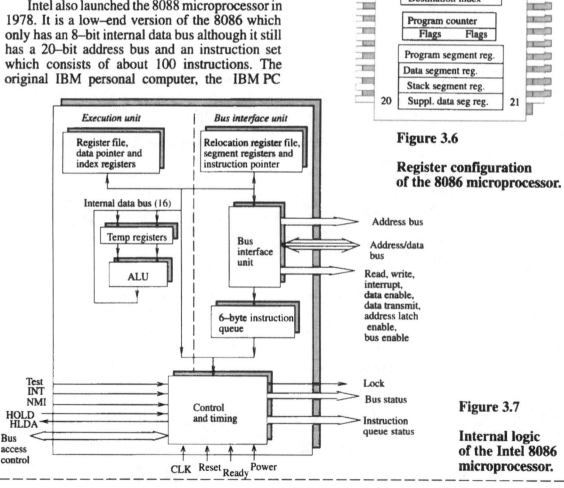

Figure 3.6

Register configuration of the 8086 microprocessor.

Figure 3.7

Internal logic of the Intel 8086 microprocessor.

MICROPROCESSOR DEVICES (16–BIT)

MOTOROLA 68000 – The Motorola 68000 (released in 1979) is an advanced 16–bit microprocessor with internal 32–bit bus. It contains 68 000 transistors etched on a die measuring 0.246 by 0.261 inches. It is packaged in a number of forms including a 64–pin dual in line package. Internally the 68000 contains eight 32–bit data registers, seven 32–bit address registers, two 32–bit stack pointers and one 32–bit program counter, uses a 24–bit address bus and offers a 16 MB direct addressing range. It runs at 8 or 12 MHz and has 56 basic instructions.

MOTOROLA 68010 – The MC 68010 is a superset of the 68000 with virtual memory and machine support built in. It appeared in 1983.

MOTOROLA 68008 – The MC 68008 is an 8- bit external bus version of the MC68000 in a 40–pin package. Its address bus was limited to 20 pins giving it a 1 MB linear address space. It provided 60% the performance of the 68000 at the same clock speed and was aimed at low cost systems. It was used in the Sinclair QL home computer.

INTEL 80186 – In 1981 Intel released the 80186 microprocessor. The architecture of the 80186 is equivalent to the 8086 except it has several added sections.
1. The DMA controllers allow block data transfer between the memory and the I/O spaces with a maximum data rate of 2 MB. Each DMA channel is configured using a channel control word which specifies the type of transfer, its internal logic, the number of bytes to be transferred, synchronization, address space selection and several other features.
2. The interrupt controller can be switched to allow the user to assign internal and external requests on a priority basis ,where the internal interrupts may be masked using the internal device control registers. Alternatively the device can be configured to act as a slave controller to an external master priority interrupt controller.
3. The programmable timers consists of two external timers/event counters and a third internal timer .
Like the 8086 the 80186 can run at 16 MHz, but has a performance which is about 25% faster then the 8086. It also has ten extra instructions in its instruction set. The 80188 is a low end version of the 80186 with an 8–bit internal data bus.

INTEL 80286 – The 80286 microprocessor from Intel (released in 1983) was manufactured in 2.5 micron technology and has 134 000 transistors. It is a sophisticated processing unit which retains compatibility with the 8086 but which also offers on chip memory management and virtual memory abilities which can facilitate multitasking, real time operations, operating system support and features to aid in high level language implementation.

The 80286 supports a 16 MB physical address space (24–bits) and the memory management facilities can provide up to 1000 MB of virtual address space. Operating system support is provided in the form of a sophisticated interrupt structure which supports two classes of context switch, one which transfers control to service routines which are in the address space of the interrupted task, while the other provides a high speed switch from the interrupted task to a special isolated interrupt service task without involving the operating system. The device had a number of different clock speeds ranging from 6 to 20 MHz. The 80286 (12 MHz) processor was chosen by IBM for the original IBM AT computer. High performance versions of the chip can run with a 25 MHz clock. The 80286 has a performance improvement of about 2.5 times compared to the 8086 for the same clock speed.

ZILOG Z8000 – The successor to the Z80 was the Z8000 which was released in 1979. It has a 32–bit address bus with on–board cache memories (256–bit data + 256–bit instruction) and a memory management unit. The Z8000 has 110 instructions and can run with either a 10, 18 or 25 MHz clock giving a maximum performance of 5 MIPS with a 2 W power consumption. The device has 64 four pins and is manufactured in 2–micron NMOS.

Micro–processor (16–bit)

The three most widely used 16–bit processors in the late 1970s and early 1980s were the Zilog Z8000, Motorola 68000 and the Intel 8086. Zilog's Z8000 found its main application in the military and industrial embedded systems, while both the Motorola 68000 family and the Intel 8086 found their way into general purpose computing systems. However, the 8086 had something of a head start and the future of this family was assured when IBM adopted the 8088 in its personal computer. Highlight 3.1 and Article 3.4 give more information on 16–bit microprocessors.

3.4.4   MICROPROCESSOR (32–bit)

The fourth generation of microprocessors was heralded in with the first 32–bit microprocessor, the Bellmac–32, which was produced by Bell Laboratories in 1979. However, it was never released commercially.

In 1981 Intel released the 32–bit Intel APX 432 processing system which incorporated over 200 000 transistors on three chips. The general data processor consisted of two chips: the iAPX 43201 with 110 000 transistors, responsible for instruction decoding, and the iAPX 43202 with 49 000 transistors which was responsible for performing the instruction execution. The third chip in the set was the iAPX 43203 I/O interface processor containing 60 000 transistors, designed to map I/O bus addresses into the main memory address space and also provide attached I/O processors with a set of interprocess communications mechanisms.

Other 32–bit processors introduced during the early 1980s were the Hewlett Packard Focus microprocessor which achieved a density of 450 000 transistors in 1981, the National 32032 microprocessor which was introduced in 1983–84, and the Intel 80386 introduced at the end of 1985.

Again compared to 16–bit devices, the present 32–bit microprocessors have more improved capabilities. These are noted in Table 3.10. All 32–bit microprocessors have to use a combination of NMOS and CMOS semiconductor technologies to give a low power dissipation and high speed.

Table 3.10

Features of 32–bit microprocessors compared to 16–bit devices.

1.	Increased instruction sets with more powerful instructions.
2.	Extra on–chip registers.
3.	32–bit arithmetic; a 32–bit adder will allow a 32–bit processor to operate at a higher speed than a 16–bit adder used twice for each addition.
4.	More addressing mode capabilities.
5.	Advanced architectural features such as data and instruction cache memories.
6.	Multistage pipelining. This allows the 32–bit machines to overlap the fetching, decoding and execution portions of a number of instructions simultaneously. Overlapping instruction execution with the fetching of the next instruction reduces the execution time of programs.
7.	Memory management hardware. The inclusion of this onto the devices means that they can handle a multiprogrammed environment by providing mechanisms for memory space protection among concurrently running tasks. The memory management's hardware can also handle program relocation and therefore permit position–independent coding and dynamically relocatable programs.
8.	Hardware and software features to support multiprocessor system organizations and distributed network configurations.

Highlight 3.2 and Article 3.5 give more information on 32–bit microprocessors.

EXAMPLE OF A 32–BIT PROCESSOR: MOTOROLA 68040

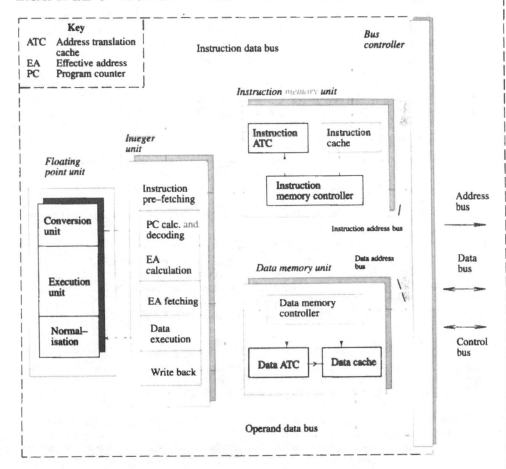

Key

ATC   Address translation
           cache
EA   Effective address
PC   Program counter

Instruction data bus

*Bus
controller*

Instruction memory unit

Instruction
ATC

Instruction
cache

*Integer
unit*

*Floating
point unit*

Instruction
pre–fetching

Instruction
memory controller

Address
bus

PC calc. and
decoding

Instruction address bus

Conversion
unit

EA
calculation

Data address
bus

Data memory unit

Data
bus

Execution
unit

EA fetching

Data memory
controller

Normal–
isation

Data
execution

Data ATC

Data cache

Control
bus

Write back

Operand data bus

Figure 3.8   **Block diagram of the Motorola 68040 microprocessor.**

   In 1990 Motorola announced the 68040, a 32–bit processor with 1.2 million transistors, almost four
times the number of the previous 68030. A block diagram of the device is shown in Figure 3.8. It is
manufactured in 0.8–micron CMOS technology. It has an integral memory management unit, floating
point unit and 4 KB cache memories for instructions and data. Both caches operate independently and
can be accessed at the same time. It uses a 25 MHz clock and comes in a 179–pin grid–array. It has an
internal Harvard architecture which gives the processor full access to both instructions and data.
   The 68040 increased the performance of the 68030 by three to ten times. It has a performance
rating of 15 to 20 MIPS and 35 MFLOPS.

MICROPROCESSOR DEVICES (32–BIT)

INTEL 80386 – The Intel 80386 microprocessor was released in 1986 with 275 000 transistors and an initial performance rating of 5 million instructions per second (MIPS). The 30386 DX device has a 32–bit data bus, a 32–bit address bus and runs with clock speeds up to 40 MHz clock. It is a true 32–bit processor providing an integrated and compatible upgrade path from the 16–bit 80286 (it is basically a 32–bit extension of the 80286 architecture which is also code compatible with the 8086). The device is capable of manipulating 32–bits of data through an address range of 4 GB (2^{32} bytes) of physical memory space. On–chip features include 32 registers and a sophisticated memory management unit which supports segmentation with up to a maximum of 4 GB per segment (compared to only 64 KB with the 80286). In a multitasking system each task has a separate segment descriptor table and task to task isolation is enforced by hardware. Extensive use of pipelining techniques within the device allow several operations to be performed at once. The co–processor interface on the device supports both the 80287 and 80387 floating point co–processors and the 82786 graphics co–processor which converts high level graphics commands into the necessary pixel updates. The 80386 processor was introduced into PCs in September 1986 with the Compaq Deskpro 386.

During 1988 an economy version of the 386 appeared called the 386SX. It has the same external connections as a 286 but with the internal architecture of the 386. This allowed 286 processor boards to be used with minimal alterations. The 386SX has a 16–bit data bus (32–bit internal), a 32–bit address bus and runs with a variety of clock speeds up to 33 MHz.

INTEL 80486 – The Intel 80486 (released in 1989) is a 32–bit processor which runs at 25 MHz. The Intel i486 CPU uses one–micron CMOS technology with 1.2 million transistors on a chip surface of about 1.7 cm^2. The 486 device contains 168 pins. The maximum current consumption of the device is 3.8 A.

The 486 is an enhanced version of the 80386 and contains four major on board units.
1. A 386 compatible processor unit with pipelined instruction execution.
2. An 80387 compatible arithmetic unit for performing floating point operations.
3. An 82385 cache controller.
4. Two 4 KB cache memory.

The device also has 342 different instructions. The i486 at a clock speed of 25MHz has a processing speed of 15 to 20 MIPS and essentially runs up to 50% faster than the 80386 at the same clock speed. The 486/40 runs at approximately twice the performance of a 386/33. Newer versions of this processor have a performance of 40 MIPS. The 486DX2 device runs at 66 MHz and has 238 pins.

INTEL i860 – The i860 microprocessor from Intel is a 32–bit CMOS microprocessor measuring 488 square mils and contains more than 1 million transistors. The i860 consists of nine units: core execution, floating point control, floating point adder, floating multiplier, graphics, paging, instruction cache, data cache, and a bus and cache controller. One third of the logic on the i860 is devoted to floating point calculations. In addition to scaler operations the device can perform vector operations using the on–chip data cache memory.

The architecture is of the Harvard type with separate instruction and data caches. Instructions feed out of a 4 KB (64–bit) instruction cache memory and can drive both the central processing unit and the floating point unit simultaneously through independent 32–bit instruction buses. Data feeds out of an 8 KB (128–bit wide) data cache memory and can drive two long real arguments at a time at the adder and the multiplier, or graphics unit. The processor also has 32 integer registers and 32 floating point registers, each 32–bits wide. The adder and multiplier are both constructed as three–stage pipelines and the two units can hook together in a number of ways.

The newer i860XP device uses 2.55 million transistors with a 50 MHz clock. It has a 32 by 32–bit register file, a 16 KB instruction cache and a 16 KB data cache. It also has a built in graphics functional unit for pixel and z–buffer manipulation.

The performance results for the i860 indicate that at 40 MHz it can attain:
1. 80 million floating point operations per second (MFLOPS) for single–precision floating point operations.
2. 60 MFLOPS for double precision operations.
3. 80 MFLOPS for matrix multiplication.
4. 67 MFLOPS for fast Fourier transforms (FFTs).
5. 10 times the speed of a 486 and twice the speed of a Cray 1F supercomputer.

(Continued)

(continued)

INTEL i860 (continued)

The problems with superscaler processors (i.e. can operate on more than one instruction at a time) such as the i860, involves identifying which instructions are independent and which must be executed in a particular sequence. The i860 therefore requires very sophisticated high level language compilers to utilize the hardware efficiently.

MOTOROLA 68020 – The Motorola MC68020 was launched in April 1984 and contains 200 000 transistors fabricated in 2.25–micron HCMOS technology and packaged in a 114–pin grid array. The 32–bit 68020 microprocessor uses a full 32–bit nonmultiplexed data bus and includes, on–chip, a high speed instruction cache memory and a bus control unit. The 68020 has 65 instructions in its instruction set and, as it has been designed to be object–code compatible with other 68000 series processors, upgrading to higher performance processors still preserves earlier investments in software developments. The device is rated at eight MIPS peak and two to three MIPS sustained when running with a 16 MHz clock.

The Motorola 68020 has a number of functional units.

1. The instruction prefetch controller loads instructions from the data bus into the decode unit and the instruction cache.
2. The sequencer control unit provides overall chip control managing the internal buses, registers and the functions of the execution unit.
3. A number of register blocks including 16 32–bit general purpose registers, a 32–bit program counter, a 16–bit status register, a 32–bit vector base register, two 32–bit alternate function code registers and a 32–bit cache address register.
4. An internal 256 byte cache holds approximately 100 instructions and is organized as a 64 by 4–byte array.

The asynchronous external bus is nonmultiplexed with a 32–bit address bus and data bus. The MC68020 can run at clock rates up to 24 MHz with a power consumption of 1.5 W. The Motorola MC68020 requires a clock generator, MC68881 floating point coprocessor, interrupt controller and an MC68851 memory management unit to complete the microcomputer functions.

MOTOROLA 68030 – The MC68030 is an upgraded version of the 68020 which has 300 000 transistors and several features additional to the 68020. It was released in 1987 and essentially combines the 68020 and the 68881 floating point co–processor into the one chip. The device has separate 256–byte data and instruction caches, each with its own path to the execution unit. Through its Harvard style architecture, the 68030 can use its three bus architecture, it has a separate path from its external bus to the execution unit, to fetch an instruction and two data operands simultaneously. The latter technique improves the data processing rate of the device. It can run with clock speeds up to 50 MHz.

3.4.5 MICROPROCESSOR (64–bit)

The fifth generation of microprocessors has seen the introduction in 1992 of 64–bit devices. The first of the true 64–bit micros are the MIPS R4000 reduced instruction set device and the DEC Alpha chip.

It is claimed that 64–bit arithmetic should allow programs which would normally have used 32–bit arithmetic to run twice as fast. The development of 64–bit microprocessors should also allow the extension of memory addressing above 32–bits. At present with a 32–bit address bus a microprocessor can address up to 4 GB (2^{32} bytes). With the increasing sophistication of microprocessor applications this can be limiting, for example in graphics manipulation. It is envisaged that 64–bit microprocessors will be in widespread use by 1995. Article 3.6 gives more information on 64–bit microprocessors.

MICROPROCESSOR DEVICES (64–BIT)

DEC ALPHA – The DEC Alpha device uses either a 150 MHz or a 200 MHz clock and crams 1.689 million transistors into a 1.4 cm by 1.7 cm die and uses a 431 pin grid array IC package. It is manufactured in 0.75–micron CMOS, dissipates 23 W of power and claims a performance rating of 400 MIPS with a 6.65 ns clock (150 MHz).

INTEL PENTIUM – The Pentium device was released in 1993. Table 3.11 notes the evolution of the device in terms of previous devices. The Intel Pentium integrates 3.1 million transistors on a chip using 0.8–micron CMOS IC technology, with a claimed performance of over 100 million instructions per second.

Processor	Release date	Performance (MIPS)	Number of transistors (millions)
8086	June 1978	0.33	0.029
80286	Feb 1982	1.2	0.134
80386	Oct 1985	5	0.275
80486	Aug 1989	20	1.2
Pentium	March 1993	112	3.1

Table 3.11

Evolution of the Intel microprocessors.

The Pentium is clocked at 66 MHz and uses a 2.16 inch square 273 pin grid array package, at 66 MHz the device draws about 13 watts. The Pentium contains two 486–like processors and is such that the chip automatically divides the work between them to keep them fully occupied. It has two 8 KB on–chip cache memories for the instructions and the data. It also has two integer pipelines using five–stage pipelining. The floating point unit is also pipelined with dedicated addition, division and multiplication circuitry.

POWER PC – The 64–bit Power PC 601 microprocessor is the first of a range of processors which are being developed by IBM, Motorola and Apple. It is designed in 0.65–micron CMOS and contains a 32 KB cache memory and floating point unit and runs at 66 MHz. Although the original 601 only has a 32–bit data bus the later MPC620 has a 64–bit data bus. The 601 can execute three instructions per clock cycle. It contains 2.8 million transistors on a die only 11 mm^2. It is a superscaler design that processes multiple instructions per clock cycle. It also contains multiprocessor support with on–chip logic to maintain cache coherency.

MOTOROLA 68060 – There is no 68050 but instead Motorola decided to introduce the MC68060 (released 1994). The Motorola 68060 is a superscaler design (i.e. uses multiple execution units) with dual on–chip cache memories, fully independent demand–page memory management units for both instruction and data, dual integer execution pipelines, on–chip floating point unit and a branch cache. It achieves a high degree of instruction–execution parallelism by using independent execution units and dual instruction issue within the instruction execution controller. It is implemented with 2 million transistors in 0.5 micron triple layer metal technology. 50 and 66 MHz versions are available. It is claimed that the 68060 has three times the performance of the 25MHz 68040 using existing compilers.

3.5 CO–PROCESSOR

A co–processor or slave processing unit is a separate microprocessor which can be added to a computer system in order to provide an enhanced performance from the combination of the standard micro plus its co–processor. Intel was the first to introduce this idea with the 8089 I/O microprocessor and the 8087 numeric microprocessor to interface to the 8086 device. Both the 8089 and 8087 devices are complex integrated circuits which perform their allotted function without any form of bus control structure, relying instead on the associated

CO–PROCESSOR

MOTOROLA 68881/68882 – The 68881 floating point co–processor provides hardware support for calculations using the full IEEE floating point standard (P754 Rev 10) and transcendental functions (sin, cos etc.). It contains 155 000 transistors and is rated at 0.4 MFLOPS. It also includes eight 80–bit floating point data registers, a microcode processor with 67–bit ALU. The co–processor interface is memory mapped into the 68020s address space and the co–processor is controlled by a sequence of writes and reads of these co–processor interface registers. All calculations are performed to 80–bit precision. It performs an addition in 2.2 μs, a multiply in 3.2 μs and a divide in 4.6 μs at a co–processor speed of 25 MHz. The MC68882 is an upgrade to the 68881 running at 33 Hz to 40MHz and was designed to support the MC68030. It can run up to twice as fast as the 68881.

microprocessor for the bus control. The advantage of using a co–processor is that it will execute its instruction in parallel with the main central processor unit operation and therefore can be arranged to relieve the CPU of some of its loading. This has the effect of increasing the throughput and processing speed of a microprocessor system.

Features available within co–processing chips are, for example,

1. floating point hardware
2. independent high speed I/O channels
3. memory management and virtual memory support

Co–processor　4. silicon operating system.

The 8087 co–processor implements the IEEE floating point standard for floating point arithmetic. Typical times for the 8087 gives a floating point addition at 17 μs and a square root operation at 36 μs. The 8087 can run at speeds up to 10 MHz. Intel also manufactures the 80287, a co–processor specifically designed to be attached to the 80286 microprocessor, thereby enhancing the performance of a 80286 system. The floating point co–processor has 68 instructions and can use 64–bit precision. The 80287 can also run at speeds up to 10 MHz. The Intel 8089 I/O processor can handle fast input/output operations, independent of the main CPU. The co–processor usually organizes the I/O operation into streams or channels so that the I/O devices are transparent to the CPU. The device has two I/O channels which can be configured as direct memory access pathways (also see Article 3.7).

3.6 MICROCONTROLLER

Several manufactures have developed single–chip microcomputers based on 8–bit architectures which predominantly differ from the microprocessor chip in the provision of on–chip memory with both random access memory (RAM) and read–only memory (ROM), I/O circuitry, display driving capability and analog to digital circuitry. Examples of the more modern microcomputer chips are the Intel 8051, Motorola 68HC11, NEC 7811, Hitachi 647180, National COP 800 and the Zilog Z8.

One of the most popular of the microcontrollers is the 8048 /8051 series of devices from Intel. The 8041 family has a split memory architecture with 1 to 4 KB of program ROM or EPROM (erasable programmable ROM) on–chip and 64 to 256 bytes of on–chip RAM. I/O within the devices has its own memory space and instructions can operate directly on the I/O ports. Family members execute their one to two cycle instructions at cycle times ranging from 1.36 to 15 μs. The devices were built using NMOS 5 V technology in a 40–pin DIP and 44–pin chip carrier.

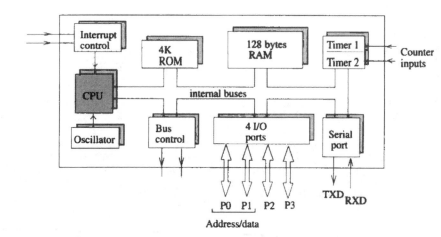

Figure 3.9

**Block diagram
of the 8051
microcomputer IC.**

*Micro–
controller*

The 8051/8052 series of devices are expandable single chip controllers, which are essentially enhanced versions of the same supplier's widely used 8048 family. A block diagram of the Intel 8051 microcomputer integrated circuit is given in Figure 3.9. It contains parallel and serial I/O, counter/timers, RAM and ROM. The different members of the family have a variety of combinations of ROM and EPROM: the 8031, for example, contains no ROM, the 8751 contains EPROM, while the 8051 contain 4 KB of ROM.

Most 8051 instructions are one or two bytes long, with all instructions except the multiply and divide executing in 1 or 2 instruction cycles (an instruction cycle is 1 μs). The multiply or divide instruction requires 4 μs.

3.7 CUSTOM LOGIC DEVICES

A microprocessor is a programmable device in that it executes a program from memory to perform its desired function. With advances in semiconductor technology it is now possible to get an extensive variety of large scale integrated programmable peripherals and control ICs which can be hardwire programmed, or indeed specially constructed, to perform particular functions. They can be used to implement unavailable hardware functions in an electronic system, from minor logic operations to complex special functions. In a microprocessor system, they are often used to replace the glue chips that are required to implement specific logic functions. In the past such things as buffers, decoders and latches have required a series of 14– and 16–pin small scale integrated (SSI) chips. With the use of custom ICs these can be replaced with a few special purpose devices.

Custom logic devices can be either of the following:

Programmable Logic Device – This is programmed through internal connections between a series of logic gates.

Application Specific IC – This is in fact a complete IC designed and built to perform a specific application.

We will consider each in more detail.

3.7.1   PROGRAMMABLE LOGIC INTEGRATED CIRCUIT

A programmable logic device (PLD) is an integrated circuit containing logic components which can be interconnected internally in various restricted ways by fusible links or other selectable interconnection methods. Some PLDs are purely combinational circuits while others also have flip–flops. The variety of PLDs is shown in Figure 3.10.

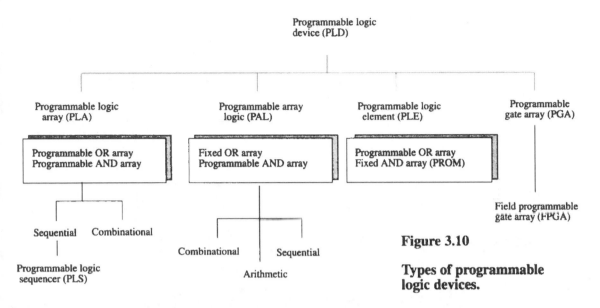

Figure 3.10

Types of programmable logic devices.

1. **PLAs** – contain a set of AND gates connecting to a set of OR gates. There are selectable links from the device input to the AND gates and between the AND gates and the OR gates so that combinations of sum of products expressions can be obtained.

2. **PALs** – these are similar to the PLAs except that the links between the AND and OR gates is fixed with only the connections into the AND gates being fixed. Hence the PAL device has a fixed predefined OR (sum) structure and a programmable AND (product) structure. Additions to the PAL device are through added registers and flexible I/O structures thereby allowing state machines to be developed.

Programm-able logic ICs

3. **PLEs** – have fixed links into the AND gates and programmable links to the OR gates. A PLE corresponds to the ROM family since there will be a particular binary output pattern for each input combination. The ROM family contains masked programmed ROMs, PROMs, EPROMs, EEPROMs and EAROMs.

4. **PGAs** – here each gate (of various types) can be programmed to connect to nearby tracks. The most popular form is the **field programmable gate array (FPGA)** where the programming of the interconnections between gates is done by the user.

By using these four forms of hardware programmed ICs it is possible to generate functions that would otherwise have to be implemented by a large number of SSI and MSI (medium scale integrated) chips. Instead of then using a large number of standard function ICs, computer designers can now employ a small number of PLD chips designed specifically to allow complex functions to be concisely implemented. With PLDs the parts count, i.e. the number of ICs, the power used, the cost, the failure rate and the board area requirements can all be reduced and an improved performance can be achieved because of the increased signal speed. An additional advantage with the presence of these programmable and specific ICs in a circuit is that it makes a printed circuit board or computer much harder to copy.

1   Programmable Logic Array (PLA) – The simplest of the programmable logic chips is the programmable logic array. A PLA is a regular array of memory or logic elements which can be used to implement combinational logic functions. It is a chip with a selection of standard SSI gates on it that can be programmed in much the same way as a read only memory. A simple PLA is shown in Figure 3.11.

Figure 3.11

**Block diagram
of a simple PLA.**

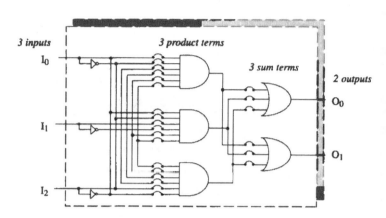

With the PLA the links are added during manufacture while with a FPLA (field programmable logic array) the links are blown, similar to the fusing of links in a PROM (programmable ROM). Built–in fuses are blown by a special programming machine such that the pattern of blown fuses leaves a particular web of logic on the chip. Because they can fit a number of logic functions into a single package, PLAs save board space and power consumption and improve the reliability of a system.

PLA

Determination of the required internal interconnections of the logic devices in the PLA can be accomplished using conventional digital design techniques. However, computer aided design systems are becoming more important to translate design specifications in the form of Boolean equations, truth tables and state diagrams into information required to program the PLA on a device programmer. The actual hardware programming of the device is done before the PLA is placed in the circuit in which it is to operate.

3.7.2   APPLICATION–SPECIFIC INTEGRATED CIRCUIT (ASIC)

An ASIC is an IC usually designed to perform a single specific function within a more complex processing system, which in most cases cannot be produced with a commercially available IC. The use of ASICs involves designing an integrated chip from scratch. This is a very expensive route to follow and is fraught with difficulties, not only in the testing of the device, but also in making the corrections if any errors are found in the design. The use of powerful computers and specific software is, however, making the job much easier and has led to a number of manufacturers adopting this approach for particular products.

Most ASICs are designed using standard cells. A **standard cell** is a small block of digital logic circuitry which can be implemented on a custom–designed IC. In designing with standard cells the designer essentially faces a blank chip area onto which he or she can place a selection of small standard cells. The manufacturer of the standard cell usually provides the designer of the custom IC with a library of standard cells which he or she can use to lay out their custom chip. The cells can be interfaced together using software. The cells within the library have usually been tuned to high performance and minimum area compared to implementing the same function from gates in a gate–array design, hence the entire chip should exhibit better characteristics than a gate–array design. However because each standard cell chip must be fabricated from scratch once the design is complete, they usually take longer to complete than gate arrays.

ASICs

The use of **silicon compilers** to produce full custom chips is now evolving. The silicon compiler behaves in much the same way as a high level language compiler in that it is a program that translates general statements of purpose into low level descriptions of particular processes to achieve that purpose. Rather than producing a low level assembly language for a computer, however, the silicon compiler produces a hardware description of a custom IC chip.

Silicon compilers are now available which allow the designer to either enter statements about the chip's desired behaviour, called a behavioural compiler, or to specify the desired data–bus width, pin arrangements, arithmetic functions etc. of the integrated circuit. These are called structural compilers. The chip produced by a silicon compiler has several advantages over a more conventional design process. It does not require as much IC design experience as a full custom chip and it can be produced in a matter of weeks or months instead of years. The disadvantages are that it does not use the chip area as efficiently as a hand designed chip would, and some functions that a designer might want are not yet available in many silicon compilers.

3.8 KEY TERMS

Accumulator	Microprocessor architecture
Address	Microprocessor development
Address bus	Microprocessor: external buses
Application–specific integrated circuit (ASIC)	Microprocessor: internal buses
Arithmetic logic unit	Microprocessor hardware
Control unit	Microprocessor software
Control bus	Microprocessor (4–bit)
Co–processor	Microprocessor (8–bit)
Custom logic devices	Microprocessor (16–bit)
Data bus	Microprocessor (32–bit)
DEC Alpha	Microprocessor (64–bit)
Field programmable gate array	Microprogram memory
Flag register	MOS 6502
Index registers	Motorola
Intel	Motorola 6800
Intel 4004	Motorola 6809
Intel 8086	Motorola 68000
Intel 80186	Motorola 68010
Intel 80286	Motorola 68008
Intel 80386	Motorola 68020
Intel 80486	Motorola 68030
Intel i860	Motorola 68040
Instruction cycle	Motorola 68060
Instruction set	Motorola 68881/68882
Microcontroller	Opcode
Microprocessor	Operand
	Pentium *(Continued)*

Power PC	Programmable logic element
Processing section	(PLE)
Program	Registers
Program Counter	Silicon compiler
Programmable array logic (PAL)	Software
Programmable gate array (PGA)	Stack pointer
Programmable logic device (PLD)	Standard cell
Programmable logic integrated	Zilog Z80
circuit	Zilog Z8000

3.9 REVIEW QUESTIONS

1. What is a microprocessor?
2. Within a microprocessor which components are involved with the data processing?
3. What do CPU and ALU stand for?
4. Briefly describe the functions of the control unit within a microprocessor.
5. What do the following terms refer to in a microprocessor: program counter, stack pointer, accumulator, index register?
6. What is the purpose of a clock in a microprocessor system?
7. Within a microprocessor what are registers used for?
8. Define what the term 'address' refers to in a microprocessor system.
9. How is a computer's main memory addressed? Why is addressing memory important?
10. What does it mean to 'interpret' an instruction?
11. Why might one microprocessor have more instructions in its instruction set compared to another?
12. Define the term operand.
13. The microprocessor is a limited operation device, explain from what it derives its usefulness.
14. Describe briefly how a microprocessor appears to make decisions.
15. Explain the difference between data and information and describe how the microprocessor processes data and converts it into useful information.
16. Approximately how long did it take to double the complexity of a microprocessor from 1971?
17. What does it mean when the Motorola 6800 is described as an 8–bit processor?
18. In a 16–bit microprocessor what size would the data bus and address bus be?
19. What are the added features that the Intel 80286 microprocessor has, compared to the 8086?
20. Why have more functions been incorporated into 32–bit microprocessors compared to their 16–bit predecessors?
21. What advantages would the addition of a co–processor bring to a personal computer system?
22. Explain the difference between a microprocessor and a microcomputer.
23. List four types of programmable logic devices.
24. Where are the main uses of programmable logic chips?
25. Contrast the properties of gate arrays, PLAs and ASICs.

3.10 PROJECT QUESTIONS

1. List four applications for microprocessors.
2. The control unit within a microprocessor can either be hardwired or micro-programmed. Explain the difference.
3. Find out more about each of the functions of the pins on the Intel 8086 micro-processor.
4. The Intel 8086 microprocessor has a multiplexed address and data bus. What does this mean?
5. How does microprogramming ensure compatibility between microprocessors?
6. By suitable reference find out as much as you can about the Intel iAPX 432 32-bit microprocessor.
7. Describe the application of microprocessors in domestic products.
8. Describe products which at present do not use microprocessors but which in the future might use them. How would the products change?
9. Choose one of the 64-bit microprocessors and investigate its architecture.
10. In what application areas are single-chip microcomputers and microcontrollers most likely to be used?

3.11 FURTHER READING

Most publishers have a wide range of books on microprocessors, all of which provide in-depth reading on particular devices. A short list of some of these books is given. Journals such as *IEEE Micro*, *Byte* and *Computer Design* carry up to date articles on modern microprocessors and their application.

Introduction to Microprocessors (Books)

1. *8086/8088 Microprocessor: Architecture, Programming and Interfacing*, B. B. Brey, Merrill/Macmillan, 1987.
2. *The Z80 Microprocessor Architecture, Interfacing, Programming and Design*, R. Gaonkar, Merrill/Macmillan, 1988.
3. *Microprocessor Architecture, Programming, and Applications with the 8085/8080a*, 2nd Edn, R. Gaonkar, Merrill/Macmillan, 1989.
4. *Digital Systems Logic, and Applications*, T. A. Adamson, Delmar, 1989.
5. *Structured Computer Organization*, 3rd Edn, A. S. Tanenhaum, Prentice Hall, 1990.
6. *Programming and Interfacing the 8086/8088 Microprocessor a Product-Development Laboratory Process*, R. Goody, Merrill Macmillan USA. 1992.
7. *The 8051 Microcontroller*, S. MacKenzie, Macmillan USA, 1992.
8. *The 68000 Microprocessor Family: Architecture, Programming, and Applications*, 2nd Edn, M. A. Miller, Macmillan USA, 1992.
9. *Digital system design*, 2nd Edn, B. Wilkinson, Prentice Hall, 1992.
10. *68000 Microcomputer Organization and in Programming*, P. Stenstrom, Prentice Hall, 1992.
11. *The 80386DX Microprocessor Hardware, Software and Interfacing*, W. Triebei, Prentice Hall, 1992.
12. *The 8086/8088 Family Design, Programming and Interfacing*, J. Uffenheck, Prentice Hall, 1987.

Introduction to Microprocessors (Books)

13. *8085 Microprocessors and Programmed Logic*, 2nd Edn, K. Short, Prentice Hall 1987.
14. *The Z80 Microprocessor*, B. B. Brey, Prentice Hall, 1988.
15. *Microprocessors Theory and Applications (Intel and Motorola)*, revised Edn M. Rafiquzzaman, Prentice Hall, 1992.
16. *16–bit and 32–bit Microprocessors Architecture, Software and Interfacing Techniques*, A. S. Singh and W. A. Triebel, Prentice Hall, 1991.
17. *Introduction to Microprocessors using the MC6809 and the MC68000*, R Horvath, McGraw–Hill, 1992.
18. *Microprocessors and Interfacing: Programming and Hardware, 68000 Family* D.V. Hall and A. Rood, McGraw–Hill, 1992.
19. *Microprocessor Support Chips Sourcebook*, A. Clements, McGraw–Hill, 1991
20. *Microprocessors and Microcomputers the 8080, 8085 and Z 80*, 2nd Edn, J Uffenbeck, Prentice Hall, 1991.
21. *16 bit Microprocessors*, I. R Whitworth, Collins, 1984.
22. *Systems Design with Advanced Microprocessors*, J. R. Freer, Pitman, 1987.

Microprocessors (Articles)

1. The birth of the microprocessor, F. Faggin, *Byte*, March 1992, pp. 145–150.
2. The Spirit of 86s, F. Hayes, *Byte*, March 1990, pp. 266–270.

Microcontrollers (Articles)

1. Why microcontrollers, S. Ciarcia, *Byte*, August 1988, pp. 239–245.

32–bit Micros (Articles)

1. Mainframe architectures and compiler techniques power 32–bit micros, K. E. Marrin, *Computer Design*, February 1987, pp. 57–76.

64–bit Micros (Articles)

1. Architecture of the Pentium microprocessor, D. Alpert and D. Avnon, *IEEE Micro*, June 1993, pp. 11–21.
2. The Alpha AXP architecture and 21064 processor, E. McLellan, *IEEE Micro*, June 1993, pp. 36–47.
3. PowerPC, *Byte*, August 1993, pp. 56–74.

Further reading

Chapter 4

Memory Systems

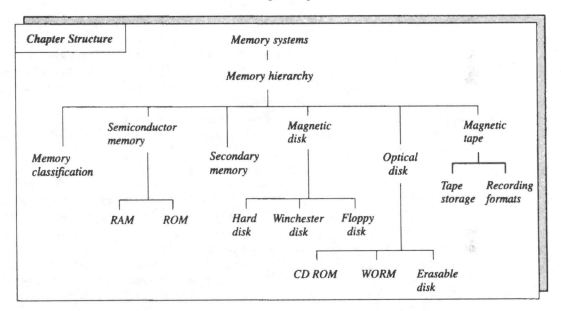

Chapter Structure

Learning Objectives

In this chapter you will learn about:

1. The different types of memory and their use in the memory hierarchy within a computer.
2. The forms of memory devices used in past generations of computers.
3. The characteristics of the various types of semiconductor memory.
4. The different backing storage media, their purposes within a computer and their relative performance.
5. The difference between direct access storage and sequential access storage mechanisms.

'The best uses of personal computers haven't been invented. To disagree with that is simply a failure of imagination.'

Mitch Kapor.

Chapter 4

Memory Systems

A computer may be seen as an information and storage manipulation machine. In order to perform this task the device requires some type of memory to be able to store and retrieve the information. The **memory** section within a computer is that part of the system's hardware that serves as storage for both program instructions and data. To be applicable to the computer's world the memory should be digital, i.e. works on a binary system of 1s and 0s, and must contain a series of discrete physical storage cells which are capable of being placed (or set) by an external signal into one of two distinct states. Each cell must also remain in this set state, ideally for an indefinite amount of time or until it is changed to the other state by another external signal.

The two distinct states of a storage cell can be naturally occurring states which require no external energy sources to be maintained. This is true for ferromagnetic, ferroelectric and superconducting cells, all of which have the property that the quantity defining the state, magnetic induction, electric polarization, or induced super current respectively, has stable states corresponding to zero energization. Typical examples of devices using this type of cell are thin magnetic films, magnetic tapes and disks, and optical disks. Alternatively, it is also possible to use storage elements which require external energization to maintain the stored state, such as the bistable circuits formed by semiconductor electronics. Table 4.1 presents the characteristics of several types of memory indicating the type of storage element used and the method of access.

Table 4.1

Memory characteristics.

Storage method	Using integrated semiconductor storage elements			Using tiny permanent magnetic fields	
Method of access	Flip–flops	Charge storage	Permanent connections	Fixed medium	Moving medium
Serial access	Static shift registers	Dynamic shift registers		Magnetic bubbles	Tape
Direct or random access	Static RAM	Dynamic RAM	ROMs	Magnetic cores	Disks and drums

4.1 MEMORY HIERARCHY

The memory of a processing system is constructed in a hierarchical structure, as shown in Figure 4.2, since it is not economically nor as yet technically feasible to use one type of storage element for all the memory requirements in a digital computer system. The variety of storage devices is used in an effort to achieve the best performance and largest capacity at a reasonable cost, i.e. optimum speed and capacity at minimum cost.

Glossary 4.1

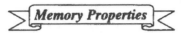
Memory Properties

ACCESS TIME – When used to describe memory devices, the access time is the delay between the time the memory device receives an address and the time when the data from that address is available at the output of the memory. This is sometimes also referred to as the read time. The access time is usually constant for a particular random access memory (RAM) device and may be regarded as a specification of the speed of the RAM device.

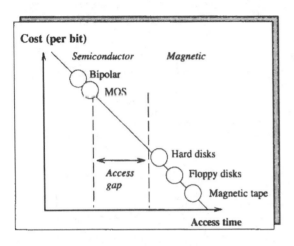

Figure 4.1 shows a graph of relative cost versus access time for typical memory devices ranging from the fast semiconductor bipolar RAM to the very much slower magnetic tapes. It can be noted from Figure 4.1 that there is a gap in access time between random access devices such as semiconductor RAM and the direct access devices typified by magnetic disks.

Figure 4.1

Relative cost versus access time for various memory types.

For memory other than RAM, the access time is often regarded as the time from when an instruction is decoded asking for a memory location until the desired information is found but not read. In this case the access time is a function of the location of the data on the medium with reference to the position of the read–write transducer(s). It can therefore vary substantially depending on the location of the information being sought and as a consequence the average access time over a short period depends on the pattern of memory references.

BANDWIDTH – Memory bandwidth is the number of bytes per second that the memory can deliver to the processor. The bandwidth of a memory is also the amount of information made available to the processor by a single memory access.

CAPACITY – The capacity of a memory module or system is simply the maximum number of bits, bytes, or words within the module or system. For example a 2K x 4 memory can store 2K ($K = 1024 = 2^{10}$) words each containing 4 bits of data or a total of 2 x 1024 x 4 bits = 8192 bits.

CYCLE TIME – In general, the cycle time is the time interval for which a set of operations is repeated regularly in the same sequence. In the area of computers, the cycle time is the total time for a program instruction to reference a memory location, read from or write to it, and then return to the next instruction.

The cycle time is also used as a measure of how often a memory can be accessed per unit time. For a static random access semiconductor memory (SRAM) the cycle time is equal to the access time; however, for a dynamic random access memory (DRAM) the cycle time is greater than the access time since a DRAM requires a frequent restore period. In this latter case the cycle time then consists of the access time plus any additional time after the completion of a memory access until the memory is available again.

It can be noted that the cycle time for a magnetic disk drive to deliver the contents of its disk is much greater than that of a silicon random access memory.

DATA RATE – The data rate is the rate (usually bits per second, bytes per second or words per second), at which data can be read out of a storage device. It can be calculated by determining the product of the reciprocal of the access time and the number of bits in the unit of data (data word) being read. The term data rate is usually associated with nonrandom access memory, where large pieces of information are stored and read serially and has little significance in random access memory since in such memories an entire word is normally read out in parallel. In most cases the data rate is constant for a given memory module.

DENSITY – In the area of magnetic storage, density is the number of characters or bits that can be stored on either one inch of magnetic tape, or per track on a magnetic disk.

ERASABLE STORAGE – The term erasable storage is given to any memory device whose contents can be modified, e.g. random access memory (RAM). This is in contrast to read–only storage.

(Continued)

Glossary 4.1   *(continued)*

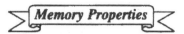

NONVOLATILE STORAGE – Nonvolatile memory is any storage medium (or circuitry) which does not lose its data contents if the power is removed. In this way it can retain its contents for an almost indefinite time.

PRIMARY MEMORY – The primary (or main) memory is the part of the total memory within a computer that is directly accessible by the central processing unit (CPU). It is usually a large block of semiconductor random access memory (RAM) which the processor directly works with, using it to access its program instructions and also to store data values and results. This statement is not wholly true since, if the computer has a cache memory, the processor will typically access the cache more than the main memory to get at its instructions and data.

READ WRITE MEMORY (RWM) – A RWM is any form of memory which may be both written to and read from. It has become synonymous with random access memory (RAM).

SCRATCHPAD MEMORY – A scratchpad memory is a small memory, or a section of a memory, used for the storage of intermediate results during the execution of a program.

STACK MEMORY – The stack may be regarded as an area of contiguous memory locations or registers which can be located anywhere in a computer system's random access memory (RAM). It is organized as a last in first out (LIFO) file (in other words the last data entered is the first to be removed). Data bytes can be pushed onto the stack from specific central processing unit (CPU) registers or popped off the stack into specific CPU registers. The stack is often used for holding the return address and processor status during an interrupt or subroutine operation but can also be used to allow simplification of many types of data manipulation.

STATIC MEMORY – A static memory is one that does not change its contents without external causes. It therefore does not need the refreshing operations characteristic of dynamic memory. A static memory may still require a power supply to maintain its information.

VOLATILE STORAGE – Volatile storage is any storage device whose contents are lost when the power to the storage device is removed. Most static and dynamic random access memories are volatile.

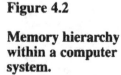

Memory hierarchy

The memory hierarchy shown in Figure 4.2 also notes the three types of memory. **Cache Memory** – Modern microprocessors are too fast for the current generation of RAM chips. It is possible to deal with slow memory by introducing wait states in which the microprocessor waits for the memory to respond after a memory request has been generated. Alternatively it is possible to use a small, fast buffer memory or a cache memory.

Figure 4.2

Memory hierarchy within a computer system.

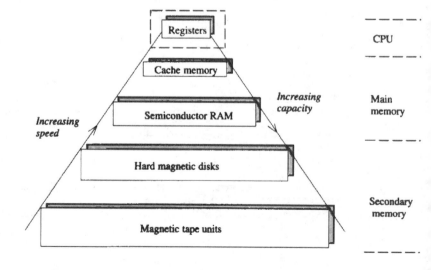

CACHE MEMORY

Within most computer programs it has been found that the central processing unit tends to spend its time executing a few main routines. This means the sequence of addresses generated by the CPU is not uniformly distributed throughout the memory, but tends to cluster in one part of the memory for a while and then possibly moves to another part of the memory. Hence certain instructions tend to be used over and over again. This phenomenon is known as a 'locality of reference'. Now if it can be arranged to have these repeating segments of a program in a fast memory, then the total execution time of the program can be significantly reduced.

Such a memory is called a cache (or buffer) memory. A cache memory is specially designed to allow a computer to load and reference information that it anticipates it will need in the future. With the correct control of the cache it is usually possible to make sure that the active portions of a program are always to be found in the cache such that once a cache is loaded with information from the main memory, it is used more than once before new information is required from the main memory. The computer can then be made to operate at the speed of the cache memory rather than the slower speed of the main memory. By adding this special form of control, the system can be made more economical than if the entire main memory was to be constructed of devices as fast as the cache. Cache memory was first used as an explicit part of a computer system design in the IBM System 360/85 mainframe computer.

When the CPU needs to make a request of memory it generates the address of the desired word. If the quantity is available in the cache a hit occurs and a copy of the object is quickly passed back to the CPU without making a request to the main memory. If the desired object is not in the cache then a miss or cache fault occurs, and the address must then be passed to the main memory for access. When the data returns, a copy of it is stored in the cache and the desired object is relayed to the CPU. Cache performance depends on access time and the hit ratio which is dependent on the size of the cache and the number of bytes brought into the cache from the main memory. Increasing the number of bytes brought to the cache increases the chance that there will be a cache hit on the next memory reference.

A diagram illustrating this mechanism is shown in Figure 4.3. There are a number of cache accessing schemes, the one shown illustrates a direct mapping scheme. It can be seen from the diagram that the memory address from the processor contains two parts – the tag and the index. The tag is compared with the tag of the data word within the cache memory using the index as a pointer to the location in the cache. If the tag is the same as the data word in the cache the word is read from the cache, if it is not a cache miss occurs and the word is sought in the main (or primary) memory.

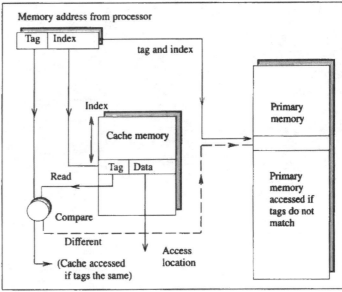

Figure 4.3   Cache memory direct mapping.

Physically the cache is a small block of random access memory (RAM) which works at high speed, typically matched to the speed of the central processing unit (CPU). It is situated between the CPU and the primary (or main) memory. In operation frequently used instructions or data are stored in the cache memory for fast access by the processor, hence rather than having a processor wait for the slower RAM the information can be read from or written to the cache. In a cache memory the data transfers are carried out automatically in hardware. Items are brought into a cache when they are referenced, while any changes to values in a cache are automatically written when they are no longer needed, when the cache becomes full or when some other process attempts to access them. See Article 4.1 for more information.

Memory hierarchy

Main Memory – The main memory is that part of the total system memory where the program instructions and associated data must be stored for direct access by the processor. It is usually a large block of random access memory.

Secondary Memory – Secondary memory devices, such as magnetic disks and tapes, store programs and data that cannot be directly accessed by the central processing unit. They allow a large amount of memory space to be incorporated into systems but have a slower access time than the main memory.

Figure 4.4 notes the three types of memory and their position within a processing system. Table 4.2 shows typical values for several components in the memory hierarchy. It gives figures for the access time for the types of memory, and an indication of the maximum size of such a memory in a high performance personal computer.

Figure 4.4

Position of the different types of memory in a computer system.

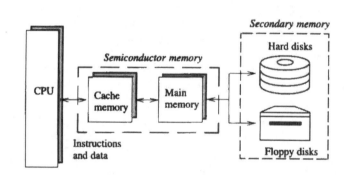

Table 4.2

Characteristics of computer memory systems.

Characteristic	Cache	Main	Secondary
Technology	TTL semiconductor	MOS semiconductor	Magnetic disks, tapes
Access time (ns)	10	70	10 000
Memory capacity (bytes)	Small to 16 K	Medium to 8 M	Large to 1 G
Usage	Processor data buffer	On–line dynamic temporary storage	On–line bulk storage

In order to control the memory structure requires the use of a memory management integrated circuit and/or sophisticated memory management routines in the operating system of the processor. Using memory management hardware and software it is possible, by providing memories of various speeds and sizes at different points in the structure of a system, to make the overall memory appear to operate at the speed of the fastest and most expensive, while offering the capacity of the slowest and least expensive. With hierarchical control it is also possible to reduce the potential conflicts between the processor and I/O devices at each level, by minimizing the time a processor must wait for the completion of a memory function.

4.2 MEMORY CLASSIFICATION

The memory can be classified into three main groups, these are semiconductor, magnetic and optical. While the main memory is solely constructed from semiconductor memory the secondary memory is usually made up from magnetic and optical disk systems.

Figure 4.5 presents the different classifications of memory used within a computer system. Article 4.2 gives information on older forms of computer memory.

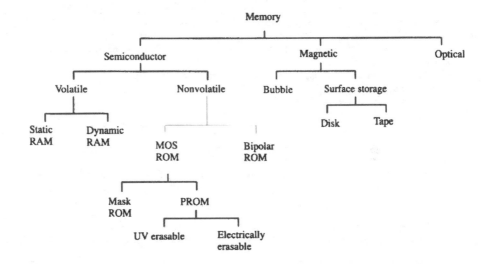

Figure 4.5

Classification for memory systems in a computer.

4.3 SEMICONDUCTOR MEMORY

Figure 4.6 shows a taxonomy of the different semiconductor memories differentiating between the random access memory and the sequential access memory varieties.

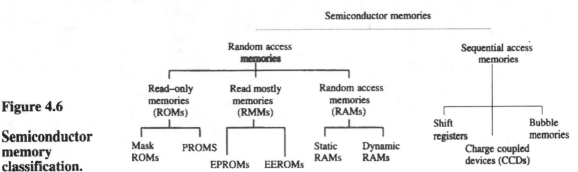

Figure 4.6

Semiconductor memory classification.

OLD FORMS OF COMPUTER MEMORY

DRUM STORAGE

Magnetic drum memory was a form of direct access memory used in most early computers. Magnetic drum storage was pioneered by A.D Booth in London in 1947. The magnetic drum consisted of a precision machined and balanced drum (or cylinder) of non–magnetic material, the outer surface of which was coated with a thin magnetic material. The magnetic surface was then divided into a number of parallel tracks, with a maximum of about 30 per inch, and each track could then be configured into a number of sections with each section recording one word or character in binary form. The density of the binary bits was up to 100 per inch along the length of each track. The information storage capacity of this type of system was determined by the number of bits per inch along a data track (bit density) and the number of tracks per inch along the drum axis (track density). The product of these two factors gave the data storage density per square inch of the recording surface.

In operation the drum was rotated at high speeds, of the order of 1200 to 3600 revs/min, causing the series of tracks to pass circumferentially beneath the special magnetic recording heads located one above each track. Since it was impossible to mount up to 30 heads in the space of one inch, they were usually staggered around the periphery of the drum. Magnetic recording heads magnetized spots along each track to indicate the binary bits, with one magnetic spot equaling a binary 1 while no spot equaled 0.

The drum is no longer used on modern computer systems as it has been replaced by the use of disks.

CORE MEMORY

Ferrite core memory is the oldest form of random access memory. It was invented at the Massechuttis Institute of Technology (MIT) in the USA in 1950 and was a nonvolatile memory consisting of a matrix of wires with a small doughnut–shaped iron alloy ring (or core) at each intersection. These rings could be magnetized in two directions, storing a 0 or 1 data bit dependent on the way they were magnetized.

During the 1960s the dominant main memory technology was magnetic core. In the early 1960s one megabyte of 1 microsecond core memory cost approximately $1 million. Typical cycle times for magnetic cores were 1 to 2 milliseconds.

Ferrite core memory is no longer used in any computer system as there are better alternatives, the most popular of these is semiconductor memory.

CHARGED COUPLED DEVICE (CCD)

A CCD is an integrated circuit storage device in which information is stored by means of packets of small electrical charges. In CCD arrays, sets of charge packets represent words, where the presence of each charge packet indicates a 1 bit (the bits are actually stored as charges on a number of small capacitors created within the IC). This is similar to the dynamic random access memories (DRAMs), except that the storage in a CCD is arranged in a shift register configuration with the charge packets being shifted from cell to cell under clock control. Hence unlike RAMs where fixed circuits hold words, charge packets do not remain stationary and the cells pass them along to neighbouring cells with each clock tick. The CCD devices are organized in such a way that the cells are formed into tracks with one track for each bit position in a set of words. Circuitry is also placed both at the beginning of each track to generate (write) the charge packets and at the end of each track to detect (read) the charge packets. Logically these tracks can be regarded as loops (in a similar way to magnetic bubble memories). Since the data is read/written serially the access time is slower than that of a semiconductor random access memory.

(Continued)

(continued)

☆ *Article 4.2* ☆

OLD FORMS OF COMPUTER MEMORY

BUBBLE MEMORY

Magnetic bubble memory (MBM) was invented at Bell Laboratories in 1969. It is an older form of solid state data storage device in which the data bits are stored by tiny magnetic bubbles (or magnetic domains) which rest on a thin film of semiconductor material.

A typical MBM's components are a thin film memory chip, constructed from a garnet crystal with a magnetic film deposited onto it. Permalloy control structures are also deposited onto the film in order to determine the paths the magnetic domains move along. The paths are controlled by the action of a rotating magnetic field, this field being created by a coil that surrounds the crystal.

The garnet chip is constructed from a large slice of the material and cut into individual chips in just the same way as the chips of standard integrated circuits. The other components in the system such as the drive coils, the permanent magnets and the control electronics are then assembled with the garnet chip and placed into a case. The case also serves as a shield to protect the device from disruptive external magnetic fields.

The data in this type of system is stored as microscopic cylindrical regions (small bubbles approximately 4 μm across) that can be moved around in the thin film of garnet magnetic crystalline material. The presence of a bubble at one of the magnetic iron–nickel strips indicates a logic 1 and the absence means a logic 0.

Two important features of this type of memory are its large capacity and its ability to retain its contents even when the power supply is switched off, .ie. bubble memory is nonvolatile. The main disadvantage is that it takes much longer to get the data into and out of an MBM, compared to a semiconductor memory, primarily because a MBM requires serial entry and read out of its information rather than the randomly accessed semiconductor memory.

Bubble memory is still being produced. Intel have produced a 4 Mbit device using a 10 mm by 10 mm chip in which magnetic bubbles are 3 micrometers. Bubble memory is rarely now used in commercial products.

Semi-conductor memory

Under the RAM branch we have distinctions of read–only, read mostly and read write. The first two can be grouped together under the ROM heading. When dealing with computers most commentators simply refer to the read–write memory within the system as RAM.

Semiconductor memory began to be widely used in computer systems after the release in 1971 of the 1 Kbit dynamic RAM. The 4 Kbit device was released in 1975 with the 16 Kbit and the 64 Kbit being released in 1981 and 1984 respectively. The 1 Mbit dynamic RAM (DRAM) was produced in 1987.

A comparison between the different semiconductor technologies used to construct semiconductor memory is given in Table 4.3.

Table 4.3

Comparison of semiconductor memory technologies.

Technology	Speed	Power required	Capacity	Noise immunity	Cost
TTL – TTL/LS	Fast	High	Low	Low	High
ECL	Fastest	Highest	Lowest	Lowest	Highest
NMOS	Medium	Low	Highest	High	Lowest
CMOS	Slowest	Lowest	High	Highest	Low

4.3.1   RANDOM ACCESS MEMORY (RAM)

Random access memory is a type of memory for which any location (word, bit, byte, record) has a unique, physically wired addressing mechanism and which can be retrieved by the central processing unit in one memory cycle time interval. It is randomly accessible in the sense that the time required to read from or write into the memory cells is independent of the previous memory location accessed. It is possible to identify three types of memory structure within a computer system that are usually composed of random access memories, these are: the internal registers of the CPU, the cache memory and the main memory.

The development of very large scale integrated semiconductor techniques has brought about the possibility of manufacturing large arrays of storage elements, or bit cells, on a single chip of silicon and configuring them as RAMs. Semiconductor RAM is the fastest and most freely available of the different memory types. Confident predictions have been made about the increasing low cost and higher densities of these active memories such that they may ultimately replace all other forms of storage within a large computer. Figure 4.7 notes the types of semiconductor technologies used to construct RAMs.

From Figure 4.7 it can also be seen that there are essentially two variations to this type of semiconductor memory, depending on the method of construction, these are static or dynamic semiconductor RAM. Although slower in operation than the bipolar device the MOS device has been particularly exploited in RAM since it takes up less substrate area, thereby increasing the packing density, requires less power and needs fewer process steps to manufacture.

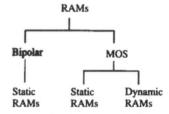

Figure 4.7

Random access memory types.

Figure 4.8 shows the evolution of semiconductor RAM. It can be seen that there are two curves, one for dynamic RAM (DRAM) and a second for static RAM (SRAM). The construction of SRAM involves a more complex integrated circuit procedure, hence the density of memory cells is not as great as that for DRAM.

Figure 4.8

Evolution of RAM size with time.

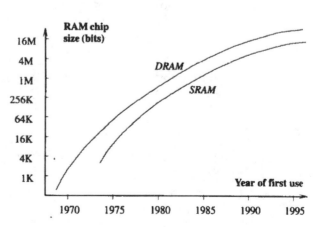

Though little is known about the long term reliability of semiconductor memories, it would appear that they are inherently more reliable than other forms of storage, moreover it is also comparatively easy to include error detection and correction circuits in the basic modules. Unfortunately, RAM storage in general is inherently volatile, since the bistable *RAM* elements (in an SRAM) or the storage capacitors (in a DRAM) require a constant source of power to maintain the stored information. This can be a severe disadvantage in some applications where it is essential that the data should not be irretrievably lost due to a power failure. This disadvantage may be overcome by incorporating power standby circuits which detect when the main power supply is about to fail and switch over to a battery source; in this way the contents of the store can be retained.

1. Dynamic Random Access Memory (DRAM) – is a form of semiconductor memory in which the binary data are stored by charging the gate capacitances of metal oxide semiconductor (MOS) transistors. A DRAM cell then simply consists of a storage capacitor and a transistor buffer/amplifier with the logic levels of 0 and 1 being represented by the state of charge or discharge. Since each of the storage capacitors is not perfect, the charge stored will tend to leak away and so cause a loss of the data, typically in a few milliseconds. In order to avoid this, the charge on the capacitor must be periodically regenerated by means of a refresh operation (see Glossary 4.2).

Figure 4.9 shows a block diagram of a DRAM memory module in which the memory control circuitry is incorporated into the unit. The buffers act as an interface to the system bus allowing the data address and control information to be exchanged with the CPU. The on–board control logic selects the particular memory cell in the DRAM that is required allowing the data to be read into or written out of the cell.

Figure 4.9

DRAM memory module.

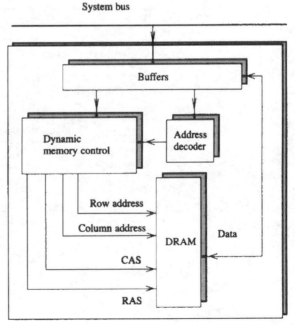

CAS – Column address select
RAS – Row address select

Memory module

Figure 4.10 is an expansion of Figure 4.9 highlighting the DRAM device. The chip contains an address decoder to select the desired memory cell word. This consists of both row and column decoders. It also has refresh circuitry fed through a multiplexor and a chip select to enable the addressing and input/output currents.

Figure 4.10

DRAM internal logic.

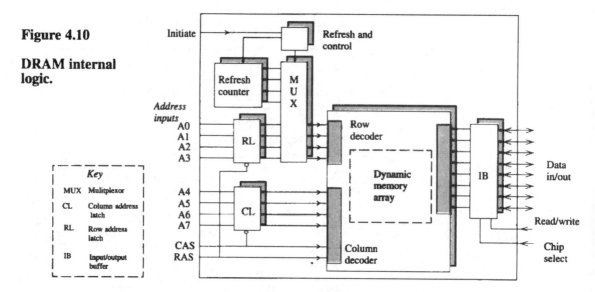

Looking at the actual dynamic memory array in more detail we would see a structure such as represented in Figure 4.11. This shows an expansion of a section of the dynamic RAM shown in Figure 4.10 and illustrates its internal construction. It can be seen that the array consists of a series of transistors with capacitors connected to them in order to store

DRAM

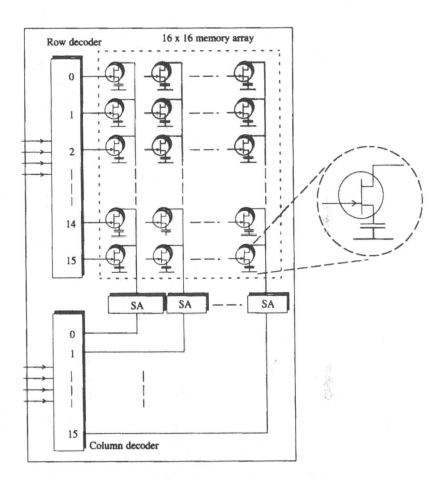

Figure 4.11

DRAM memory matrix array.

the charge. Each of the transistors essentially behaves like a switch allowing charge to be written into or read out of the capacitor. Onto this memory array are connected the sense amplifiers (SA) which detect and amplify the data coming from the memory and, finally, the row and column decoders, which allow the system to select individual transistor storage cells
DRAM   to either write information into or read information out of them.

The 1103 1–Kbit DRAM chip was introduced by Intel in 1970; however, since then the density has increased dramatically until present DRAMS offer megabits per IC.

2. Static Random Access Memory (SRAM) – SRAM is so named because of its ability to indefinitely retain the logic levels memorized, as long as the power on the memory is maintained. Static should not be confused with nonvolatile, as SRAMs are volatile to the power being switched off.

The block layout of a static RAM is shown in Figure 4.12. It is similar to the block layout of the DRAM. It can be seen that the integrated circuit has 256 different memory locations each of which can hold 8 bits of data at a time. The individual locations are selected through the address lines. Each time a location is selected, 8–bits of data can either be written into the memory or read out from the memory.

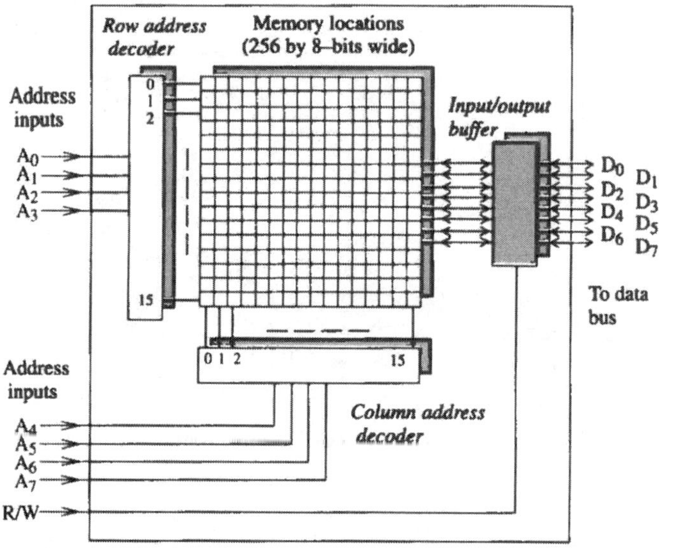

Figure 4.12

Block layout of a 256 by 8 static RAM.

SRAM

Compared to DRAM, SRAM is usually preferred within most integrated circuit boards. This is because the static RAM cell is based on a logical flip–flop arrangement (a typical SRAM cell uses six transistors in a flip flop circuit) rather than the storage capacitor found in a DRAM; the information held is stable and requires no clocking or refresh cycles to sustain it. Interfacing to SRAM systems is also simpler then DRAMs since static memory devices are usually fully decoded for interfacing directly to address and data lines in a system. The development of CMOS static RAM has been particularly important in that it enables the power consumption in a circuit to be kept down. SRAMs can achieve lower access times than DRAMs and CMOS SRAM may be used easily for battery backup storage which retains the memory contents when the computer power supply is removed.

The first semiconductor SRAM was developed by the Intel corporation. It was a 64–bit bipolar device called the 3101 which was introduced in 1969. The 256–bit MOS SRAM was introduced later in 1971.

3. Associative Memory – An associative memory is a special type of random access memory that is content addressable, i.e. the individual memory cells are addressed by their contents rather than by the use of an address. In addition to having a conventional wired–in addressing mechanism, the memory also has wired–in logic that enables a comparison to be made between an input address and a series of desired bit locations for a specified match, and is able to do this for all words simultaneously during one memory cycle time. In operation, the incoming memory address is compared with all the stored addresses simultaneously, using the internal logic of the associative memory and, if a match is found, the corresponding data is read out. In this way every memory register that contains a referenced string of data symbols can be accessed, rather than the single register whose location, or address, is specified. This allows a user to specify part of a pattern, or key, and retrieve the complete data item associated with the key from the memory. The parallel bit–slice associative memory is often referred to as a **content addressable memory (CAM)**, parallel search memory or catalogue memory.

4.3.2   READ–ONLY MEMORY (ROM)

A ROM is a randomly accessible semiconductor memory, the contents of which cannot be altered, except under certain circumstances. There are many different types of ROM devices. Figure 4.13 shows the different members of the ROM family with the distinction in manufacturing separating them into MOS and bipolar types.

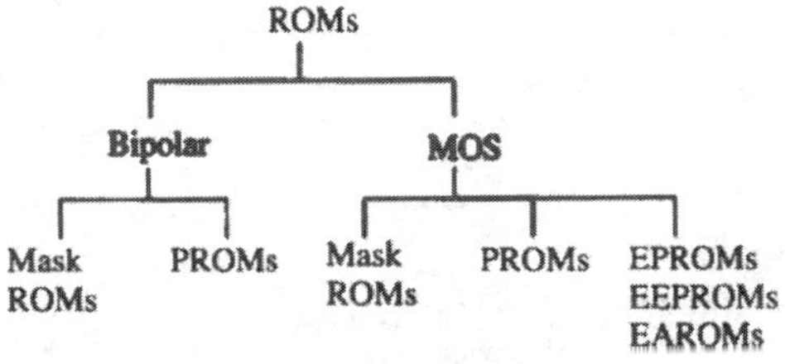

Figure 4.13

ROM family.

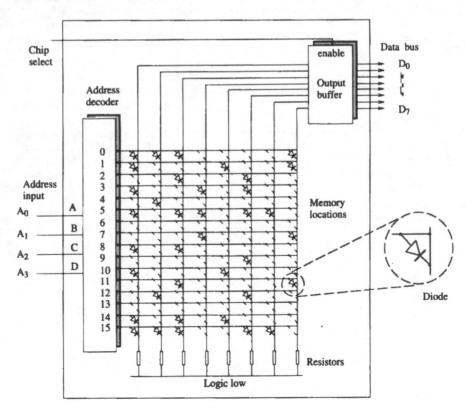

Read–only memories are simpler and less costly than random access memories and are also nonvolatile in that they retain their data regardless of the presence of power. They have the disadvantage that once they are programmed and inserted into the computer system, they are difficult to change.

ROMs are extensively used when there is no need for modification or replacement of data, for example in programs or tables of data that remain fixed within the lifetime of a system. They are also used for implementing modules of a computer's operating system (i.e. for system programs) or where a program is required to be permanently implanted into the computer, such as a translator or an emulator.

Figure 4.14

Internal logic of a programmed 16 by 8 ROM.

READ–ONLY MEMORY

PROM – A programmable read–only memory (PROM) is part of the read–only memory (ROM) family. It is a programmable chip which once written cannot be erased and rewritten, i.e. it is a read only after written memory. To implant programs, or data, into a PROM a programming machine (called a PROM programmer) is used to apply the correct voltage for the proper time to the appropriate addresses selected by the programmer. As the PROM is simply an array of fusible links the programming machine essentially blows the various unwanted links within the PROM leaving the correct data patterns, a process which clearly cannot be reversed. Like the ROM, the PROM is normally used as the component within a computer used to carry any permanent instructions that the system may require.

EPROM – An erasable programmable read–only memory (EPROM) is a special form of semiconductor read only memory that can be completely erased by exposure to ultraviolet light. The device is programmed in a similar way to the programmable read–only memory (PROM); however, it does not depend on a permanent fusible link to store information, but instead relies on charges stored on capacitors in the memory array. The capacitors determine the on/off state of transistors, which in turn determine the presence of 1s or 0s in the array.

The EPROM is so arranged that the information programmed into it can be erased, if required, by exposing the top surface of the package to ultraviolet radiation. This brings about an ionizing action within the package, which causes each memory cell to be discharged. EPROMs are easily identified physically by the clear window that covers the chip to admit the ultraviolet light. Once an EPROM has been erased, it can be reprogrammed with the matrix being used again to store new information. The user can then completely erase and reprogram the contents of the memory as many times as desired.

Intel first introduced the EPROM in 1971 ; however, the storage capacity has increased dramatically with improving IC technology. Current EPROMs can store multiple megabytes of information.

EAROM – An electrically alterable read–only memory (EAROM) is a nonvolatile read–only memory device in which the data can be altered electrically, but will be retained in the memory when the power is removed. The device can be easily read from, but requires a high voltage in order to clear individual memory cells before they can be written to. A more accurate description of an EAROM would probably be a mainly readable memory, rather than read–only. This is because, unlike the ROM, the EAROM can be written on; however, the writing operation is slower than the reading process, (milliseconds compared to microseconds). The device also uses complex internal techniques (thereby allowing only low integration levels) and demands multiple power supplies to achieve the writing process (hence an EAROM cannot be used as a normal read/write memory).

EEPROM – An electrically erasable programmable ROM (EEPROM) is a closely related device to the erasable programmable ROM (EPROM) in that it is programmed in a similar way, but the program is erased not with ultraviolet light but by the use of electricity. Erasure of the device is achieved by applying a strong current pulse, which removes the entire program, thus leaving the device ready to be reprogrammed. The voltages necessary to erase the EEPROM can either be applied to the device outwith or (more often) from within the host system, thereby allowing systems to be reprogrammed regularly without disturbing the EEPROM chips. In this way electrical erasability does yield certain benefits; however, this comes at the cost of fewer memory cells per chip and lower density, than on a standard ROM or EPROM.

FLASH MEMORY – Flash memory is a type of nonvolatile storage that can be erased and rewritten in situ like a magnetic disk. Flash memory is an intermediate form of memory between EPROM and EEPROM, essentially a cheap dense version of the EEPROM. One of the major difficulties with flash memory is that it can only be erased in blocks and not byte by byte, with erase blocks for various flash ICs ranging from 512 bytes to 64 KB. One of the results of this is that the read speed for flash memories is good but it is poor for writing. Furthermore, the flash memory cells in most flash ICs can withstand only about 100 000 erase/write cycles. The original flash memory used 5 V for reading and 12 V for erasing and writing, current devices operate with only a single power supply of 5V or even 3.3V. Flash memory ICs are usually incorporated into a card form.

The properties of a typical flash memory card are shown in Table 4.4. 16 Mbit flash memory devices were developed in 1993 using a 0.5 micron integrated circuit process, while it is envisaged that 64 Mbit devices will be available in 1995. With such densities the device could be incorporated into a computer system producing so called 'solid state disks' and replacing the use of magnetic hard disks.

(Continued)

Article 4.3

◆ READ–ONLY MEMORY ◆

(continued)

FLASH MEMORY

Table 4.4

Properties of a typical flash memory card.

Typical capacity		20 MB
Thickness		3.3 mm
Weight		35 g
Power:	Sleep	< 2 mW
	Operating	125 to 350 mW
Speed	Flash turn on time	3 to 20 ms
	Flash random access time	200 to 250 ms
Transfer rate	Burst read rate	4.5 MBs^{-1}
	Sustained read rate	0.6 to 4.5 MBs^{-1}
	Burst write rate	4 to 5 MBs^{-1}
	Sustained write rate	30 to 775 KBs^{-1}

An example of the internal block diagram of a ROM is shown in Figure 4.14 with the central diode matrix already having been programmed; this can be seen from the fact that some of the diode links have been removed. The matrix is arranged so that where the diode is intact it corresponds to a logic 1, while if the link is removed the output is a 0. Table 4.5 notes the data bus output for the ROM example in Figure 4.14.

Table 4.5

ROM data output for example in Figure 4.14.

Address A_0..A_4	Decoder output	Data bus output D_0........D_7
0000	line 0	11100001
0001	line 1	10001001
⋮	⋮	⋮
1111	line 15	11100110

Read–only memory

The **mask–programmable ROM** is a semiconductor type of read–only memory whose stored information is programmed during the manufacturing process. In essence this type of ROM is an array of possible storage cells with a final layer of metal forming the interconnections to determine which of the cells will hold 0s and 1s. Although the contents may be read they may not be written. It is programmed by the manufacturer at the time of production using a specially constructed template. All masked ROMs are therefore very similar, since most of their layers are the same with the differences only being introduced in the final layer metal mask.

Masked ROMs are typically manufactured in high volumes in order to minimize costs. They can cram a lot of data onto a relatively small chip area; unfortunately, because the information is programmed by the design of a mask which is used during the manufacturing process, it cannot be erased. Also, as the final metal layer is deposited at the chip factory, any repair of a masked ROM requires a long turnaround time. The process of repair consists of identifying the error(s), notifying the chip firm, and altering the mask before finally fabricating new chips. Any detected bug also has the unwanted effect of leaving the user with lots of worthless parts. Most commentators usually refer to mask–programmable ROMs as simply ROMs. Article 4.3 gives details on other types of ROM.

4.4 SECONDARY MEMORY

Secondary memory storage devices (such as magnetic disks and tapes) store programs and data that cannot be directly accessed by the central processing unit. They allow a large amount of memory space to be incorporated into systems but have a slower access time than the main memory. The characteristics of a number of different types of secondary storage are given in Table 4.6 and Glossary 4.3.

Table 4.6

Characteristics of several types of secondary memory devices.

Characteristic	TAPE			DISK	
	7 in reel	Phillips cassette	3M type cartridge	Floppy disk	Large hard disk
Capacity (MB)	40	1.44	11.5	2.4	571
Transfer rate (kbits s^{-1})	180	9.6	48	500	10 000
Number of tracks	9	2	4	200	600
Density (bits per in)	880	880	1600	3200	1600

The memory system within a computer consists of both primary memory, usually semiconductor memory, and secondary memory, usually **magnetic memory**. Figure 4.15 shows the different devices that can be found within this latter type of memory.

The magnetic disk family of devices are usually referred to as being directly accessible, while the tape units are often referred to as sequential access devices.

Figure 4.15

Magnetic memory storage.

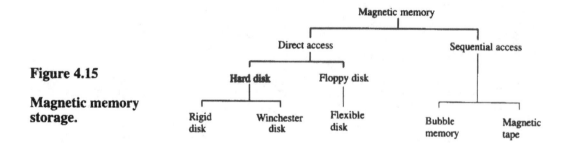

4.5 MAGNETIC DISK

A magnetic disk is a nonvolatile mass storage device which allows fast access to its data and the ability to both readily modify this data and to transfer it rapidly to the processing unit. Magnetic disk units come in different sizes and types with the most obvious distinction being that some are rigid (or hard) and some are flexible (or floppy). Both types basically consist of a rotatable disk, similar to phonograph record, that has a coating of magnetic material on one or both sides.

A typical modern processing system may have a variety of disk storage units with diverse speeds, interconnection details, capacities, and physical structures. Disks are also shareable objects in a processing system, in contrast to terminals which tend to be non–shared character devices.

The main advantage of magnetic disk storage over magnetic tape storage is that the time taken to find and read a piece of data, access time, is considerably faster. Disks achieve this speed increase due to their ability to refer to the data they hold by a key (or an identifier) rather than simply by a position within a sequence of data records. This key is held in the system during a search for a particular block of information, called a record, and compared against the keys of the different records that rotate underneath the read head during the search. Since magnetic tape allows only sequential access to find a piece of data, all other previously recorded data must be run through. Hence the access time for tape systems varies with the record being accessed. At an extreme it may require searching of every record in the tape before the correct one is located.

Magnetic disk

4.5.1   HARD DISK

A hard disk is a rigid magnetically coated disk which is used for the mass storage of data and programs. It provides faster access and can store more data than comparably sized floppy disks. The hard disk drive was developed in 1956 by IBM who introduced the Rampac 350 which stored 5 MB of data on 50 24–in disk platters. The diameter of the later disks was typically 14 in and often several disks would be mounted onto one spindle.

For example, the disk store shown in Figure 4.16 consists of 11 aluminium disks, coated on both sides with a ferric oxide material, which have then been mounted on a vertical shaft. Hard disks can either be fixed, that is permanently mounted in their drive units, or they can be removable, meaning that the disks and the information recorded on them can be changed whenever necessary.

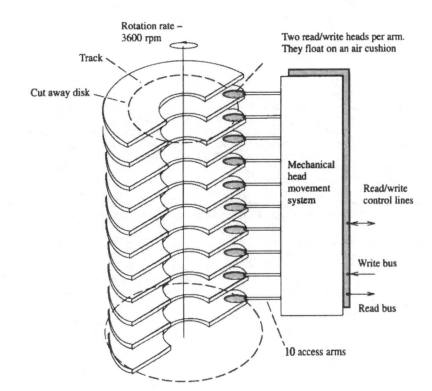

Figure 4.16

A hard disk pack.

Glossary 4.3

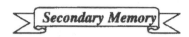

AUXILIARY STORAGE – The auxiliary storage in a computer system is usually taken to be the storage located on the magnetic disk and tape units or on optical disks. It is also known as external storage (or secondary storage) and supplements the primary (or main) memory storage. In most cases it is much larger than primary memory but operates at a slower speed.

BERNOULLI DISK – With removable hard disk drives for computer systems, it is normal to transport the entire hard disk and head drive mechanism, including the motor and the read/write heads. This is not necessary for removable drives based on a Bernoulli disk. The first commercial Bernoulli Box was introduced in 1983 by the American company Iomega. A Bernoulli cartridge uses a 5.25 inch disk rotating at 1800 rpm. Bernoulli cartridges come in 20, 44 and 90 MB capacities.

The magnetic surface on the Bernoulli system is on a flexible disk which hangs loosely in its resting state. Above this movable disk is a second disk at a fixed location. If the movable disk is accelerated by the drive motor to 2000 rpm it becomes flat and straight. An air cushion which is exactly 0.127 mm thick is created between it and the fixed disk. The read/write head oscillates at a height of 0.13 microns within the air cushion and can read/write data from or to the flexible disk. Since the rotation of the flexible disk automatically carries dust particles to the outside, the disk surface does not have to be hermetically sealed and hence it can be inserted or removed into a suitable disk drive. Access times of 22 ms are available.

CYLINDER – A cylinder is a set of tracks on a series of magnetic disks within a large disk drive that are under the various read/write heads of the drive at any one time. All such tracks are then accessible by the read/write heads with one movement, or positioning of the access mechanism.

Figure 4.17 shows eleven magnetic disks within a multiple disk drive. Each disk with the exception of the top and bottom disks has two surfaces to read from or write to. This makes a total of 20 recording surfaces. The tracks on each disk in the disk drive form a cylinder, hence in the example of Figure 4.16 we have 20 tracks per cylinder.

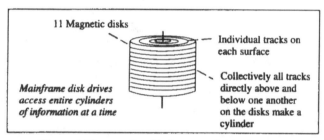

Figure 4.17   A cylinder on multiple disks.

DIRECT ACCESS MEMORY – In a computer system, direct access memory is made up from a variety of memory devices most of which all rely on the use of magnetization to retain information, the one exception is the optical disk. The difference between the various magnetic storage devices simply relates to the way in which the magnetic material is contained – be it on a plastic tape or disk or a metal disk. Each of these media provides a thin layer of magnetic material deposited on a supporting medium. Bits are stored by magnetizing small regions of the magnetizable medium in one direction for a logic 0 and in the other direction for a logic 1. The great benefits of this bulk storage are that it substantially brings down the cost per bit for the memory and that it provides a nonvolatile form of retention.

To address a particular word in direct storage requires a combination of jumping to a general vicinity, direct access, plus a subsequent sequential search to find the exact location asked for. The access time for such devices depends on the physical location of the requested data block (or record) at any given time, thus access time can vary considerably both from record to record, as well as to a given record when accessed at a different time. Direct access devices require a complex control system to allow them to operate and also require that the storage media itself must contain a certain amount of information to assist the controller in the location of the desired data.

DISK – A disk is a circular direct access memory device resembling a phonograph record which can be used for the mass storage of data in a computer system. Disks may be either magnetic or optical. Magnetic disks may be of the hard type, such as the Winchester disk, or may be made of flexible plastic in a protective envelope, as in the case of the floppy disk. Disks may also be removable from the computer system (floppy disks) or fixed into it in the case of the Winchester disk. The binary data on the magnetic disk is stored in the form of magnetic spots where the information may be stored and retrieved by read/write heads within a disk drive.

DISK DRIVE – The disk drive is the mechanical device used to rotate a magnetic or optical disk during data transfer. The circuitry in the disk drive also controls the position of the read/write heads over the surface of the disk in order to read/write the data from/to the disk. .

DISK PACK – A disk pack is a stack of magnetic disks which can be loaded and removed from a disk drive as a single unit. It is usually used in large mainframe computers. An older form of disk had 11 hard platters, each with 200 tracks.

(Continued)

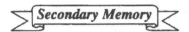

Glossary 4.3   *(continued)*

Secondary Memory

EXCHANGEABLE DISK – A magnetic hard disk system is usually designed such that the disks are fixed into the unit. However it may also be designed so that the read/write head(s) retract sufficiently for the disk(s) to be removed from the drive and replaced with other disk(s). Such a system is often known as a cartridge magnetic disk drive since the disk pack is contained in a plastic cartridge case. In large mainframe computers, the most common disk storage employs these exchangeable or removable disk cartridges. The standard form uses 10 read/write heads which are mounted on five movable arms. The arms can move in and out to enable the heads to cover all the disk's usable area.

MAGNETIC MEDIUM – A magnetic medium is any data storage medium and related technology, including disks, diskettes and tapes, in which different patterns of magnetization are used to represent bit values.

MASS MEMORY – In a computing system, the term mass memory is given to all forms of memory that are peripheral (or auxiliary) to the central processing unit (CPU) and the main random access memory. This type of nonvolatile memory is used to store large quantities of data as well as libraries of programs. Commercially available direct access devices are confined to magnetic and optical disks, while sequential access devices include magnetic tape on reels, magnetic tape cassettes and magnetic tape cartridges.

READ/WRITE HEAD – A read/write head is a type of electromagnet used as a component of a magnetic tape or disk drive. When the device is used to read from a particular magnetic medium it detects the individual magnetized areas and translates them into electrical pulses. In performing the write operation the device magnetizes the appropriate areas on the magnetic medium, thereby erasing previously stored data.

SEQUENTIAL ACCESS STORAGE – Sequential access storage is a form of memory for which the stored words or records do not have a unique address, but are stored and retrieved entirely sequentially. The main type of sequential access storage is magnetic tape.

In hard disk memories, information is stored on the concentric tracks of the disk(s) with access by read/write head(s). The data bits are recorded onto each disk by the write mechanism which induces magnetic dipoles in the rotating magnetic surface of the disk. To read the data from a disk, the magnetic surface passes under a read head (within the disk drive unit) where the relative motion of the magnetized areas on the disk, past the read–out transducer and induces an output voltage which is then amplified to produce the output data.

Hard disk

On some large capacity multiple disk stores a single scanning head is used, which not only scans the tracks, but can also be positioned vertically to select a particular disk. Other units have multiple read/write heads. For the example shown in Figure 4.17 it can be seen that each side of the disks, except for the top and bottom disk, is able to be scanned in a horizontal direction by a separate read/write head. The disk surface is physically formatted into concentric tracks which may also be subdivided into addressable sectors.

In the hard disk pack shown in Figure 4.16 all the heads are positioned together so that one track per disk side is available to the read/write head at any given position, such a set of tracks is called a cylinder. There are 20 read/write heads and therefore 20 tracks can be accessed at any one time. Older disks also had up to 2000 tracks per disk and therefore 2000 cylinders each containing 20 tracks. The typical rotational speed of the disk is 3600 rpm (revolutions per minute). Table 4.7 presents typical characteristics of a hard disk system.

Table 4.7

Characteristics of a large fixed disk drive.

Characteristic	Specification
Number of spindles per cabinet	2
Capacity per spindle	400 MB
Data rate	5 Mbits s^{-1}
Average access time	15 ms
Rotational speed	3600 rpm

The tracks on a surface are typically separated by between five and ten mils and are less than 3 mils in width, on both sides of the disks. The storage capacity would then be in the range of several hundred million bits.

To integrate a hard disk drive into a computer system requires that the activities of the asynchronously analog device, the disk, be co-ordinated with the synchronous digital interface (the computer system). This necessitates the use of **a hard disk controller**, i.e. special analog circuitry to synchronize and encode/decode the data pulses going to/coming from the disk during a write/read operation, with attention having to be paid to data recovery methods as well as to error detection and correction. In addition, software support and system interface overheads must be carefully considered in order to ensure effective communications with the computer system.

With disk systems in general, the electronic circuitry connecting the disk to the computer can be divided into two physical parts, the drive electronics and the controller logic. It is usual in small disk systems to provide the minimum of electronics within the disk drive itself and to concentrate most of the control in a separate control unit. This unit often takes the form of a circuit board within the computer system. In this way the disk drive need only contain the analog circuitry required to position the read/write heads as well certain circuitry to interface to the controller. There is then very little intelligence built into the drive electronics, which are largely analog in nature, and simply provide an analog/digital interface.

The controller, on the other hand, may well contain a large amount of intelligence such that it is able to take over the complete control of the reading/writing process of the hard disk system, even to the level of stepping the head actuator. The controller then provides the required sector and track to the drive to enable it to find the correct part of the disk. In this way the disk controller acts as an interface between the computer and the hard disk drive, receiving commands and data from the microprocessor before passing them on to the disk drive, and on the return journey, taking back data and status signals from the drive to the microprocessor. The commands from the processor are interpreted by the disk controller to provide signals to the disk drive to control the reading and writing of data.

Hard disk

The use large scale integrated (LSI) technology allows a programmable LSI disk controller to be incorporated on a single integrated circuit. Such a part easily interfaces to the microprocessor bus appearing as simply an input/output peripheral device with several internal addressable read/write registers.

To read data from or write data to the disks, the computer's operating system must issue commands to the disk controller. It must also transmit the track and sector addresses from a table it maintains in memory. The controller then positions the read/write head(s) over the appropriate track(s) on a particular disk(s), based on the starting address for the desired data. The disk controller reads the address of the sectors that pass under the head and when it finds the appropriate position on the disk it either reads the contents of the appropriate number of sectors into memory or writes data from memory into empty or unused sectors. The transfer of these bytes can be via programmed input/output instructions or via direct memory access (DMA). The operation requires both an access time and a read/write time.

1. Access Time – The access time is the sum of both the seek time and the latency time.
(a) Seek Time: As a hard disk has multiple tracks along its surface, but there is usually only one read/write head per disk, a read or write operation is normally preceded by a seek operation to locate the head above the track containing the required data. The seek operation is essentially a start/stop movement.

(b) Latency Time: In addition to the seek time, a disk operation also has to wait until the start of the block required is located under the read/write head; this waiting time is called the latency time.

2.   Read/Write Time – Compared to the access time the read/write operation is very rapid. Since the read/write operation time is related to the rate of disk rotation, reading a single item of data can take nearly as long as reading a whole track of data. Hence, in order to be time efficient, data transfers to and from a disk are arranged in large blocks. However, too large a block size would be space inefficient, since it may not always be possible to fill every block with data. Fixed sizes of 128, 256 or 512 bytes per block are accepted as compromise standards. It can be seen from this description that to access a particular part of the disk surface takes a variable amount of time.

Hard disk

Built–in circuitry within the disk controller, and software in the disk I/O routine are usually arranged to detect common errors, and to repeat any failed operations until their successful completion. The error detection circuits incorporate such things as a Hamming code check character for each record transferred. Other functions within the controller may entail it carrying out the formatting of the disk and running a cyclic redundancy check (CRC) for data errors.

4.5.2   WINCHESTER DISK

Fixed magnetic hard disks are employed within processing systems when fast access is required to a large amount of data. The most widely used fixed hard disk drive is now the Winchester type of system. This is a recent form of the flying head type of disk system introduced by IBM in 1973 (the model 3340) in which the disk was fixed into the disk drive unlike the removable hard disk units of the time. A design team member Ken Haughton called the 3340 a 'Winchester' after its two 30 MB modules which reminded him of his Winchester 30–30 rifle. Figure 4.18 shows a block layout of a Winchester hard disk system.

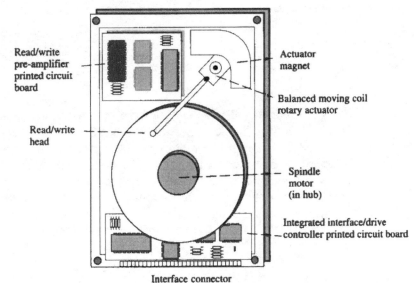

Figure 4.18

Winchester type of hard disk (block layout).

Within modern hard disk systems, such as the Winchester type of hard disk shown in Figure 4.19, when the disk is rotating, it carries along with it a thin layer of air caused by the viscous friction between the air and the disk. The read/write heads are spring loaded and are shaped so that the moving air layer forces the head away from the disk, so keeping the head separated from the disk by a small distance less then 20 μm. In this way the read/write heads are kept close to, but not touching, the hard disk as it revolves at high speed. This separation prevents wear on both the disk and the head. Due to the nature of the operation of the read/write head, the system may also be termed a flying head hard disk system. Figure 4.19(a) shows the flying read/write head.

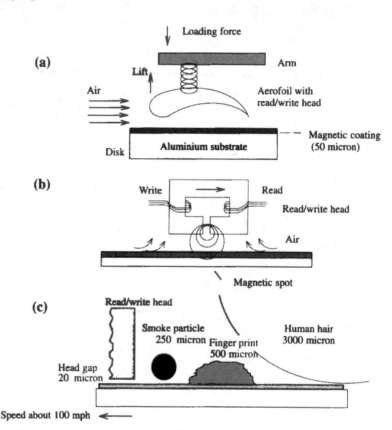

Figure 4.19

(a) Operation of a flying read/write head
(b) Read/write head
(c) Relative size of the head gap to various hazards

This technique allows greater precision in the head actuator and disk assembly and is such that with modern drives a reduced flying distance of twenty μm or less can be achieved (the head of the original 3340 flew on a thin film of air 18 millionths of an inch above the magnetic disk surface). Since the head can fly closer to the disk surface, compared to the previous type of hard disk, then the individual magnetic spots can be made smaller, therefore greatly increasing the bit density. Figure 4.19(b) notes the creation of the magnetic spots on the disk. It can be seen that the closer the head flies to the surface of the disk then the smaller the spots will be. However, to achieve the small flying distance the head assembly must be very lightly loaded, compared to older flying head systems.

Winchester disk

It is important in a Winchester system to exclude airborne particles. For example, a smoke particle may be about 250 μm in diameter, many times the head to media separation, and might easily cause a head crash if sucked under the head.

Figure 4.19(c) shows the relative size of the head gap to potential hazards in a disk system. This means the complete system is sealed in a dustproof package, except for a highly efficient air filter. The air filter is required since some particles are generated internally through the action of the head landing on the disk surface and dislodging oxide. These particles must be removed from the unit as soon as possible. As a further point, because of the possibility of head crash, most manufacturers have endeavoured to ensure that the whole system is made very reliable and in most cases is even able to withstand the occasional head crash.

Winchester disk

In a Winchester system is also necessary to either retract the heads away from the surface completely when the system and disk motor are turned off or to allow the head to land on the outer track of the disk. Unlike the older type of hard disk drive, the Winchester system is designed such that the head actually lands on the surface of the disk on the outermost track. This special outer track is not used for recording but has been arranged to be the head's landing zone. In order to prevent damage of the recording head during take off and landing, this outer disk surface is specially lubricated. When the disk is stationary, the head rests on the landing zone; however, when the disk motor is turned on, the disk rapidly accelerates and the head experiences a sliding contact with the surface of the disk for about 1/4 a revolution, taking off at about 400 rpm (revs per minute). On switching the system off, a brake is applied to the disk spindle within the disk drive unit and the head lands on the outer track, as the speed reduces below 400 rpm.

Although Winchester development began as a replacement technology to large flying head disk systems, it soon became apparent that it could be developed as a low–cost mass storage system suitable for microprocessor systems using 8 in and 5.25 in diameter disks. The capacity of the first 5.25 in Winchester disk delivered in 1980 from Seagate Technology was 5 MB. The storage capacity of a 200 MB Winchester hard disk is roughly equivalent to 140 000 book pages.

The characteristics of a typical Winchester disk are given in Table 4.8. Since their introduction, disk drive dimensions have steadily fallen and their storage capacity has risen greatly.

Table 4.8

Typical characteristics of a Winchester disk system.

Characteristic	Specification
Number of disks	1 to 4
Data surfaces	1 to 7
Bit density	34,768 bits in^{-1}
Track density	500 tracks in^{-1}
Tracks per surface	1200
Surface capacity per disk	> 50 MB
Rotational speed	3600 rpm
Data transfer rate	10 Mbits s^{-1}

Low end drives, those storing less than 100 MB of data, use designs and materials that lend themselves to high volume, low cost production. On the other hand the higher capacity drives require much more costly approaches. There are several ways to raise the capacity of the disk drive without necessarily increasing its physical size.

1. One of the more obvious is to put more platters, or disks, into the case. To do this for a Winchester drive the designers have moved the motor from its previous location underneath the stack of disk platters to inside the disk's hub.

2. The capacity also rises when more data can be put into each platter. Manufacturers do this in two ways.

(a) By using a larger area of each platter's surface. This is difficult without resorting to an increase in the physical size of the Winchester drive.

Winchester disk

(b) By increasing the density of bit storage on the disk platter. The number of bits that can be put onto a given area depends on the magnetic material that coats the disk surface, on the head size and on the configuration. The recording surface of a hard disk used to be a plastic binder sprinkled with slivers of gamma ferric oxide. The problem with ferric oxide is that it is not coercive enough to allow a large number of bits and tracks to be packed together (coercitivity is a measure of the field required to reverse the direction of magnetization of a bit on the magnetic medium). As more and more bits are packed closer together very high coercitivity materials are required to ensure that a bit will not be demagnetized or have its magnetization reversed by neighbouring bits. Coating the gamma ferric oxide with splinters of cobalt doubles this coercitivity. Modern disks are made from aluminium, coated with a film of a cobalt and nickel alloy. This allows high packing densities up to 34 768 bits per inch.

There are now a large number of different types and sizes of hard disks such as 1.3, 1.8, 2.5, 3.5 and 5.25 in. **Mini hard disks** measuring less than 45 mm by 50 mm with a thickness less than 15 mm have also been constructed for use in portable computers. The properties of a typical 2.5 inch disk drive are noted in Table 4.9.

Table 4.9

Properties of a 2.5 in mini–disk drive.

Capacity		40 MB
Thickness		10.5 mm
Weight		60 to 80 g
Power	Sleep	10 – 15 mW
	Operating	1.5 W
	Disk spin–up	2 to 3 W
Speed	Disk spin–up time	1.0 to 1.5 s
	Disk seek time	18 ms
Transfer rate	Burst–read rate	1.2 to 5 MB s^{-1}
	Sustained read rate	0.9 to 2.2 MB s^{-1}
	Burst write rate	1.2 to 5 MB s^{-1}
	Sustained write rate	0.9 to 2.2 MB s^{-1}

4.5.3   FLOPPY DISK SYSTEM

The floppy disk storage system was originally designed by IBM in the mid 1960s as a semirandom access memory that could combine the head positioning technology of hard disks with the recording surface technology of magnetic drums. It was initially used to store diagnostic programs for fault detection in the hard disk system and was subsequently introduced as part of the 3300 system in 1970. By 1973 the floppy disk drive had been sufficiently refined that the IBM 3740 commercial data entry system came with an 8 in diameter floppy disk drive for data storage.

The floppy disk system proved an immediate success and was used to replace punched cards as a medium for data entry and also to store programs and data files. In a floppy disk system, the disk and head make mechanical contact just like the tape and head in a magnetic tape system. Therefore the disk need not be particularly rigid and can be thin enough to be slightly flexible. Originally a single read/write head was used on the floppy disk, with a track density of 48 tracks per inch but by 1977 a double head system, recording on both surfaces, had been perfected. An example of a 5.25 in floppy disk system is shown in Figure 4.20.

Floppy disk system

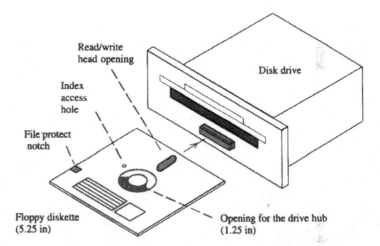

Figure 4.20

5.25 in floppy disk and disk drive.

Table 4.10 indicates the floppy disk systems that have been incorporated into different personal computers. The capacity and a brief description of the different types of disk systems currently available is given in Table 4.11.

Table 4.10

Different disk systems within personal computers.

Disk	Disk density	Size (KB)	Personal computer
5.25 in	Single	140	Apple 2
	Double	360	IBM personal computer
	Double (high)	1200	IBM AT
3.5 in	Double	400	Apple Macintosh
	Double	800	Apple MacIntosh SE
	Double	720	IBM PS/2Model 30
	Double (high)	1440	IBM PS/2Model 50

Table 4.11

Different floppy disk systems.

Disk type	Capacity (KB)	Description	
5.25 in	360	2S 2D	Double density
	1200	2S HD	High density
3.5 in	720	2S 2D	Double density
	1440	2S HD	High density
	2880	2S ED	Extra high density
		(where S = sided ; D = density)	

SECTORING – Before a magnetic disk can be used it must be divided into tracks and sectors. A track on a magnetic disk or tape is the part on which the data can be recorded. It is a horizontal row stretching the length of the magnetic tape or one of a series of concentric circles on the surface of a disk onto which the information is written. Before a magnetic disk can be used in a computer system the tracks must be divided up into sectors. There are two form of sectoring.

HARD SECTORING – With hard sector formatting, the disk is provided with a small hole at the beginning of each sector, punched near the spindle hole; for example, with 10 sectors there would be 10 sector holes around the disk. A photosensor is used to detect these sector markers as they spin by the index hole opening and a count is kept in hardware within the disk drive. The first sector is identified by an additional hole known as the index hole, punched near the first sector hole which can be used to reset the sector count.

SOFT SECTORING – The more common sectoring technique is to use software to write sector identifications (IDs) at the beginning of each sector. The disk is provided with an index hole to identify the first sector while each sector is subsequently identified by a prerecorded address which includes a special address mark written at the beginning of the sector. This requires that a special format program be used on each new disk before it can be used for program or data storage. Figure 4.21 shows an example of sectors on a five–track disk with eight sectors per track.

It can be noted that due to their different construction, hard and soft sectored disks are not interchangeable. Most disks are soft sectored and although this is more complex than hard sectoring it provides more flexibility.

Figure 4.21   Example of a five–track disk with eight sectors per track.

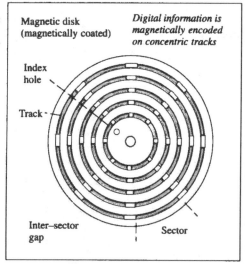

The 3.5 in disk system is becoming the most widely used of the different sizes of floppy disks. Table 4.12 notes the properties of three types of 3.5 in disk systems.

Before a floppy disk can be used it must be sectored (Article 4.4).

Table 4.12

Properties of several types of 3.5 inch disk systems.

Property	Low density	High density	Toshiba 2.88 MB
Cylinders	80	80	80
Sectors per track	9	18	36
Formatted capacity (MB)	0.72	1.44	2.88
Unformatted capacity (MB)	1	2	4
Rotation speed (rpm)	300	300	300
Recording density (bits in^{-1})	8717	1432	34 868
Track density (tracks in^{-1})	135	135	135
Transfer rate (KB s^{-1})	22.5	45	90

FLOPPY DISK HANDLING

1. A disk must not be hand written on with a hard device. It is best to write out labels before placing on the disk.
2. Any magnetic part of a disk must not be touched.
3. A disk must not be forced into the disk drive.
4. A disk must not be removed from microcomputer disk drive while the disk drive is running.
5. A disk must not be exposed to direct heat or cold.
6. A disk must not be near any magnetic fields such as those generated by television sets and electric motors.

Floppy disk system

The **floppy disk** itself is a magnetically coated flexible plastic disk which is mounted in a protective envelope and can be used as a mass data storage medium. The eight in floppy diskette was introduced in 1970 for loading microprograms into disk controllers for the IBM System 370 mainframes. Further developments have included the introduction of the 5.25 in diameter floppy disk by Shugart in 1976 and more recently the 3 in and 3.5 in diameter floppy disk. The smaller 3 and 3.5 in floppy disks are enclosed in rigid plastic cases with sliding openings for the read/write heads. These smaller disks are less prone to any thermal changes of the media and hence allow a greater track density to be achieved. Figure 4.22 presents the different parts of a 3.5 in disk. Article 4.5 gives information on how to take care of a floppy disk.

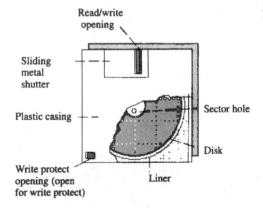

Figure 4.22

A 3.5 inch floppy disk.

1. **Floppy Disk Drive** – In operation the floppy disk is inserted into the disk drive unit which clamps the centre of the disk and rotates it inside its stationary jacket. Three holes are cut in the protective sleeve to allow access to the floppy disk. One is the spindle hole through which the spindle within the floppy disk drive protrudes when the disk is loaded into the drive. The spindle turns the floppy disk inside its plastic jacket. The second is a rectangular aperture in the jacket which allows the read/write head of the disk drive access to the circular tracks, in order that the head may record or play back data on the disk. The third hole is used to sense the start of the various tracks and is known as the index hole. The index hole in the jacket allows the index hole in the floppy disk to be optically sensed for the purpose of generating a synchronization signal.

There are three main tasks that must be performed by a floppy disk drive. These are to spin the disk at a constant speed, position the read/write head over the desired track and transfer data to and from the media.

Figure 4.23 shows a block diagram of the interior of a floppy disk drive indicating the position of the different component parts. It can be seen that the system contains a stepper drive motor to rotate the disk, a read/write assembly to control the position of the read/write head, and various control and interface integrated circuits mounted on a printed circuit board.

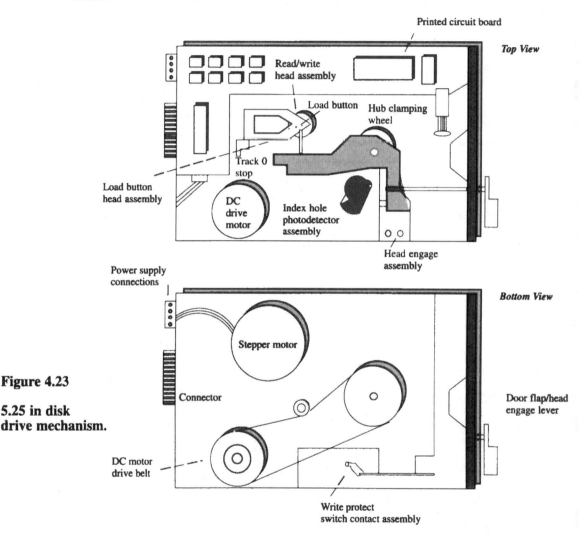

Figure 4.23

5.25 in disk drive mechanism.

Floppy disk drive

The read/write head of a floppy disk is three heads in one. The data head is in the centre and a tunnel erase head is mounted on each side. The tunnel erase heads are to clear the intertrack region of any flux transitions. As we can see from Figure 4.23 the floppy disk drive has a door that covers a slot into which the floppy disk is placed. When the door is closed, the disk is clamped to the rotating spindle by the action of a solenoid forcing a pressure pad against the opposite side. For reading and writing, the disk is rotated by the drive spindle at a fairly slow speed (normally 300 rpm for the 5.25 in disk). The rotation of the disk is maintained at a constant rate by a precision servo controller motor. This compares to 3600 rpm for a hard disk system.

The read/write head is moved radially across the disk above the head access hole by using a stepper motor. One pulse applied to the motor steps the head one track in or out. In order to access a particular track on a disk, the head is brought to the starting track (track 00) and its position detected by a limit switch. The head is then stepped the number of required tracks to its destination by pulsing the stepper motor, this process is called the seek operation.

Floppy disk drive

When the head is over the desired track, a control mechanism in the drive lowers the head, causing it to come into direct contact with the floppy disk. The read/write head then either writes the 0/1 magnetization dots serially along a given track or, alternatively, plays back the data by reading the magnetic state of the dots through the magnetic read head.

(a) Floppy Disk Read/Writing – The recording technique used within a floppy disk drive is such that a logic 1 is written onto the disk by reversing the direction of the current through the read/write head, which in turn causes a reversal of the magnetic flux stored in the iron oxide coating on the disk itself. No flux reversal stores a 0. Reading the disk consists of sampling the disk at regular intervals and detecting any current reversal, indicating a stored logic 1. In this way information on a floppy disk is stored as long strips (i.e. tracks) of magnetic dots. Tracks are numbered from the outside towards the centre, with the outermost track being numbered as track 00. This track is usually kept as a catalogue or directory of the contents of the disk. The tracks are further divided into a number of sectors with a typical sector being divided into a series of fields.

After writing or reading, the head is lifted from the disk to minimize the wear on both the disk and the head. In this way the heads are only in contact with the disk surface when a read or write operation is taking place. A few seconds after a read/write operation has been completed the disk drive motor is automatically turned off.

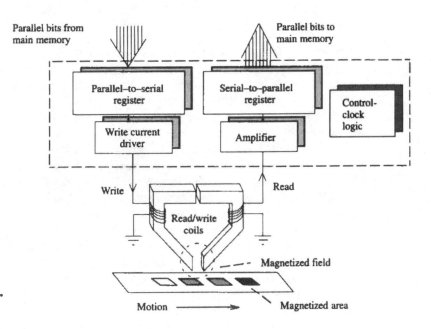

Figure 4.24

Floppy disk read and write operation.

The read/write operation is shown in more detail in Figure 4.24 where it can be seen that the parallel data from the microprocessor must first be converted to a serial form before being sent to the read/write head. Here the signals form magnetized areas on the magnetic medium of the disk. Reading from the disk is the opposite procedure.

Read/Write Time: Each read/write operation in a floppy disk system takes at least the sum of the seek time, the latency time and the read/write time. The average access time for a 5.25 in disk is of the order of 80 ms compared to just over 10 ms for a rigid disk system. 5.25 in disks using double density mode can achieve transfer rates of 500 kbits s^{-1}. The track to track stepping time for 3.5 in disks is of the order of 3 ms for an 80 track disk.

*Floppy
disk
read/write*

Recording Format: Data is recorded on a floppy disk either using FM (frequency modulation) format or MFM (modified frequency modulation) format. Using the FM format, both data bits and clock pulses are recorded onto the disk, whereas using MFM only the data bits are recorded. This last technique doubles the amount of data which can be stored using the same number of flux changes in the magnetic layer of the disk, i.e. double density disks as opposed to single density. One limiting factor is the stability of the recording mylar media, which is subject to deformities by mechanical and thermal stress.

Storage Capacity: A typical 5.25 in disk may have 80 tracks per side with 15 sectors per track, each sector containing 512 bytes. 3.5 in disks have track densities greater than 135 tracks per inch. Track density for a hard disk is 200 per in, meaning the total storage capacity of a floppy disk may typically be over 30 times less than a hard disk.

It can be noted that not all the disk surface is utilized for recording and the linear recording density increases towards the middle of the disk. This latter effect means that the individual bits tend to be spaced out more along the tracks at the outer tracks compared to those nearer the centre. This is necessary since towards the centre of the disk the tracks become physically shorter and it is important to ensure the same number of bits are recorded on each track. In order to assist in the reading and writing process each disk has information on it other than simply data values or program instructions. This special information is recorded during a programmed formatting procedure.

(b) Floppy Disk Track Format : Figure 4.25(a) presents a diagram of the position of tracks on a magnetic disk. Each track is formatted in a particular way. The format for an IBM 3740 soft sectored disk is shown in Figure 4.25(b), each sector on the track has an identification (ID) section and a data section.

To assist in the operation of a floppy disk system it is necessary to be able to identify each sector. This can be accomplished by recording special identification synchronizing patterns between each sector. The format divides each sector into four fields: the ID record, ID gap, data field record and data gap. The data address mark is one byte to indicate the start of the data record. It can be seen that the data record is 128 bytes long plus two cyclic redundant check (CRC) bytes. These last two checksum bytes are used to check that the data contains no errors. A buffer of 33 bytes after the data record allows the read/write head to prepare for the next sector by switching to read mode if the previous sector involved writing data. At the beginning of each sector is an address mark, or ID field, which passes the read/ write head first and identifies the upcoming areas of the sector.

The information recorded on each sector of a soft sectored disk then consists of two parts, an identification part and a data part.

Identification field: The ID field identifies the data field by sector and track number while the data mark indicates whether the upcoming data field contains a good record or a deleted record. The length of the sector and the head number in the case of double sided systems

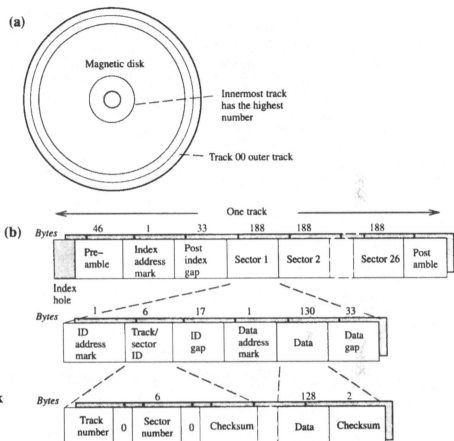

Figure 4.25

(a) **Tracks on a magnetic disk**
(b) **IBM 3740 track format.**

Floppy disk track format

may also be incorporated in the identification information. The identification field concludes with two CRC error detection bytes in order to ensure the correctness of the identification bytes. The CRC code is required since a floppy disk is significantly less reliable than a floating–head hard disk, due mainly to the head to disk contact wear. Disk operations are also prone to errors because of surface imperfections, dust and nonconsistent rotation.

Data field: After the identification field comes the data field. The data field begins with either a data, or a deleted data, address mark (special recording patterns) followed by the data bytes and terminated with two CRC bytes – again for error checking.

4.6 OPTICAL DISK

The biggest challenge to magnetic mass storage comes from optical technologies such as the CD ROM (compact disk read–only memory), the WORM (write once read many time memory) and the erasable optical disk. Although optical storage is slower than magnetic media, primarily because of the greater mass of the optical read/write heads, it offers greater capacity in a removable cartridge.

Figure 4.26 shows a block diagram of an optical disk system. An optical disk drive uses a laser to scan an optical disk to transfer information from the optical bubbles on the disk into the computer. The optical disk contains a series of small lands and pits in the substrate, the optical bubbles, which the laser system can read as large or small signals. The lands give the largest signal which is normally interpreted as a 0, while the pits may be interpreted as a 1. The optical bubbles which store the information are typically less than a micron in diameter. The reflected light from the pits and lands is detected and passed via an interface unit to the computer.

Optical disk

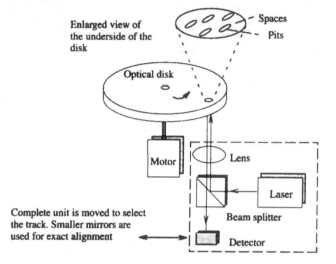

Figure 4.26

Block diagram of an optical disk system.

Optical storage comes in a number of different forms.

(a) **CD** – The 12 cm compact disk audio system was released in 1983. It is a nonerasable disk with 60 min of playing time.

(b) **CD ROM** – this can store 550 MB on 12 cm disks.

(c) **CD–I** – this describes methods for providing audio, video, graphics and text.

(d) **WORM** – 5.25 in disks which can hold 200 to 800 MB of data.

(e) **Erasable Disks** – 3.25 in and 5.25 in disks with typical capacity of 650 MB.

4.6.1   CD ROM

CD technology was developed in 1976, the result of a joint effort between the Philips and Sony Corporations. The potential application of CD ROM technology as a high capacity, low cost medium for read–only data storage appeared in 1984. A CD ROM measures 120 mm (4.72 in) across, has a 15 mm (0.6 in) spindle hole and a thickness of 1.2 mm. With CD ROMs the information is fixed into the disk during manufacture.

CD ROM disks have a single spiral track, as a record does, that starts near the centre and spirals outward, this is different from magnetic floppy disks which have concentric circular tracks divided into sectors. The three–mile long track on the CD ROM is divided into equal length sectors or blocks. The track width is about 600 nanometers (nm) wide and adjacent turns of the spiral track are about 1600 nm apart producing a density of about 25 000 pits per inch. Data on the spiral track is in the form of small variable length, 120 nm deep depressions called pits and intervening flat areas called lands. The CD ROM's read head is an optical assembly with a low power gallium arsenide laser and a photodetector which is able to read the pits and the space. The assembly directs the laser beam through a one way reflective mirror to the disk surface where the lands reflect the light from the laser while the pits disperse it. The reflective mirror redirects the returning light to the photodiode.

CD ROM

The CD ROM has emerged as a means to distribute large amounts of fixed information between/to large databases. Using a 4.72 in optical disk it is possible to store 540 MB. This is equivalent to 200 000 pages of text or 12 000 scanned images, equivalent to 1500 low density floppy disks. CD ROM has a typical average seek time of 300 to 500 ms.

4.6.2   WORM

WORM disks can store information in real time. The early WORM drives could store over 2 GB on 12 in disks and have proved effective in archival storage applications. These optical disks can be written on once, but the data is subsequently fixed and can only then be read. A laser is used to write onto the disk by increasing its power by 20–fold. Present 5.25 in WORM disk drives have access time less than 98 ms and can store up to 600 to 650 MB, some up to 1 GB. 12 in disks can store about 5 to 6 GB per side.

4.6.3   ERASABLE DISK

A new technology is the introduction of 5.25 in optical disks which are erasable. Two types are available, magneto–optic and phase change.

1. Magneto–optic Disks – The magneto–optic erasable disk consists of a thin vertically magnetized film of a rare earth transition metal, such as terbium or gadolinium, alloyed with iron and/or cobalt. At room temperature the film's resistance to magnetic change is very high; however, when part of the film is heated by a laser pulse the resistance changes to a point where it is susceptible to magnetic reversal by a small bias field. These marks can be used to store binary 1s and 0s. Marks are erased by applying another laser pulse and reversing the bias field. Since the material is initially amorphous the cumulative effect of temperature fluctuations on the film over numerous read/write cycles can change the film's atomic structure. At present disks come in two sizes: the 5.25 in disk can store 600 MB per side with 1 GB expected, the 3.5 in disk can store 128 MB per side. The fastest MO drives have seek times in the 25–35 ms range. They are rewritable and have a rotational speed of 3600 rpm with a data rate of 950 kbps. They have a claimed 25 year data retention life–time.

2. Phase Change Technology – Phase change technology depends on the ability of a thin film of a chalcogenide alloy, usually materials built around tellurium, to be switched from a crystalline to an amorphous state and back again. When a spot on a phase change disk is heated by a brief laser pulse, the film loses its crystalline structure and becomes amorphous. As the spot cools, the amorphous state is locked in. A longer less intense pulse reverses the change in state. Phase change technology therefore involves the continual rearrangement of the atoms on the thin alloy film on the disk, unlike magneto–optic technology which only changes the direction of the electrons. Concerns have been raised regarding how many times the write/erase process can be performed in this type of erasable medium before the phase change disk suffers degradation. Laboratory tests indicate that problems in disk reliability occur beyond about 10 000 read/write cycles. There is also a question about the archivability of phase change disks because tellurium is used. This alloy is attractive for optical disks in that it heats quickly and allows rapid change of states; however, it has the unfortunate property that it oxidizes rapidly in a humid atmosphere. If phase change techniques can overcome these problems, optical disks using such a technology do offer the advantage of higher bit densities and better data transfer rates than magnetic disks.

The data transfer rate performance of an optical disk drive can be of the order of 1.4 MBs^{-1}. Various techniques are being applied to increase this.

1. The spinning speed of the disk can be increased. Speeds have been increased from the 30 revolutions per second (rps) for the 12 in disk up to 60 rps for the 5.25 in disks
2. The linear density of the media can be increased.
3. Lightweight optical heads can be used in the drive. These allow rapid acceleration and deceleration in the heads, resulting in a faster seek time.

4. Semiconductor laser technologies are also evolving such that laser diode arrays will be available soon. This would allow designs to multiply the data rate by using parallel track on the disk. The concept of combining laser diodes arrays with a fixed drive holding multiple erasable platters could lead to an optical drive capable of out–performing a Winchester disk drive. With the inclusion of multiple platters it is now possible to build 3 GB optical disk drives.

Optical disk drives require sophisticated error correction techniques This is required since in optical recording systems problems such as insufficient write power, defects in the media, or noise in the read channel can contribute to generating errors. This is particularly acute in optical disks because of the high storage density in optical recording (10^8 bits per cm^2) hence pinholes and particles as small as 1 μm can cause errors. As standard error rates of 1 in 10^{12} or better have become the target for most optical drives sophisticated error correction codes are required.

4.7 MAGNETIC TAPE

Tape storage is a form of secondary memory storage which stores data on magnetic tape. It uses a sequential access method to read or write the data. A block diagram of a magnetic tape system is shown in Figure 4.27

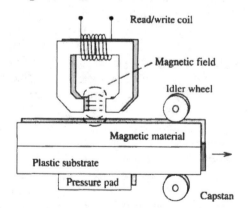

Figure 4.27

Magnetic tape operation.

In a **magnetic tape** the magnetic coating is supported on a long flexible plastic tape. To write onto the tape, a current is injected into the read/write coil on the recording/playback head in either of two directions, thereby magnetizing the magnetic coating in one direction or the other. As the tape moves across the head, the digital bits are recorded onto the tape adjacent to each other, along the complete tape length. To read, the tape is moved across the head and at each change in direction of magnetization a voltage is induced in the head coil, the voltage polarity depending on the direction of change in magnetization. Address information (in the form of simple interrecord gaps) is stored on the tape in order to separate individual records and allow the different data sets to be distinguished. The interrecord gaps also allow the tape to reach its correct operating speed before writing/reading of the data takes place.

Magnetic tape

Tape storage cannot be used for reading and writing simultaneously. If the user wishes to change the contents of a tape, the tape must be read into the memory in blocks where the change can then be made, a block at a time, and recopied back onto the tape. Hence tape systems are not a convenient backing store for frequent accessing because of the long time involved in winding and rewinding the tape to select the records. Tapes are much more suited as a back-up or archival form of storage, i.e. data that is stored and rarely changed.

4.7.1   TAPE STORAGE

There are a number of different types of tape systems.

1.   Open Reel Tape – Open reel magnetic tape was first introduced in 1953. The first tape subsystems could store over 2 MB on 2400 ft. Present open reel magnetic tape is typically 0.25 in wide and wound in lengths of 600 ft, 1200 ft or 2400 ft in spools up to twelve inches in diameter.

The most common method of representing data on the tape uses a nine track coding scheme, although other coding schemes such as seven track are also available. Using the nine track coding scheme allows as many as eight bits to be written side by side across the width of the tape at a time, eight data bits and one parity bit, using nine read/write heads. Along the length of the tape, bits can usually be written with a density of about 1600 bits per inch, hence for a 2400 ft tape the total number of bits per track is of the order of 46 million. A typical speed for a drive unit using open reel tape is 45 in s^{-1}, where, at this rate, bits are read or written from each track at a rate of 72 kbits s^{-1}. The most common computer I/O device which utilizes open reel tape is the vacuum tape mechanism.

A **vacuum tape mechanism** is shown in Figure 4.28, while Table 4.13 presents the specifications of a typical vacuum column digital tape transport.

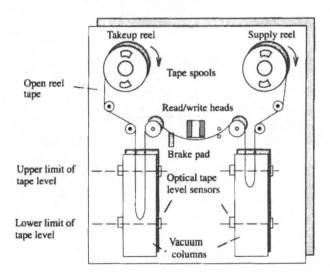

Figure 4.28

Vacuum tape mechanism.

Characteristic	Specifications
Data density	7 or 9 tracks, 1600 or 6250 characters per in
Tape velocity	200 in s^{-1}
Start–stop time	3 ms at 200 in s^{-1}
Reel diameter	10.5 in
Tape length	2400 ft
Tape width	0.5 in
Tape thickness	1.5 mils
Rewind speed	300 in s^{-1}

Table 4.13

Specifications of a vacuum column digital tape transport.

2. Cartridge (Magnetic Tape) – The magnetic tape cartridge unit is a compact enclosed package of magnetic tape mostly used for storing computer programs or data values. The most pervasive magnetic tape technology is the IBM 3840 cartridge. The cartridge tape is 0.5 in wide, 2400 ft long, is fully enclosed in a plastic case measuring 6 in by 4 in, and can hold 400 MB. Data is recorded serially at 1600 bits per inch and at a speed of 30 in s^{-1} on the tracks (from 1 to 6 tracks).

Figure 4.29 shows a cartridge tape unit. It can be seen that there are several rollers to guide the path of the tape and two reels to supply and take up the tape. The specifications of a 3M tape cartridge are given in Table 4.14.

A mini cartridge is a smaller version of the cartridge which is about the size of a standard tape cassette and contains 140 ft, 300 ft or 450 ft of tape.

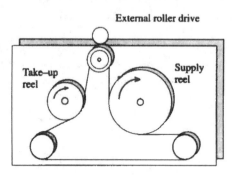

External roller drive

Take–up reel

Supply reel

Figure 4.29

A cartridge tape drive unit.

Characteristic	Specification
Operating Speed	
Read/write	30 in s^{-1} forward and reverse
Fast forward, rewind	90 in s^{-1} forward and reverse
Packing density	6400 bits per in.
Transfer rate	48 kbits s^{-} maximum
Interrecord gap	1.33 in typical
Speed variation	+/– 4%
Interface logic	TTL compatible
Power	5 V DC

Table 4.14

The characteristics of a 3M tape cartridge.

In workstations and computer networks cartridge tapes are often used as a back–up for Winchester disks. Protection against lost data due to disk crashes or otherwise can be accomplished by simply dumping the contents of the Winchester disk(s) to a cartridge tape at the end of each day after all transactions and computations have been done. Taking a copy of a hard disk for safety is done using a tape operation called streaming. **Streaming** was first introduced on open reel tape systems to back up disks, an idea now extended to cartridge tape systems.

Tape cartridge

3. <u>**Cassette (Magnetic Tape**)</u> – A magnetic tape cassette is an enclosed package of 0.25 in magnetic tape which is usually housed in a plastic container measuring approximately 4 in by 2.25 in. Data is recorded serially on each track. An exploded view of the tape mechanism is shown in Figure 4.30.

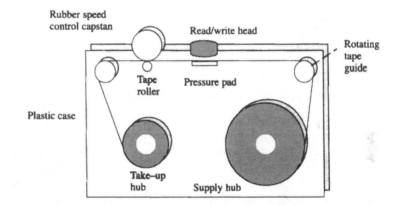

Figure 4.30

A cassette tape unit.

The read/write heads are contained in the tape recorder as is the speed control capstan. The standard cassette unit is the Philips type cartridge, containing 282 ft of 0.0015 in thick magnetic tape which is intended to be phase encoded at 800 bits in^{-1}. Tape systems based on 8 mm cassette format can store up to 2.3 GB.

4.7.2   RECORDING FORMATS

The recording format for magnetic tape is given in Figure 4.31. Individual records are written into blocks on the tape with interrecord gaps separating them. The data within the record is preceded by identification bytes with a checksum added to the end of the data record. The checksum is used to check for errors in the reading or writing process. In order to record data onto a magnetic tape, various formats are often used. The following are two of the most widely used tape recording formats.

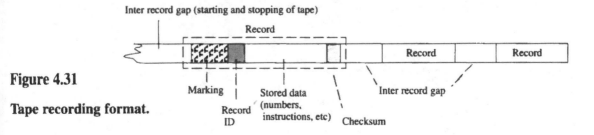

Figure 4.31

Tape recording format.

Tape recording format

1. **Manchester encoding** – This is a digital encoding technique in which the period assigned to each bit in the digital message to be sent is divided into two complementary halves. In this scheme a negative to positive transition of the voltage in the middle of the bit period designates a logic 1 while a positive to negative transition represents a logic 0. With this form of transmission no extra clock signal need be sent to enable synchronization between communicating parties, i.e. the transmission method is self clocking.

2. **The Kansas city standard** – This is an audio cassette recording format which uses four cycles of a 1200 Hz tone for a logic 0 and eight cycles of a 2400 Hz tone for a logic 1. An example of this is shown in Figure 4.32.

Figure 4.32

Kansas City recording.

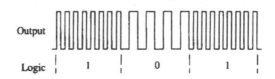

4.8 KEY TERMS

Access time	Electrically erasable
Associative memory	programmable read–only
Auxiliary storage	memory (EEPROM)
Bandwidth	Erasable optical disk
Bernoulli disk	Erasable programmable
Bubble memory	read only memory (EPROM)
Cache memory	Erasable storage
Capacity	Exchangeable disk
Cartridge (magnetic tape)	Flash memory
Cassette (magnetic tape)	Floppy disk
Charged coupled device (CCD)	Floppy disk drive
Compact disk read–only	Floppy disk handling
memory (CD ROM)	Floppy disk reading/writing
Content addressable memory	Floppy disk system
(CAM)	Floppy disk track format
Core memory	Hard disk
Cycle time	Hard disk controller
Cylinder	Hard disk (mini)
Data Rate	Hard sectoring
Density	Kansas City standard
Direct access memory	Latency time
Disk	Magnetic disk
Disk drive	Magnetic medium
Disk pack	Magnetic memory
Disk sectoring	Magnetic tape recording
Dynamic random access	formats
memory (DRAM)	Magnetic tape
Drum storage	Magneto–optic disks
Electrically alterable read–only	Main memory
memory (EAROM)	*(Continued)*

Manchester encoding	Refresh
Mask programmable read–only	Scratchpad memory
memory (ROM)	Secondary memory
Mass memory	Sectoring
Memory	Seek time
Memory classification	Semiconductor memory
Memory hierarchy	Sequential access storage
Nonvolatile storage	Soft sectoring
Open reel tape	Static random access
Optical disk	memory (SRAM)
Phase change technology	Stack memory
Primary memory	Static memory
Programmable read–only	Streaming
memory (PROM)	Tape storage
Random access memory (RAM)	Vacuum tape mechanism
Read–only memory (ROM)	Volatile storage
Read/write head	Winchester disk
Read/write time	Write once read many times
Read write memory (RWM)	(WORM)

Key questions

4.9 REVIEW QUESTIONS

1. What is the purpose of storing data in a microprocessor system?
2. What is a memory storage cell?
3. What properties of materials can be used to store information?
4. List as many different types of memory device as you can.
5. What is the main purpose for primary storage and secondary storage?
6. What does the term 'volatile' mean when applied to memory system? What types of memory are volatile?
7. What is the central memory of a computer used for?
8. Distinguish between reading and writing memory.
9. What is meant by the term 'contiguous' memory?
10. What is a megabyte?
11. Detail how a cache memory improves the performance of a computer system.
12. Describe the differences between static and dynamic RAM.
13. Outline the operational characteristics of
    (a) ferrite core memory,
    (b) magnetic drum,
    (c) bubble memory,
    (d) CCD memory.
14. Explain the difference between RAMs and ROMs and suggest two possible uses for a ROM chip.
15. Does a ROM employ flip–flops?
16. What do the following terms refer to when used to describe computer memory: RAM, ROM, PROM and EPROM?
17. Why is an EEPROM called a 'read mostly memory'?

18. Which of the two, RAM or ROM is volatile?
19. Explain the term 'access time' as applied to an auxiliary storage device.
20. Why is secondary storage necessary in a personal computer system?
21. How does a hard disk drive work?
22. What is the advantage of ensuring the read/write heads of a hard disk fly close to the disks surface? What are the disadvantages?
23. What is so disastrous about a head crash in a hard disk drive?
24. What do the following terms mean when used to describe disk systems: access time, seek time, rotational delay, data transfer?
25. What is the meaning of the term 'direct access'?
26. Summarize the types of backing storage available which are used on micros, minis and mainframes.
27. What advantages are there in Winchester disk technology compared to the older fixed head hard disks?
28. How could the data rate from a disk system be increased?
29. What advantages in terms of memory density are obtained by sealing a Winchester disk?

*Review
questions*

30. Define the following terms in respect of magnetic disk memories: exchangeable disk storage, Winchester disks and floppy disk.
31. What were the main reasons for the development of the floppy disk?
32. What are the advantages and disadvantages of floppy disk as a means of storage compared to a hard disk?
33. There are three standards for modern floppy disk drives. What are they?
34. Why must a disk be formatted before its use?
35. What are the advantages/disadvantages of a soft sectored disk compared to a hard sectored disk?
36. State the recommendations for taking care of floppy disks.
37. Describe the process of loading a piece of data from a magnetic disk in terms of the read/write head movement and the times required at each stage.
38. If a floppy disk rotates at 360 rpm how many data bytes can be transferred from the disk per second (ignoring all the address and preamble bytes)?
39. Why is it bad practice to leave a disk in a disk drive when the power to the drive is switched off?
40. Why is it necessary to use error detecting codes on data coming from a disk system?
41. Define each of the following with respect to disk processing: track, sector, cylinder, address and density.
42. List three different types of optical disks.
43. What are the advantages of using optical disks over magnetic hard disks?
44. Distinguish between sequential and direct access memory types.
45. Outline the general characteristics of magnetic tape.
46. What are the advantages and disadvantages of magnetic tape for storage of data?
47. Why are tape systems slower than disk systems for accessing data?
48. What are the factors which have led to magnetic tape becoming the principal form of back up storage?
49. Explain what 'streaming' is used for.
50. What is meant by the following terms used in connection with storage devices, access time, transfer rate, random access and serial access?

4.10 PROJECT QUESTIONS

1. Investigate how a PROM is programmed.
2. Find out what different forms and of memory are available in a small personal computer.
3. Why do you think bubble memory is no longer used?
4. With suitable reference what are thin–film read/write heads?
5. What would a notebook computer require in terms of memory systems?
6. Investigate vertical recording techniques for improving the memory capacity of hard disks.
7. Why can the disks used in the IBM AT computer not be read by the Apple MacIntosh computer?
8. Investigate the manufacturing process for optical disks.
9. What will be the main application areas of optical disks?
10. Describe how magnetic tape storage is used in mainframe computers.

4.11 FURTHER READING

A number of articles, providing a range of information on the different types of memory are listed below.

Memory Hierarchy (Articles)

1. Scaling the memory pyramid, B. Ryan, *Byte*, March 1992, pp. 161–170.
2. Semiconductor–memories the frontiers of production, *Electronics Industry*, October 1986, pp. 11–17.
3. Performance trade–offs for microprocessor cache memories, D. Alpert and M. Flynn, *IEEE Micro*, August 1988, pp. 26–44.

Flash Memory (Articles)

1. Flash memory challenges disk drives, G. Legg, *EDN*, February 18th 1993, pp. 99–104.

Disk Drives (Articles)

1. Megafloppies, S. Satchell, *Byte*, October 1990, pp. 301–309.
2. Winchester drives reach for a gigabyte, J. Voelcker, *IEEE Spectrum*, February 1987, pp. 64–67.
3. The new wave of removable storage, R. G. Cote and S. Wszola, *Byte*, October 1992, pp. 198–210.

Optical Disks (Articles)

1. Optical technology what's mature and what's on the horizon, M. De Haan, *ESD*, September 1987, pp. 41–49.
2. Parallel optical memories, D. Psali, *Byte*, September 1992, pp. 179–182.
3. The new wave of removable storage, R. G. Cote, S. Wszola, *Byte*, October 1992, pp. 198–210.
4. Entering a new phase, B. Ryan, *Byte*, November 1990, pp. 289–296.

5. The once and future king, B. Ryan, *Byte*, November 1990, pp. 301–306.
6. State of the media, D. A. Harvey, *Byte*, November 1990, pp. 275–281.
7. Optical provides direct access storage at a reasonable price, J. Donavan, *EDN*, 22nd July 1993, pp 92–98.
8. CD–ROM inside and out, R. C. Alford, *Byte*, March 1993, pp. 197–206.

Magnetic Tape (Articles)

1. Digital audio tape for data storage, E. Tan and B. Vermeulen, *IEEE Spectrum*, October 1989, pp. 34–38.

I/O Systems

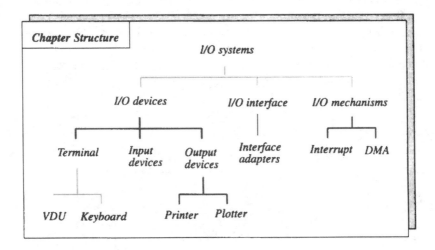

Chapter Structure

I/O systems
- I/O devices
  - Terminal
    - VDU
    - Keyboard
  - Input devices
  - Output devices
    - Printer
    - Plotter
- I/O interface
  - Interface adapters
- I/O mechanisms
  - Interrupt
  - DMA

Learning Objectives

In this chapter you will learn about:

1. The different types of input/output hardware – their uses, advantages and disadvantages.
2. The construction of a terminal for a computer system.
3. The types and capabilities of impact and nonimpact printers.
4. The computer hardware interfaces required to enable input and output of data from and to a computer.
5. The use of DMA in I/O transfers.

> *'Engineering is the application of science for profit.'*
> *Anon*

Chapter 5

I/O Systems

The Input and Output (I/O) section of a computer handles the transmission of information into and out of a computer (Glossary 5.1). The hardware units that are used for input and output are called I/O devices, or peripherals. The **I/O devices**, make up the third of the architectural sections within a processing system. After the microprocessor and the memory system they constitute what may be regarded as the most important aspect of the overall architecture of a computer. I/O devices both record information in a machine readable form and display information in a form that is easily understood by an operator, e.g. through alphanumeric displays. We will consider a number of I/O devices starting with the computer terminal.

5.1 TERMINAL

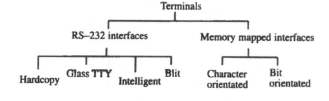

Figure 5.1

Different types of terminals.

A terminal is a form of I/O device which is usually equipped with a keyboard, a visual display unit and a transmitter and receiver capable of sending and receiving data over a communications link to the main host computer (Glossary 5.2). Terminals come in two basic types as shown in Figure 5.1, the RS–232 device or the memory mapped device.

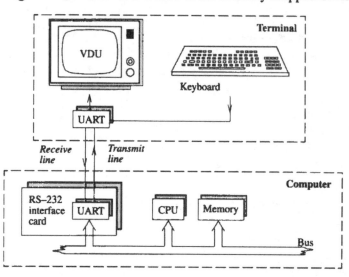

Figure 5.2

An RS–232 VDU terminal.

Glossary 5.1

EXTERNAL I/O – External I/O devices are devices such as terminals, local printers, network interfaces and tape drives, which are effectively accessed by only one user at a time. These devices essentially provide a method for entering new work into a computer system and outputing work from the system. Most are character devices in that they deliver or accepts a stream of characters without regard to any block structure. They are not addressable and do not have any seek operation.

INTERNAL I/O – The term internal I/O device within a computer system tends to refer to magnetic disk or tape drives. Internal I/O devices tend to be block devices in that they store information in fixed size blocks, each with its own address. Common block sizes range from 128 to 1024 bytes. The essential property of a block device is that it is possible to read or write each block independently.

PERIPHERAL DEVICE – Within a computer system, a peripheral device is any piece of external equipment, such as a printer, floppy disk unit or visual display unit, that provides the processor with outside communications. The peripheral device communicates with, and is usually controlled by, the central processing unit to perform such tasks as the storing and displaying of data and the conversion of data to a form which is usable by the computer.

RS–232 Terminals – RS–232 terminals are devices containing a serial interface which the computer can output data to or receive data from, one bit at a time. The host computer uses a 25–point connector connected to a special interface unit, or card, within the computer to communicate with the display and keyboard, see Figure 5.2.

Terminal

The interface makes use of a special I/O integrated circuit device called a universal asynchronous receiver/transmitter (UART) to handle the communications. The RS–232 is the oldest and most popular of the interface standards. There are different types of RS–232 interfaces for different applications; some are listed in Figure 5.1. The hardcopy RS–232 is reserved for devices such as printers while the glass TTY is used in displays.

Memory Mapped Terminals – Memory mapped terminals consist of a monitor display and a separate keyboard. The keyboard is typically connected to an I/O port on the computer while the display is interfaced to the computer via a special memory called a video random access memory (VRAM).

Figure 5.3

A memory mapped VDU terminal.

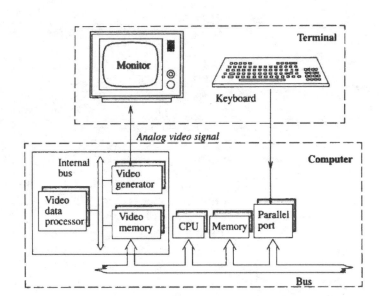

The VRAM forms part of the computer's address space and is addressed by the central processing unit (CPU) in the same way as the rest of the memory, see Figure 5.3. A video controller chip or video data processor subsequently accesses the VRAM to generate the signals to control the display monitor.

Memory mapped terminals allow for extremely fast interaction, although most of the low cost visual display units (VDUs) operate at the standard rate of 30 characters per second, to avoid non standard interfacing requirements.

Terminal

There are two types of memory mapped interfaces, those that are **character orientated** in which individual characters are mapped onto the screen, or **bit mapped terminals** in which every bit in the VRAM controls an individual pixel on the screen. With bit mapped terminals the screen may also be used to form pictures as well as letters and numbers. The bit mapped approach is used in graphics terminals to enable high resolution displays.

5.1.1   VISUAL DISPLAY UNIT (VDU)

A VDU is the part of a computer terminal where the computer produced image is projected onto the face of a cathode ray tube. The display has become virtually an obligatory feature within modern computer systems to visually communicate data held in the computer's store. The data from the computer may either appear in the form of numbers and letters, in diagrammatic or picture form, or as a combination of both. It can be noted that the early computers did not have any form of VDU and instead the output was simply channelled through printers. The VDU is a much more useful form of output since human users tend to think in images and thus a screen that can display text and graphic information is extremely convenient. Although it is a very fast way of communicating information it does not form a permanent record of computer output.

VDUs make use of a cathode ray tube to form the characters. A **cathode ray tube (CRT)** is constructed from an electron gun, electron lens and fluorescent glass screen. A block diagram of a cathode ray tube is shown in Figure 5.4.

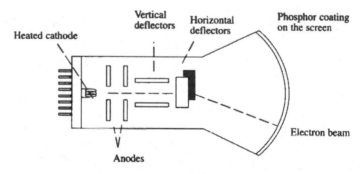

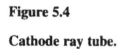

Figure 5.4

Cathode ray tube.

The electron gun produces a stream of electrons which is focused by an electric field into a beam. The electron beam is deflected horizontally and vertically by either an electric field formed between two sets of metal plates or a magnetic field formed by two sets of coils carrying current. In the case of the metal plates, they are fixed at right angles to each other inside the neck of the tube. This form of electrostatic tube is mostly used for oscilloscopes. In CRTs using two two sets of coils, the coils are fitted at right angles to each other outside the neck of the tube.

Glossary 5.2

⟩⟩⟩ ⟨ **Terminal Description** ⟩ ⟨⟨⟨

CONSOLE – The console on a computer refers to the front panel of the device. The console normally has certain control keys and special switches, e.g. a start key, stop key and power key, located on it for manual operation of the unit. It may also contain various lights that can display the information located in certain registers within the computer.

CURSOR – A cursor is a position on a visual display unit (VDU) screen which normally indicates the next screen position for an input. The input can be an alphanumeric character, a graphic symbol or the erasing of a character or symbol. On a typical computer keyboard there are four keys to control the movement of the cursor on any part of the screen.

ICON – An icon is a display image which is presented in place of words to the user through the VDU screen. The icon is usually a simple picture showing the control options, for example instead of the phrase 'delete file' the user may be shown a picture of a file and one of a wastepaper basket.

PROGRAMMABLE TERMINAL – A programmable or intelligent user terminal is one that has some form of computational capability which allows it to perform complex I/O functions.

RASTER – A raster is a scanning pattern used in generating, recording or reproducing graphic images on a screen.

REMOTE TERMINAL – A remote terminal is a computer peripheral unit, connected at a point in a computer network at which data can either enter or leave the host computer. A user can then communicate with the main (or host) computer in order to enter and run his programs on the computer. Remote terminals may be either intelligent or dumb.
1.   An intelligent, or programmable, terminal is a stand alone facility that has both internal and external storage facilities.
2.   A a dumb terminal does not have its own central processing unit but is an input/output device only and is not capable of processing information.

WINDOW – A window is a portion of a display screen that can be used to display information. In the 'Windows' graphical interface, the windows are also used to display the execution of different tasks, thereby allowing a computer to appear to be doing more than one task at once.

These electromagnetic tubes are used for VDUs and television receivers. The screen for both types of tube is coated with phosphorus which emits light when bombarded by electrons.

The display process within a VDU consists of generating an electron beam which creates a dot of light when the beam strikes the phosphor coated surface of the cathode ray tube. The VDU builds characters and numbers from a series of individual points on the screen (pixels) forming the different symbols on the television type screen from a matrix of dots. Most VDUs use a matrix of 5 by 7 dots (pixels) to form the characters, with additional dots being used to separate the characters horizontally and vertically.

Visual display unit

The character images needed for the display of text are stored as patterns of bits in a special read only memory (ROM) called a character generator. The clarity of the text depends on the number of dots employed in forming each character. A typical monitor can display 24 lines of text, each line of which has 80 to 130 characters, hence VDUs can display well over 2000 characters on the screen at any one time. The displays are refreshed at a rate of 50 to 60 Hz.

VDUs are used in **graphics display devices** which is a piece of visual display equipment that is able to project its output visually, in the form of graphs and line drawings. The device may also be able to accept inputs from a keyboard and/or light pen to manipulate the presentation of its output.

The display of graphic images, whether they are engineering drawings, graphs, or moving targets in a video game, also calls for complex software and for large amounts of memory. Typically the special software routines can produce circles, curves and vectors. Each pixel on the VDU display is stored in the computer memory and requires one word of storage. Graphics–associated text may also be displayed in various sizes and angles.

The graphic controller for the display uses a video data processor which consists of arithmetic and state machine logic, implemented usually as gate arrays. The controller drives a raster scan display, allowing the control of every pixel on the screen of the cathode ray tube.

In operation the data processor transfers a string of pixel information to the video random access memory. The string consists of the pattern length, 1 to 16 bits, the co–ordinate of the pixel at which the operation will begin, the type of operation to be performed and the number of times the operation is to be performed.

There have been many different **graphics standard interfaces** for personal computers. These have tended to be set by the IBM personal computer (Table 5.1).

Graphics Standard	Description
Monochrome display adapter	The original IBM PCs only had a display that could handle text.
Hercules graphics card	This was the dominant adapter used in the early IBMs to produce graphics facilities. It had a 720 by 348 pixel display mode.
Colour graphics adapter (CGA)	In 1983 IBM produced its first colour graphics system consisting of the CGA and IBM personal computer colour display. It could only show four colours with a resolution of 320 pixels across by 200 pixels down; however, with a monochrome monitor it could display 640 by 200 pixels.
Enhanced graphics adapter (EGA)	In 1984 IBM introduced the EGA display which could display 640 by 350 with 16 colours at a time.
Multicolour graphics array (MCGA)	In 1987 IBM introduced the MCGA adapter for its personal System PS/2 model 30 computer.
Video graphics array (VGA)	For all PS/2 models above the model 30, IBM introduced a different adapter on the mother board. This was the VGA which offered a number of modes: (a) high resolution at 16 colours, (b) 640 by 480 with 256 colours, (c) a quarter million colours in CGA resolution. VGA is also compatible with EGA.
Super VGA	1024 by 768 with 256 colours.
XGA	1280 by 1024 up to 32 768 colours.

Table 5.1

Graphics standards for the IBM PC.

5.1.2   KEYBOARD

In most computer systems the keyboard is the prime input interface between the user and the operating system. Through the keyboard the operator can instruct the computer to load and run programs, change the structure of application programs, enter data into a running program, instruct the operating system to copy and transfer files as well as operate many other control functions.

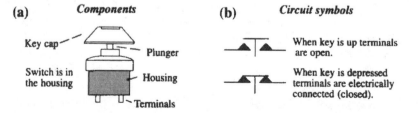

Figure 5.5

Keyboard switch.

A keyboard is a collection of **key switches**, or pressure sensitive pads, which are collected in a configuration usually based on a standard typewriter. In general a switch is a

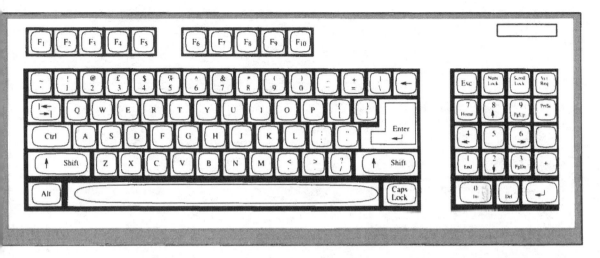

Figure 5.6   Typical layout of a computer keyboard

device which is capable of being set into more than one particular state. The circuit symbols which are used when a switch is open or closed are shown in Figure 5.5(a) while Figure 5.5(b) shows the components within a switch for a computer keyboard.

The keyboard has been styled on the standard QWERTY keyboard of the typewriter, the layout of which can be traced back to the early part of this century when, after analysis of the distribution of letters within ordinary text, an attempt was made to optimize the placement of the letters on a keyboard to prevent the jamming of the print hammers when used at high speed. A typical keyboard layout is shown in Figure 5.6. With modern computers the layout has been adopted despite the fact that there are no longer any print hammers to jam within the keyboard. Several new key layouts have been tried but so far have not proved popular. As well as the standard typewriter keys many computers have additional special function keys to the layout and have included a separate number keypad, usually located on the far right of the keyboard. Cursor keys may be included, to move the cursor pointer on the VDU screen.

Keyboard

There are two methods by which the computer can read the keys being struck on the keyboard, these are to use either an encoding method of key interpretation or a scanning method of data entry.

1. Keyboard Encoder – A keyboard encoder is a device that produces a unique output code for each possible closure of one of the keys on a keyboard. The encoder determines which key has been pressed on a keyboard through the use of a large encoder circuit, shown in Figure 5.7. The encoder has an input line for each key such that when a key is pressed the encoder transforms the decimal number of the input to a binary code. For example if the 8th key was pressed then the output from the buffer would be 0001000, while if the 9th key was pressed the output would be 0001001. The standard format adopted is the ASCII seven bit code, Article 5.1

2. Keyboard Scan – The process of examining the rows and columns of a matrix keyboard to determine which key has been pressed is known as a keyboard scan. In this mode of operation the processor continually runs a small program which scans through groups of keys on the keyboard such that when a key is pressed the processor is able to determine which key it was, by looking at the s ignal from an output buffer. In most applications using

Article 5.1

ASCII

The American Standard Code For Information Interchange (ASCII) is alphanumeric, i.e. pertaining to a set comprising letters, digits and normally associated characters (e.g. punctuation marks). It is widely used in computers and communications, being established in 1968 in an attempt to standardize the data representation among different computers in order to achieve compatibility between all data processing systems. With ASCII, each character is represented by a 7–bit data word (Table 5.2) plus a check digit. In this way the code can be used to represent the 96 displayed characters and the 32 non–displayed control characters within the character set of most modern computers. Composite symbols are often used to extend the range of symbols available.

Table 5.2

The ASCII character set.

Hex	ASCII	Hex	ASCII	Hex	ASCII
00	Null/idle (NULL)	2B	+	56	V
01	Start of header (SOH)	2C	,	57	W
02	Start of text (STX)	2D	–	58	X
03	End of text (ETX)	2E	.	59	Y
04	End of transmission	2F	/	5A	Z
05	Enquire (ENQ)	30	0	5B	[
06	Positive ack (ACK)	31	1	5C	\
07	BELL	32	2	5D	]
08	Backspace (BKSP)	33	3	5E	I
09	Horizontal tab (HT)	34	4	5F	–
0A	Line feed (LF)	35	5	60	`
0B	Vertical tab (VT)	36	6	61	a
0C	Form feed (FF)	37	7	62	b
0D	Carriage return (CR)	38	8	63	c
0E	Shift out (SO)	39	9	64	d
0F	Shift in (SI)	3A	:	65	e
10	Data link escape (DLE)	3B	;	66	f
11	Device control (DC1)	3C	<	67	g
12	Device control (DC2)	3D	=	68	h
13	Device control (DC3)	3E	>	69	i
14	Device control (DC4)	3F	?	6A	j
15	Negative ack (NAK)	40	@	6B	k
16	Synchronous idle (SYNC)	41	A	6C	l
17	End of block (EOB)	42	B	6D	m
18	Separator (S0)	43	C	6E	n
19	Separator (S1)	44	D	6F	o
1A	Separator (S2)	45	E	70	p
1B	Escape (ESC)	46	F	71	q
1C	Separator (S4)	47	G	72	r
1D	Separator (S5)	48	H	73	s
1E	Separator (S6)	49	I	74	t
1F	Separator (S7)	4A	J	75	u
20	Space (SP)	4B	K	76	v
21	!	4C	L	77	w
22	"	4D	M	78	x
23	£	4E	N	79	y
24	$	4F	O	7A	z
25	%	50	P	7B	{
26	&	51	Q	7C	I
27	'	52	R	7D	}
28	(	53	S	7E	~
29	)	54	T	7F	Delete
2A	*	55	U	80	

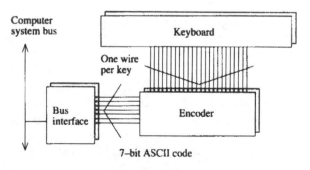

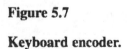

Figure 5.7

Keyboard encoder.

INPUT DEVICES

DIGITIZING TABLET – Digitizing refers to the process of converting graphic representations into digital data that can be processed by a computer system. In 1978 the company Summargraphics announced the 'Bit Pad' which is generally recognised as the first commercial digitizer.

Figure 5.8 shows a digitizing tablet. Signals from the computer are sent along the x and y direction into a grid of wires. The sensor probe picks these up to indicate its position on the grid. By activating the buttons on the probe an operator can use the digitizing tablet to transfer the details of a drawing or a map into a computer.

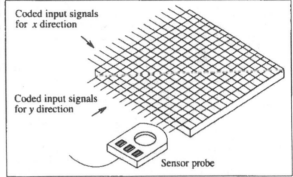

Coded input signals for x direction

Coded input signals for y direction

Sensor probe

Figure 5.8   Digitizing tablet.

GRAPHICS PAD – An engineer or designer using a graphic pad makes use of a stylus (which is a writing instrument like a pen but with a point that has no ink) in order to make a drawing on a special board. The drawing is read into a computer and can be manipulated by software in the computer to produce a display on a visual display unit. Various electrical techniques are used to sense the position or the location of the pen on the tablet.

JOYSTICK – The joystick is a computer input device, similar to a mouse, which is used to control the position of the cursor on the computer screen. The joystick may also incorporate buttons to allow an operator selection to be entered. Joysticks are frequently used to allow an operator to play certain games.

LIGHT PEN – A light pen is a pen–shaped photosensitive device with a photoelectric cell at its end; they first appeared in about 1967. A light pen is used to manually direct the stream of electrons that form the image on a cathode ray tube into any required pattern. It can be used with the aid of software within a computer, to draw lines, modify the display, or manipulate the data on the screen of the visual display unit. Alternatively, it can be used to call up standard patterns and characters for which the machine is programmed.

MOUSE – A mouse is a generic name for a hand held digital input device that translates movement across a surface to cursor position on a visual display screen. The user moves the mouse across a flat surface, causing a pointer (or highlight) to move in a corresponding way on the display screen. It is used frequently in menu–driven programs, word processing, high resolution graphics and video games. Most mice have buttons that can be depressed to provide programmed inputs. In most cases there are two, which either selects the menu choice on which the mouse rests or causes the screen to scroll to show different material. The mouse was first introduced in 1972 in the ALTO computer developed by Xerox at its Paulo Alto Research Center (PARC).

The mechanical construction of a mouse is such that it generates pulses using photo–optical sensors and optical interrupter disks. The sensors detect rotation of an internal ball that rolls when the mouse is moved across a desktop. With appropriate software the display cursor automatically tracks the mouse whenever it moves.

SCANNER – A scanner is a device for converting text or pictures into digital information which can be stored in a computer. With suitable software a scanner can enable a user to easily incorporate line art, photos and other graphical images into documents. A gray level scanner supports 256 shades of gray with 8 bits of digital data being used for each pixel. Colour scanners use a red, green and blue filter to give a 24–bit colour representation of the input picture (i.e. allowing a representation of 16.7 million different colours) with a resolution greater than 800 dots per inch.

There are two types of scanner, hand held and desktop. Hand held scanners use a four inch scan width. Desktop scanners use A4 size. The document being scanned is first illuminated by a light source. A single scan line from the document is focused onto a charge coupled device (CCD) and the output voltage of each light sensor is converted to a digital binary bit pattern.

(Continued)

INPUT DEVICES

(continued)

Scanners (Continued)

Scanners can be used with **optical character recognition** (OCR) software to read in text from a document and convert it into ASCII character information. OCR is the process of reading words in a scanned image and translating them into computer–readable codes. It was originally designed for input to typesetting machines for the newspaper industry; however, it is now commercially available for a wide range of different computers. The software uses feature analysis techniques such that by matching the shapes of the different character inputs in terms of the curves, straight lines and enclosed spaces, the software can quickly decide on the particular alphanumeric character being represented.

TOUCH SCREEN – A touch screen is a special form of I/O device which is attached to or built into the screen of a VDU. There are two approaches which can be used.

1. A resistive touch screen uses a thin transparent panel which mounts in front of the CRT display monitor. When the panel is touched two sheets of conductive material connect and an output voltage that is proportional to the position of the touch is produced. The two axes of the CRT are scanned separately to allow the x and y position of the touch to be determined.

2. A second type of screen uses an ultrasonic technique. Piezoelectric transducers mounted along the edge of the screen set up acoustic waves in the screen. Receiver transducers are mounted to pick up the echoes of any object touching the screen allowing its position to be determined. It can also be built from a series of infrared beams which, when a finger breaks the beam, activates a couple of sensors.

TRACKERBALL – A trackerball is an input device used on a computer for controlling the position on the screen of the cursor. It is similar to the mouse input device except that the ball is at the top of the unit. The operator moves his hand across the trackerball to manipulate the cursor. The trackerball translates the sphere's direction and speed of rotation into a digital signal used to control the cursor. Movement of the ball causes optical encoders to generate pulses that indicate the speed and direction of rotation. Most trackerballs produce between 50 and 550 pulses per revolution. Trackerballs are used in graphics applications in computer aided design systems, image processing and video games.

Keyboard scan

keyboard scan techniques the keyboard is a fully mechanical contact keyboard with diodeisolation which is arranged as a matrix of keys consisting of a series of rows and columns. By placing a logic low on the different columns of the matrix in turn (normally called scanning) then when a key is pressed its row is signaled at the output buffer with a logic low. Since the computer knows which column was activated when the row number is indicated, it can therefore determine the coordinate of the key that has been pressed in the matrix. It subsequently uses an ASCII look up table to determine which of the keys has been pressed. The scanning rate is kept high enough that even the fastest typist cannot enter data that the computer will miss.

5.2 INPUT DEVICES

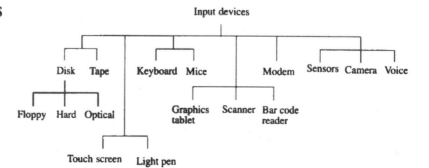

Figure 5.9

Computer input devices.

Input devices

The input devices in a computer system allow data and instructions to be read in by the central processing unit. Example devices are a keyboard, a light pen, a digitizer, an optical scanner, a punched card reader, a touch sensitive screen, a mouse, a modem or a disk drive, (Article 5.2). Figure 5.9 notes a number of these devices.

5.3 OUTPUT DEVICES

Typical output devices used in a computer system are video display unit, printer, plotter, hard copy graphics device, loudspeaker, modem and disk drive. These devices have been grouped according to various criteria and presented in Figure 5.10. We will consider the most widely used output device other than the VDU – the printer.

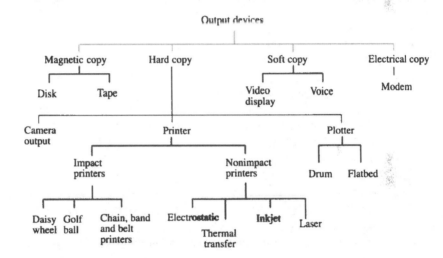

Figure 5.10

Computer output devices.

5.3.1   PRINTER

A printer is a computer output device which can be used to produce a permanent hard copy. There are essentially two types: impact printers, which use a mechanical means to produce printed text, and nonimpact printers which use electrostatic, chemical or other forms of technology to produce the text.

Table 5.3

Printer characteristics.

Mechanism	Quality	Graphics Support	Speed	Reliability
Impact printers				
Typewriter	Letter	No	15 cps	Medium
Daisy wheel	Letter	No	15–70 cps	Low
Dot matrix	Draft/correspondence	Yes	125–300 cps	High
Nonimpact printers				
Thermal	Draft	No	30–120 cps	Low
Inkjet	Correspondence	Yes	300 cps	Medium
Laser (micro)	Typeset (330 dpi)	Yes	500–600 lpm	High
Laser (main)	Typeset (1000 dpi)	Yes	18 – 21 klpm	High

cps = characters per second, lpm = lines per minute, dpi = dots per inch
Quality is such that typeset is better than letter which is better than correspondence which is better than Draft.

Glossary 5.3

CHARACTER–AT–A–TIME PRINTER – Character–at–a–time printers simply print individual characters onto the paper one at a time. The daisy wheel printer and the dot matrix device fall within this class.

FONT – A font is a set of printing or display characters in a particular type. Printers may offer a choice of fonts either by exchanging the print head (daisy wheel) or by software control as with a dot matrix or ink jet printer.

LETTER QUALITY PRINTER – The term letter quality is often applied to printers to describe the output text as being equivalent to that produced from a standard typewriter. Most impact printers can produce high quality characters with the use of either print wheels, print chains or print belts. Laser printers can also produce letter quality output while dot matrix printers using 24 pins can produce near letter quality output.

LINE–AT–A–TIME PRINTER – A line printer prints one whole line of text, known as a print line, at a time. The three types of line printer most generally used for on line computer work are the print wheel printer, the chain printer and the drum printer. Line printers, whether they are barrel or chain types, can print lines with up to 160 characters on continuous stationary at speeds from about 300 to 3000 lines per minute.

TEXT – Text is the words printed on a sheet of paper or written on a visual display unit (VDU) screen. The term also includes numbers and mathematical symbols but excludes diagrams, graphs or pictures.

Table 5.3 gives the speed and quality of several different types of printers including the impact and nonimpact varieties. Speed is normally measured in characters per second, or for high speed devices in lines per second or even pages per second. Quality depends on the clarity of the text with the highest quality being referred to as typeset, and the lowest referred to as draft. For printers, the higher the quality and the faster the speed means the higher will be the cost of the device.

1. Impact Printer – The term impact when used to describe printers relates to the manner in which the characters are formed on the paper. It implies using some form of medium which has the individual characters embossed on it and these embossed characters being pressed onto an inked ribbon which is subsequently pressed against a roll of paper. In this way an imprint of the character is left on the paper. It is possible to characterize impact printers as being either character–at–a–time (600–900 characters per minute) devices or line–at–a––time (up to 2000 characters per minute) machines (Glossary 5.3).

Figure 5.11

Daisy wheel mechanism.

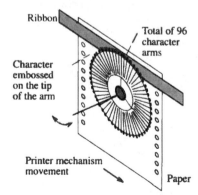

(a) Daisy Wheel Printer: A daisy wheel printer makes use of a print wheel to produce letter quality print. The print wheel has a plastic hub around which are arrayed ninety six radial spokes; with a letter, number or other symbol being moulded into the end of each spoke, see Figure 5.11. In operation, the print wheel is rotated to select the character to print, a sliding

wedge is pushed forward onto the end of the radial arm to press the inked ribbon against the paper. The cartridge and ribbon then advance as the wheel is spun to bring the next symbol into position. The device can usually print at up to one hundred characters per second.

(b) *Golf Ball Printer:* A golf ball printer is a form of impact printer which uses a golf ball head to produce the alphanumeric characters. The golf ball head can be rotated and pivoted to select the character to be printed. Figure 5.12 shows a diagram of the head.

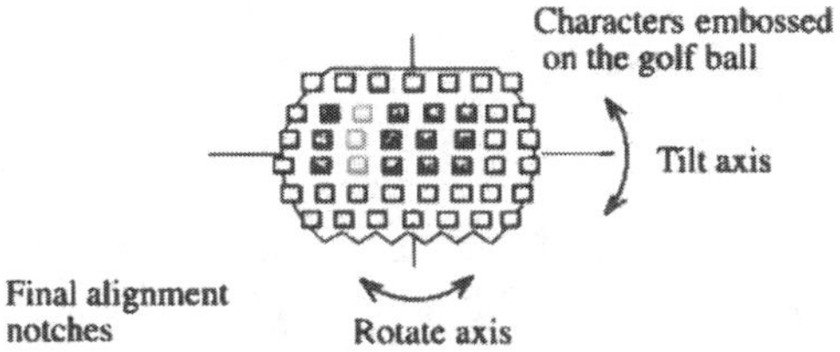

Figure 5.12

A golf ball mechanism.

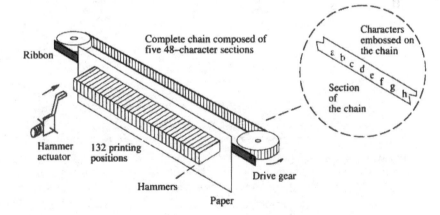

(c) *Chain Printer:* A chain printer is a computer output device which is able to convert the electrical output from the computer into type, printed on paper sheets or rolls. It is a special type of printer that has the alphanumeric character set engraved in type and assembled on a chain. A diagram illustrating the mechanisms within a typical chain printer is given in Figure 5.13. From the diagram it can be seen that the chain revolves horizontally past all the print positions on a sheet or roll of paper.

Figure 5.13

Chain printer mechanism.

The device is arranged so that it prints onto the paper when an electrically operated print hammer (one for each column of the paper) presses the paper against an inked ribbon that in turn presses against the character on the print chain.

(d) *Print Belt Printer:* The print belt printer use a belt with all the different characters on it and is similar in operation to the chain printer. A diagram of the print belt mechanism is shown in Figure 5.14.

The belt passes horizontally along the print line with hammers located at all the print positions on the line. These hammers then strike the characters in the belt as they pass the appropriate position. One line is printed for each complete revolution of the belt.

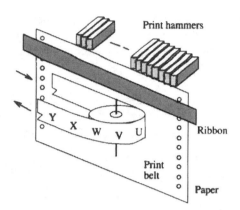

Figure 5.14

Print belt mechanism.

(e) Drum Printer: A drum printer utilizes a metal drum embossed with alphanumeric characters. Each column on the drum contains the complete character set and corresponds to one print position on the line.

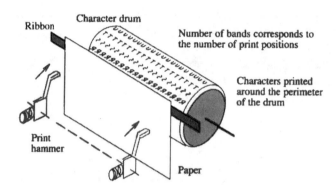

Figure 5.15

Drum printer mechanism.

On the drum printer the rows of characters are arranged with As in one row, Bs in the next and so on to a full compliment of characters. Each row contains between 120 and 160 characters so this number of characters can be printed on one line. The characters are printed onto the paper by means of electrically operated hammers pressing the paper against the characters on the drum, via an inked ribbon. This is shown in Figure 5.15. To print a line there is an electromagnetic hammer in line with each character in the row of characters. The drum rotates, paper and carbon paper are in position between the drum and the hammers and all the As are printed first then all the Bs and so on until the line is complete in one revolution of the drum. The individual hammers are activated when the appropriate character is in place, thereby printing one complete line with each rotation of the drum. Printing speed is between 500 and 3000 lines per minute.

(f) Dot Matrix Printer: A dot matrix printer is one which prints characters formed by a matrix (or array) of small dots. The dot matrix printer shown in Figure 5.16 produces its images by using a series of small pins which can be activated to form the individual characters. The print head may contain anything from 7 to 48 pins arranged in a vertical column. As the print head moves horizontally it constructs a character by repeatedly striking these pins against an inked ribbon and the paper. An example of the production of the letter B using a seven by five format is shown in Figure 5.16(b).

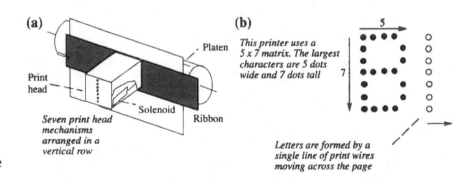

Figure 5.16

Dot matrix printer
(a) Mechanism
(b) Example of the formation of the letter B.

In a dot matrix printer with a seven by five dot matrix printing at 120 characters per second (cps) the printer must process over 400 vertical lines or dots per second or over 25,000 vertical lines per minute. If the printer is operated for a long time the print head may become very hot. An example of part of the character set produced by a 5 by 7 dot matrix printer is shown in Figure 5.17.

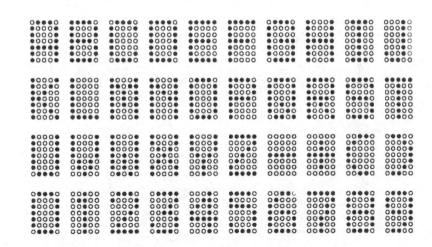

Figure 5.17

5 by 7 dot matrix printer character set.

This type of printer has proved popular for small personal computer systems and is flexible enough to also be able to generate graphic type printouts. The EPSON MX–80 dot matrix printer is generally regarded as the first of the low cost dot matrix computers released in 1978. The main disadvantage with this type of printer is that the text it produces is of a lower quality than that produced by a letter quality printer. In terms of print rate the nine–pin dot matrix devices can print at rates greater than 360 cps, while the more complex 24–pin devices offering seven colour printing can do so at greater than 240 cps in a draft mode.

Dot matrix printer

2. Nonimpact Printer – Nonimpact printers use one of a number of different non–mechanical techniques to form their characters, e.g. chemical, thermal, laser or electrostatic techniques. There are many different forms of nonimpact printers, some of which are shown in Table 5.4.

Table 5.4

Different forms of nonimpact printers.

	Printer	Description
1.	Electrostatic	This forms an image of a character on special paper using a dot matrix of wires or pins.
2.	Thermal	This type of printer uses a special printing head which heats up to form the character shapes and a special type of paper which responds to this heat to display the characters.
3.	Ink jet	This forms the character images using a thin jet of ink to essentially paint the characters onto the paper.
4.	Laser	This makes use of a low powered laser to form the character images.

(a) Electrostatic Printer: An electrostatic printer uses electromagnetic impulses which are applied to magnetized powdered ink to form the characters on the page. Heat treatment is then applied to affix the characters to the paper.

Nonimpact printer

(b) Thermal Printer: An electrothermal printer uses a special type of heat sensitive paper. The print heads convert electrical data input to heat.

(c) Ink Jet Printer: An ink jet printer forms characters or graphics with a print head containing tiny nozzles or jets that spray drops of ink onto the paper. The effect is similar to dot matrix printing. A block layout of an ink jet printer is shown in Figure 5.18. It uses a piezo crystal to force a small jet of ink out of the printer. Thermal ink jet printers (or bubble jets) are similar to those developed using piezo crystals except they work by heating a drop of ink until it forms a vapour. The bubble forces the ink out of the nozzle and onto the paper.

Figure 5.18

Block diagram of an ink jet printer.

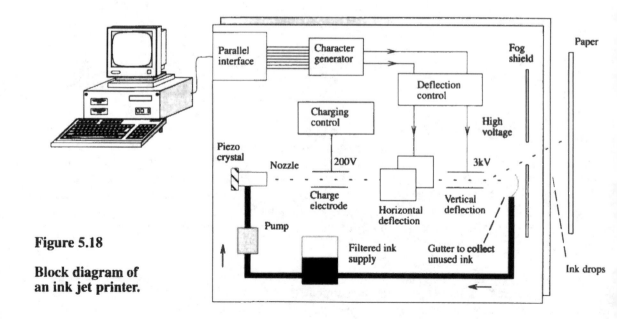

In both types, the position of the ink drops on the paper is determined by horizontal and vertical deflection plate, controlled from a character generator which builds the characters required from a series of ink droplets.

Ink jet printer

Ink jet printers can print with up to 300 dots per inch (dpi) and are also very quiet and highly reliable, due to their few moving parts. They however require specially coated paper to produce the highest resolution. An ink jet printer is capable of printing colour images. By using a separate nozzle and ink cartridge for each of the three subtractive primary colours (cyan, magenta and yellow) plus a fourth for black, up to six different hues can be printed. Ink jet technology provides the middle ground for colour output with good colour rendition and medium to high resolution output.

(d) *Laser Printer:* A laser printer combines laser and electrophotographic technology to form images on paper. High speed laser printers appeared in 1975. A laser printer works in a similar way to a photocopy machine; its operation consists of 6 phases.

Charging the drum – Before use the drum within the laser printer is charged.

Copying with the laser – Using a pattern of small dots, a laser beam conveys information from the computer to a positively charged drum inside the laser printer. Whenever an image is to be printed, the laser beam is turned on and wherever the laser beam hits the drum it causes the spot on the drum to become neutralized. In this way the charge from the nonprinting (white) areas is removed using the laser.

Developing the image – As the drum passes by a toner cartridge, toner sticks to the charged spots on the drum.

Transferring the image to paper – The toner is then transferred from the drum to a piece of paper.

Fusing the image – The drum prints the image onto the paper, and after heating, the particles of the toner are fused to form a permanent image on the paper.

Cleaning the drum – The drum is subsequently cleaned for the next pass.

Laser printers have many advantages over other printer types (Table 5.5). Laser printers can print with 600 by 600 dpi resolution for A3 and A4 mono output. The average speed of most laser printers is between ten and fifteen pages per minute.

Table 5.5

Comparison between laser printer and other printer types.

Printer type	Print graphics	Print quality	Speed	Plain paper	Noise	Colour
Laser	Yes	High	High	Yes	Low	Yes
Print wheel	No	High	Medium	Yes	Noise	No
Dot matrix	Yes	Medium	Medium	Yes	High	Yes
Band/belt	No	Medium	High	No	High	No
Ink jet	Yes	High	Medium	Yes	Low	Yes
Thermal	Yes	Low	Low	No	Low	No
Thermal transfer	Yes	Medium	Low	No	Low	Yes
Electrostatic	Yes	High	High	Yes	Low	No

5.3.2   PLOTTER

A plotter is a computer peripheral output device, similar to a printer, which converts the data emitted from the central processing unit into a hard copy graphic output. It displays the data in a two dimensional form and can produce lines, curves and complex shapes. Plotters originally appeared in the mid–1950s.

Figure 5.19 shows a diagram of a plotter. The major difference between a plotter and a graphic display device is that the plotter produces a hard copy output (paper), whereas the graphic display device produces a soft copy output (screen image).

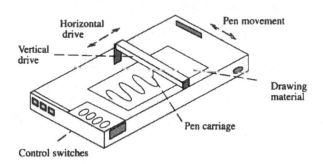

Figure 5.19

Plotter.

5.4 INPUT/OUTPUT INTERFACE

I/O units consist of a mechanical part and an electronic part. The electronic part is often called the controller while the mechanical component is the I/O device itself.

5.4.1   I/O COMPONENTS

Functionally we can identify several components in an I/O device.

1.   External Mechanism – Within an I/O device there must be some kind of mechanism which is able to interact with the external medium to perform the I/O operations.

2.   Electronic Control Circuits – To control the external I/O mechanism requires electronic control circuits. It is to be noted that because the external mechanism frequently contains electromechanical or magnetic components, which take a much greater current to operate than for example memory cells or processor components, the electronic parts needed to control the external mechanism cannot be produced in the same minute sizes as for standard semiconductor devices. The individual integrated circuit chips in the electronic control circuits, also do not have the same internal circuit density as that of memory devices. The consequence of these two facts is that for I/O devices even relatively simple functions may take several IC chips to achieve, the number increasing usually with the speed, the data capacity of the device and the procedural complexity for device control.

3.   Interface Circuits – The third main component within an I/O device are the interface circuits. Microcomputer interfaces may be classified in the following two general categories. **External Orientated** interface circuits are used to provide interfacing between the I/O controller and external devices such as switches, keyboards, VDUs, light emitting diodes and alphanumeric displays, relays, stepping motors, magnetic tape cassettes and cartridges, floppy disk memory systems and modems. **Computer Orientated** interface circuits are used to provide interfacing between an I/O controller and the internal buses in the computer.

Glossary 5.4

BUFFER – A buffer is any device designed to be inserted between two systems or program elements which can perform any of the tasks listed below.
1. Matches impedances or peripheral equipment speeds.
2. Prevents mixed interactions, i.e. the buffer isolates a signal source from another circuit being driven by that source. This prevents the driving circuit from being affected by the characteristics of the driven source.
3. Supplies additional drive or relay capability.
4. Delays the rate of information flow.
5. Acts as a temporary storage device used to compensate for a time difference in the occurrence of events. By using buffering, transmission devices are able to accommodate differences in data rates and to perform error checking and retransmission of data received in error.

CENTRONICS INTERFACE – The Centronics interface is perhaps the most popular of the parallel interfaces used within microcomputer printers. It is based on the connector first used in the printers manufactured by the Centronics Data Computer Corporation. The parallel interface allows data to be transmitted between a computer and an external device an entire byte at a time.

INTERFACE – In terms of hardware, an interface can be seen as a physical point of demarcation between two devices, where the electrical signals, connectors, timing and handshaking protocols are all defined. An interface in a computer is the connecting circuitry that forms a shared boundary to link the computer to other peripheral devices in order to allow the synchronization of the transmission of digital information between the computer and the peripheral devices. An interface may also be regarded in data communications terms as the procedures, codes and protocols that enable two entities to interact for the meaningful exchange of information.

I/O PORT – An I/O port is a special input/output register which allows the central processing unit (CPU) to communicate with various I/O peripheral units in a system. The port can usually be configured, or programmed, by the CPU to provide a data path either to or from the microprocessor and the external device.

LATCH – A latch is an electronic circuit that can be locked into a particular stable condition.

MEMORY MAPPED – Most microprocessor based systems use I/O devices that have interface circuits which are memory mapped into the microprocessor's address bus space. Here the status, control and data circuits in each I/O interface are made to resemble memory locations so that the microprocessor may generate the I/O operations by performing read/write operations on these pseudomemory locations. In this way the switching circuits within the I/O system connect the data lines of the I/O device to the data lines of the microprocessor system when the address and control signals indicate that the device is to perform an I/O operation.

PORT – A port is a physical or electrical interface into a computer, a network or other electronic device. A port is normally situated where a bus or other link connects to an I/O node, and through it a computer can gain access to its external devices such as keyboards or displays. It may also be seen as the basic addressable unit of a computer's I/O section.

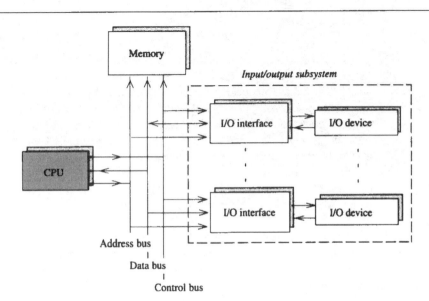

Figure 5.20

Input/output interfaces.

Hence all input/output equipment must be connected to the CPU usually through a unit called an I/O controller or interface unit. This provides for the decoding of the address instructions and commands from the CPU, as distributed on the common address highway, and the generation of the appropriate responses to the transfer control signals.

In some cases the interface unit may also be called upon to control a buffer store (Glossary 5.4) and perform assembly and coding operations. In addition the controller must develop the necessary control and timing voltages required to operate the peripheral device.

Interface circuits

Figure 5.20 shows a block diagram of an I/O subsystem within a computer. It can be seen that the various I/O devices (such as disk drives, line printers, and tape units) are normally connected onto the system address, data and control buses through I/O interfaces, one interface per I/O device is normal.

A typical I/O interface may have a number of specialized integrated circuits on a single board or module designed to handle the different types of I/O such as parallel and serial to various devices (such as disks, tape units, printers and modems). This is shown in Figure 5.21.

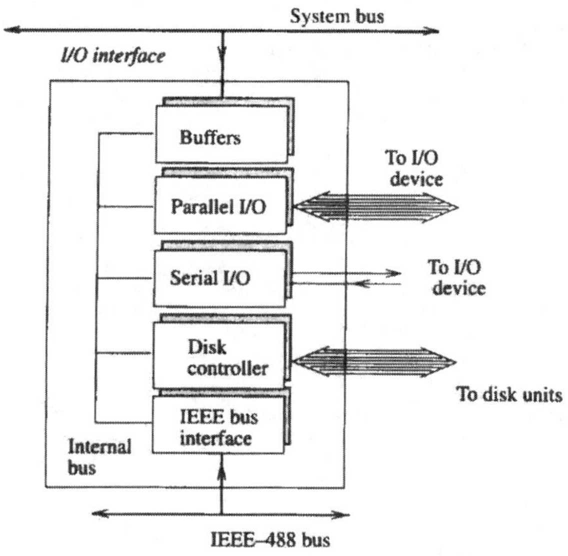

Figure 5.21

Block diagram of an input/output interface.

Some of the mechanisms that must be included within the I/O interface are as follows.
(a) *Logic Shifters and Line Terminators:* The interface unit must provide ways through which the data from each external device can be transferred properly to and from the microprocessor, without causing interference to the other devices connected to the system's buses. This requires that the data lines between the processor and the external mechanism must be correctly terminated to ensure protection against interference from noise. Logic level changers may also required within the I/O interface if equipment employing various voltage levels is used.
(b) *Interrupt Generation:* The interface circuits may be configured to produce interrupt signals to force the microprocessor to react immediately, in case some peripheral demands service.
(c) *Synchronization:* The interface unit must resolve any differences that may occur regarding the timing between the microprocessor and the peripheral I/O device. The microprocessor runs on its own internal clock, while the peripherals may or may not have their own internal clocks.

(d) *Protocol Converters*: Interface units are responsible, in the majority of cases, for the conversion of character formats between the processor and the I/O device.

(e) *Error Detection*: I/O devices, being electromechanical, are prone to incorrect information transfer. The performance of parity and sum checking operations, including parity insertion, on data transfers between the CPU and the peripherals is often handled by the interface unit. The interface circuits may also take care of the activation of an error interruption, usually on a special line, when errors are detected.

(f) *Control Signalling*: The operation of I/O devices requires a variety of control signalling information, some of which may be quite complex. To reduce the burden of the complex control of the I/O device from the microprocessor, the control signals for the external mechanism are usually not generated within the processor itself. Instead, the processor issues a limited number of signal pulses, merely to indicate what operation is required by a particular device, and the I/O interface hardware accepts the shorthand command and produces from it, the detailed control pulses required by the I/O device. The control situation within interface circuits must be a two–way communication in that there must be some form of feedback to the microprocessor from the interface, so that the microprocessor can detect the status of the I/O device itself, e.g. whether it is available, whether it has successfully performed the last operation and is therefore ready for the next one, or whether an error condition exists.

I/O interface

More recently the development of inexpensive microprocessors has allowed I/O devices to be controlled by dedicated, inexpensive microprocessors. In this way it is possible for a single interface circuit to control several peripheral units at one go, with the interface responsible for controlling the routing of data transfers between the CPU and the selected units.

(g) *Buffer Registers*: Most interface circuits make use of buffering registers to hold the data coming from or going to the microprocessor. Where buffers are used to store input data, the microprocessor is given a larger time to react to the fact that data is available at the interface circuit than if it had to respond to single data items. On the other hand if registers are used at the output, the processor can fill these registers and so it does not have to continually keep servicing the I/O device with individual data items. Moreover, whereas the processor and the memory both operate at high transmission rates, I/O devices are slower and require control, status and data pulses over extended periods. Thus, the processor would normally write control and data values into control buffer registers, which can hold their values throughout the period required by the I/O devices. When status information is generated in an I/O device, it too is stored in a register for access by the processor.

5.4.2   INTERFACE ADAPTERS

The I/O interface for computer systems makes use of specialised IC devices to produce the data interfacing and control signalling.

1. Peripheral Interface Adapter (PIA) – A PIA is an integrated circuit which can act as a parallel I/O buffer device in order to allow a processor to communicate with external peripherals. It is typically part of an I/O port and can be programmed by the central processing unit to act as an input or output. The PIA example shown in Figure 5.22 is a two–port device which in general has three registers for each port, Figure 5.22(a). The device is programmed by configuring the data direction register, Figure 5.22(b), which controls the direction of the data lines as either an input or an output. After configuration the CPU can read/write the data from/to the data register.

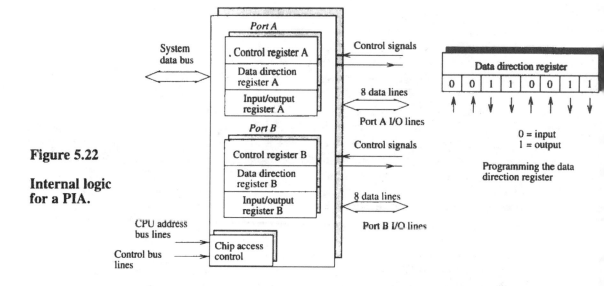

Figure 5.22

Internal logic for a PIA.

2. Asynchronous Communication Interface Adapter (ACIA) – One of the most common requirements in a computer system is to effect a communications interface between terminal equipment, such as a visual display unit, and the central processing unit. Typically this takes the form of an asynchronous serial data transmission channel. There are four basic types of interface adapters which perform the parallel to serial and serial to parallel conversions required to convert characters handled as words in the computer and the data format used for serial data transmission. These are

1. the asynchronous communications interface adapter (ACIA),
2. the universal asynchronous receiver/transmitter (UART),

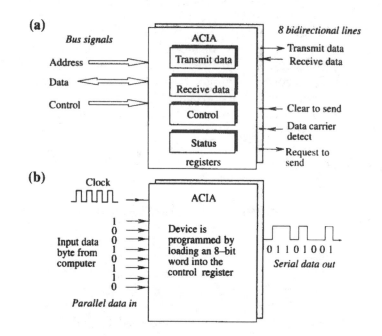

Figure 5.23

An ACIA
(a) Block diagram
(b) Example of its operation.

SERIAL COMMUNICATION ICs

UART – A universal asynchronous receiver/transmitter (UART) is a LSI device used to control the transmission of asynchronous serial data. It acts as an interface between systems that handle data in a parallel form and devices that handle data in an asynchronous serial form. A block diagram of the Motorola MC 6850 is given in Figure 5.24. It can be seen that it is composed of a number of sections dedicated to transmitting or receiving data and controlling this data transfer.

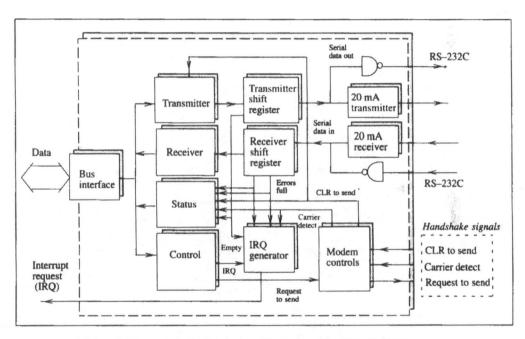

Figure 5.24   Block diagram of the Motorola 6850 UART.

USRT – A universal synchronous receiver/transmitter (USRT) is a LSI device that acts as an interface between systems that handle data in a parallel form and devices that handle data in a synchronous serial form.

USART – A universal synchronous/asynchronous receiver/transmitter (USART) is an integrated circuit common to many data communication devices which is able to convert the parallel data from the central processing unit into a serial form for transmission, either synchronously or asynchronously, across a communications network. The device can also operate in the opposite direction by taking serial data and converting it back to a parallel form.

3. the universal synchronous receiver/transmitter (USRT),
4. the universal synchronous/asynchronous receiver/transmitter (USART).

ACIA

These devices are made by almost all of the integrated circuit manufacturers and are easily interfaced to the CPU. They are single chip large scale integrated devices and although there are some differences between them, their functional operation is similar. The simplest is probably the ACIA, an example of which is shown in Figure 5.23.

Figure 5.23(a) shows a block diagram of an ACIA in a system with the various registers within the device indicated. ACIAs may have several serial I/O lines and various registers to control the I/O lines as input or output.

ACIA   The operation of an ACIA is shown in Figure 5.23(b) where the parallel input data word is transferred to serial output data. The operation of the device is shown in Figure 5.23(b), the parallel word 10010110 is transferred into a serial form. Article 5.3 gives a brief description of the other serial communication ICs.

3. Analog To Digital Converter (A/D) – Microprocessor systems, especially those employed in control situations or which have to handle signal processing applications, may well have to process data from various types of nondigital transducers. This data, for example, may come from audio, light, temperature or pressure sensors. The signal from these sensors is usually in the form of an analog voltage approximately proportional to the value of the external input. In order to be used within a computer the voltage has to be converted into a digital form, using an analog–to–digital converter. The digital data from the A/D converter can then be processed by the computer system. The process of A/D conversion applied to a time varying signal involves the representation of the signal as a sequence of ordinate values spaced out in time. Such ordinate values each define the signal at one instant of time and can be regarded as impulse values. The input sequence is then represented as a train of impulses separated in time by the sample interval.

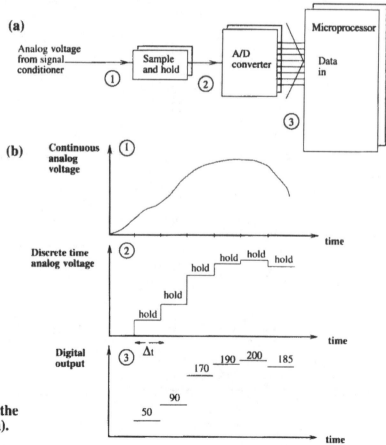

Figure 5.25

Analog to digital conversion
(a) **Block diagram of the conversion process**
(b) **Sample waveforms in the points 1 to 3 of part (a).**

A block diagram of an 8–bit conversion process is shown in Figure 5.25(a). It can be seen that before the signal is used by the A/D converter a sample and hold device must be used to hold the input voltage at a constant level during the conversion process. Figure 5.25(b) presents an example of the waveforms at different points in the conversion process, while 5.25(c) notes the hexadecimal output to the microprocessor.

The following are some important qualities of an A/D converter.

(a) **The conversion time:** This is the time required to perform the A/D conversion. From Figure 5.25(b) the conversion time is shown as Δt. The speed of the conversion depends on the chosen conversion system and the number of bits in which the signal value is expressed after conversion.

(b) **The resolution:** The number of bits determines the deviation between the real signal value and the represented binary number. The resolution is mostly expressed in bits but may also be in percentage, e.g. 8–bit resolution in Figure 5.25 implies $2^8 - 256$ different levels so each level represents 0.25% of the signal.

(c) **The overall accuracy:** There are factors other than resolution which determine the overall accuracy of the A/D, such as the stability of reference voltages and sensitivity to temperature variations.

Although A/D converters vary greatly in their range of speed, precision, sensitivity, internal construction, cost, or physical size, etc., there is little difference as far as functionality and system interfacing are concerned, and the software and hardware development for their utilization is fairly standard. This is because the operation of A/D interfaces is comparatively simple, allowing data conversion within the devices to proceed with little need for program intervention from the microprocessor.

4. Digital To Analog Converter (D/A) – In computer control situations it is often required that the results from the computer are used to generate certain environmental effects called for by the control system. However since the output from the control system is usually required in the form of an analog signal a digital to analog converter is required. D/A conversion takes the data from the processor and where necessary applies amplification and scaling, to produce an output of usable power and dynamic range, to drive such devices as audio speakers, lamps and stepping motors. A block diagram of the D/A conversion process is shown in Figure 5.26.

The digital data from the microprocessor is converted at the D/A into a quantized output signal. This is amplified and filtered to produce the output signal.

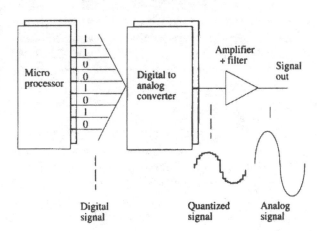

Figure 5.26

Digital to analog conversion.

5.5 I/O MECHANISMS

Most interfaces make use of interrupts and direct memory access device to control data transfer. In a processing system where the I/O data transfers are controlled by the program running on the processor, a large amount of time is required to transfer even small amounts of data. Not only does this leads to a low I/O data rate but during the I/O transfer the microprocessor cannot execute any other processing function. This can be overcome by using an interrupt–driven system where interrupts are used to indicate the start and finish of a data transfer. The use of interrupts can increase the attainable data rate, requires less software management and allows concurrent processing by the I/O device and the microprocessor.

5.5.1   INTERRUPT

An interrupt is an input signal generally initiated by an external device connected to a computer that temporarily suspends the normal sequence of operations of a running program and transfers control to a special high priority routine to deal with the interrupting device before resuming the main program sequence. The use of an interrupt is generally for the purpose of transferring information between the interrupting device and the computer. An interrupt may also be seen as a mechanism for diverting the attention of a computer when a particular event has occurred, in such a way that it allows one device to gain the attention of another so that the first unit may report a status change.

When an interrupt is received at a central processing unit it is usually converted to a particular data input combination that the processing unit can read. Once it interprets the nature of the interrupt, the processing unit initially inhibits other maskable interrupts, saves the status of the current running programs, then jumps to the appropriate interrupt service routine and carries out the set of instructions it finds there. In this way the interrupt facility enables an external signal from a peripheral I/O device, for example, to suspend or stop a currently running program and to force the processing unit to execute the interrupt procedure (or interrupt service routine). At the end of the interrupt routine an instruction is initiated to restore the suspended registers, thereby restoring control back to the main program.

Interrupts enable an I/O device to send an immediate signal to the CPU whenever it is ready to send or receive data. This signalling is accomplished by activating an interrupt request line that goes from the I/O interface directly to the CPU and usually forms part of the system bus. The CPU then responds by initiating an interrupt service request routine. The interrupt is usually initiated asynchronously either by the running program or by an external influence such as an I/O device. It is usually also necessary to establish some level of priority for each type of interrupt that can occur in order to sort out any simultaneously occurring interrupts. It can be noted that, in general, different interrupts will cause the processing unit to move to different interrupt service routines. Glossary 5.5 explains several terms related to interrupts.

5.5.2   DIRECT MEMORY ACCESS (DMA)

Interrupts can be used to improve the performance of a computer system handling data I/O. However, in applications where the required data rate is simply too high to be achieved by using interrupts or where the data rate is such that the time spent in interrupt service routines impacts the concurrent processing to an unacceptable degree, there is another method which can be used called direct memory access.

Glossary 5.5

DAISY CHAIN – Daisy chaining is a technique used in multiple interrupt inputs. It uses bus lines that are connected to the interrupting units in such a way that the interrupt signal passes from one unit to the next in a serial fashion (the daisy chain). With this method of linking units, signals pass along a single line from the high priority units to the low priority units until accepted or blocked. Lower priority units are then automatically prevented from interrupting the actions of higher priority units.

INTERRUPT HANDLER – When an interrupt occurs the first routine usually invoked is an interrupt handler. The interrupt handler controls the reception and action generated by the interrupt. Associated with each I/O device class is a location called an interrupt vector which contains the address of the interrupt service routine. The interrupt service routine starts out by saving all the registers in the process table entry for the current process. The processor then jumps to the rest of the interrupt subroutine with the registers of the processor being appropriately configured for the routine. The current process number and a pointer to its entry are kept in global variables so they can be found easily. The device handler routines are then invoked to handle the I/O transfer. After the completion of each I/O operation, the interrupt handler is again activated by the I/O to check for error conditions, to notify the processor of the outcome of the I/O.

INTERRUPT SERVICE ROUTINE – An interrupt service routine is the software routine that performs the actions required to respond to an interrupt. The routine stores the present status of the central processing unit (CPU) to the stack in order to respond to the interrupt request. The CPU then performs the interrupt service routine before restoring the saved status of the machine and resuming the operation of the interrupted program.

INTERRUPT VECTOR – An interrupt vector is an address in memory which points to the start of an interrupt service routine.

NONMASKABLE INTERRUPT (NMI) – An NMI is an overriding interrupt signal that in operation will always interrupt the microprocessor regardless of the interrupt enable status.

POLLING – Polling is a communications control procedure by which several peripheral devices may be monitored in sequence at regular intervals to see if they need attention. It may also be used to find out which of a set of peripherals caused an interrupt. In operation the master station systematically invites the slave stations on a multipoint circuit to transmit data in a round robin fashion, i.e. the highest priority device is allowed to transmit any data first, then the next highest device and so on down the line to the lowest priority device.

PROGRAM DRIVEN I/O – Program driven I/O, requires software to explicitly transfer each byte or word of information between an I/O device and a central processing unit (CPU) register's or memory. All the steps for an I/O operation therefore require the execution of instructions by the CPU. Program driven is the simplest but slowest form of I/O operation since the CPU must continually monitor the I/O device to check if it is ready or not.

VECTORED INTERRUPT – A vectored interrupt is a special form of branch operation that provides the central processing unit (CPU) with an identification code that it can use to transfer control to the start of the corresponding interrupt service routine.

Direct memory access

A DMA (or I/O) channel is an arrangement of hardware that can be added to a processing system such that data can be transferred directly to or from the primary memory and the I/O device, independent of any regular central processing unit intervention. In this way the DMA unit is allowed to take control of the address and data buses instead of the CPU, and transfer data blocks directly into or out of the system's main memory.

This greatly speeds the transfer process as it bypasses the need for the CPU to repeat a service routine for each word to be transferred. Unlike programmed I/O and interrupt driven I/O which route data through the microprocessor, DMA directly transfers data between an I/O device and the main memory. It has the additional advantage that it does not use the internal registers of the CPU for the data transfers and so allows the CPU to continue with some internal task. The internal logic of a DMA device is shown in Figure 5.27. It can be seen that the device uses a number of registers to store the information required to perform the DMA transfer.

The DMA controller is initiated by a data request from the microprocessor, after which the DMA device takes control of the system bus to perform the data move. Several DMA transfer methods and bus access techniques exist, the two most often used being:

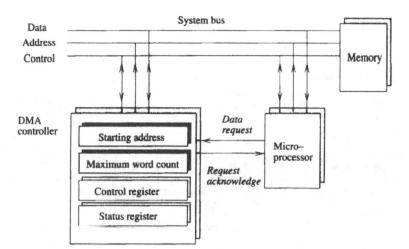

Figure 5.27

Internal logic of a DMA device.

Direct memory access

the halt, also called the suspend or idle, method and the cycle–stealing method.

1.   Halt Method – With the halt method, the operation of the microprocessor is suspended while a DMA transfer occurs. When the CPU becomes disconnected from its bus, its bus drivers are set to their high impedance state and the DMA controller is then given control of the CPU bus to manage the data transfer. By using this technique we have a situation in which either the CPU or the DMA controller is using the bus at any given time.

It may appear on first looking at this arrangement that very little has been accomplished as no form of concurrent activity has been gained. However, because the DMA controller can transfer data quicker than the CPU, there will still be a speed up of operations. This usually means that the data must be transferred in a large block to make it viable in terms of performance.

This type of transfer is often known as 'burst data transfer'. In this case the maximum transfer data rate is limited only by the read or write cycle time of the memory and by the speed of the DMA controller. Below this maximum, the data rate is limited by the rate at which the I/O device can supply or receive data. It is commonly used in transferring data to or from a floppy disk mass storage unit or for refreshing a cathode ray tube display.

2.   Cycle Steal Method – As was stated the halt method is usually used for block DMA transfers, in which a block of words is transferred between the I/O device and the memory while the CPU is deactivated. The cycle steal method (also called halt–steal or byte transfer mode) on the other hand can be used to transfer only one word at a time, between CPU cycles. With this method, memory cycles are stolen from the microprocessor for use by the external device. In this way the DMA can execute a memory access at any time the CPU is not using the memory, hence allowing the CPU to continue with its normal functions. The DMA data can then be transferred concurrently with other processing being carried out by the microprocessor. In effect the memory may be regarded as a separate subsystem to which the CPU and the DMA channels send requests to for service. This means that rather than interrupting the microprocessor, as the halt method does, the cycle steal method simply slows the processor down (since the processor must continually contend for the memory with the DMA device). In practice the processor is only delayed when an instruction fetch or an operand store or fetch is to be made to or from a memory bank that is the target or source of an I/O transfer. The execution steps are similar to those for burst DMA except that the controller issues a DMA request for every single word of data that is to be transferred.

Compared to the halt method the cycle steal method usually involves a somewhat higher hardware cost. However it is considered the best on a cost/performance basis. It also presents the lowest single–word transfer delay.

5.6 KEY TERMS

Analog to digital (A/D)	Keyboard scan
ASCII	Keyboard switch
Asynchronous communication interface adapter (ACIA)	Laser printer
	Latch
Buffer	Letter quality printer
Cathode ray tube (CRT)	Light pen
Centronics interface	Line–at–a–time printer
Chain printer	Memory mapped
Character–at–a–time printer	Memory mapped terminals
Console	Mouse
Cursor	Nonimpact printer
Daisy chain	Non maskable interrupt (NMI)
Daisy wheel printer	Optical character recognition
Digitizing tablet	Output device
Direct memory access (DMA)	Peripheral device
Dot matrix printer	Peripheral interface adapter (PIA)
Drum printer	Plotter
Digital to analog (D/A)	Polling
Electrostatic printer	Port
External I/O	Print belt printer
Font	Printer
Golf ball printer	Program driven I/O
Graphics display device	Programmable terminal
Graphics pad	Raster
Graphics standard interface	Remote terminal
Icon	RS–232 terminals
Impact printer	Scanner
Ink jet printer	Serial communication ICs
Input device	Terminal
Input/output devices	Text
Input/output interface	Thermal printer
Input/output mechanisms	Touch screen
Interface	Trackerball
Interface adapters	Universal asynchronous receiver/ transmitter (UART)
Internal I/O	
Interrupt	Universal synchronous/ asynchronous receiver/transmitter (USART)
Interrupt handler	
Interrupt service routine	
Interrupt vector	Universal synchronous receiver /transmitter (USRT)
I/O Port	
Joystick	Vectored interrupt
Keyboard	Visual display unit (VDU)
Keyboard encoder	Window

5.7 REVIEW QUESTIONS

1. What is a monitor?
2. What is the difference between an alphanumeric display terminal and a graphics display terminal?
3. What creates the images on a CRT?
4. Draw a cathode ray tube and explain how it operates.
5. What are pixels? How do they relate to the screen resolution of a VDU?
6. In colour displays what are the three primary colours used?
7. Note four of the graphic standards which have been produced for personal computers.
8. What is the purpose of a keyboard on a computer system?
9. Why has the layout of a computer keyboard been designed the way it is?
10. Why is a standard keyboard layout referred to as a QWERTY keyboard?
11. Which pointing input device can replace the cursor keys on the keyboard?
12. Why is a mouse a useful input device? What is it usually used in conjunction with?
13. What is the similarity between a trackerball and a mouse?
14. Define the operating characteristics of a light pen, graphics tablet and a mouse.
15. What are the uses of ASCII?
16. What is OCR and where is its major application area?
17. Give two examples of I/O devices which convert computer readable code into human readable code.
18. In what ways would a light pen be used for input into a computer?
19. List several input devices which could be used with a graphics system.
30. Summarize the different methods of producing a computer output.
21. Name three types of computer output which are not intended for people.
22. Define the term 'hard copy' when used to describe an I/O device.
23. Which of the different computer output devices produces a hard copy output?
24. Compare the properties of soft and hard computer output.
25. Which of the different forms of computer output is the most widely used?
26. What is the major advantage associated with the use of hard copy?
27. Why do most businesses require a hard copy of all their computer output?
28. How does an impact printer form characters on paper?
29. Name and describe the different degrees of print quality obtainable from printers.
30. List the main features of a dot matrix and a daisy wheel printer.
31. Which type of printer produces an output similar to an electric typewriter?
32. The best quality of print is produced by a matrix printer, true/false?
33. Outline the features of a thermal printer and a line printer.
34. How does a nonimpact printer form characters on paper?
35. What would be the advantages of a laser printer compared to a dot–matrix printer?
36. Which of the different types of printers can produce a wide range of images, provide they are programmed to do so?
37. If a company wishes a printer to print 3000 characters per second which type should be chosen?
38. Which of the two printers the dot matrix or the inkjet can print colour? Which is the quietest?

39. In a legal firm what type of printer would be selected?
40. If you were to start a business to sell printers, what type of technology would you select? Why?
41. Detail the differences between a printer and a plotter.
42. List three hard copy output devices which could be used with a graphics system.
43. Contrast the jobs of an I/O device's electronic and mechanical parts.
44. What is the purpose of an interface in a microprocessor based system?
45. Why is error checking needed for many types of input?
46. Note the differences between a PIA and an ACIA interface I/O device.
47. What is the purpose of an interrupt?
48. Define each of the terms when used to refer to I/O systems: polling, buffer, and port.
49. Detail the fundamental nature of programmed I/O.
50. Why are DMA devices important for handling data to and from a disk drive?

5.8 PROJECT QUESTIONS

1. Compare bit mapping and character mapping techniques within a VDU for producing images on the screen.
2. What factors determine the resolution of a computer monitor?
3. Find out about electroluminescence displays for computers.
4. What are gas plasma display?
5. What additions could be added to a computer to enhance its ability to deal with complex graphic displays?
6. In which application fields are high resolution displays required.
7. What is an LCD?
8. Of the four types of flat panel displays – LCD, dot matrix display, electro-luminescent and VCR which offers the best potential for advanced computer displays?
9. Find out what type of technology is used for displays on portable computers.
10. What prevents the acceptance of newer arrangements of keyboard layouts taking place?
11. Contrast the advantages/disadvantages of the scanning and encoding methods of keyboard monitoring.
12. Which keys on the computer keyboard are often used in conjunction with the control and ALT keys?
13. What are the advantages and disadvantages of hand–written recognition as a form of computer input?
14. A bar code reader has found many applications in retailing. For what?
15. What are the advantages and disadvantages of optical character recognition?
16. Investigate the current status of speaker–independent voice recognition.
17. What applications would there be for computer generated speech synthesis?
18. Why is paper tape no longer used for computer input?
19. How does an image scanner work? What are its advantages as an input device?
20. Investigate the range of ASCII symbols available outwith the standard character set.
21. Which of the four types of plotters – drum, cylinder, tablet or flat bed – is the most popular?

22. Through suitable reference find out about data–gloves as input devices.
23. Discover as much information on the operation of A/D converters and detail their characteristics and where they are chiefly used?
24. Microfiche and microfilm are two forms of output which a computer can produce, investigate where they are used.

25. What does the term memory mapped mean when describing an I/O peripheral?
26. What is the main advantage of 'memory mapped' I/O?
27. Select one of the parallel/serial converters, such as a UART or USRT, and from its data sheets note some of its features.
28. Explain what the difference is between an I/O device being described as an on–line or off–line device.
29. Explain the reason why I/O in a computer can be a bottleneck to a system's performance. What measures can be taken to minimize this problem?
30. Describe what a virtual environment is and what features it requires from a computer.

5.9 FURTHER READING

The popular computer magazines give a large number of articles on modern I/O devices, noting their features and properties. A short list of books and articles is presented below.

I/O Interface (Books)

1. *Microcomputer Interfacing a Practical Guide for Scientists and Engineers*, J. J. Car, Prentice Hall, 1991.
2. *Microcomputer Interfacing an Experimental Approach*, M. C. Cavenor and J. F. Arnoid, Prentice Hall, 1989.
3. *An Introduction to Microcomputer Systems Architecture and Interfacing*, J. Fuicher, Addison–Wesley, 1989.

User Interface (Articles)

1. Colour and resolution, M. Lurie, *Byte*, June 1992, pp. 171–176.
2. Of mice and menus: designing the user–friendly interface, T. S. Perry and J. Voelcker, *IEEE Spectrum*, June 1992, pp. 46–51.

Scanners (Articles)

1. How scanners work, R.C. Alford, Byte, June 1992, pp. 347–350.

Printers (Articles)

1. Ink jet takes off, A. J. Rodgers, *Byte*, October 1991, pp. 163–168.
2. Smart printing, M. Riezenman, *Byte*, October 1991, pp. 139–146.
3. Hot colour, M. D. Nelson, *Byte*, October 1991, pp. 177–182.
4. Printing with electrons, B. Smith, *Byte*, October 1991, pp. 185–192.

A/D Conversion (Articles)

1. Signal computing, E. C. Anderson, S. Shepard and P. Sohn, *Byte*, November 1992, pp 155–164.

Chapter 6

Computer Buses

Chapter Structure

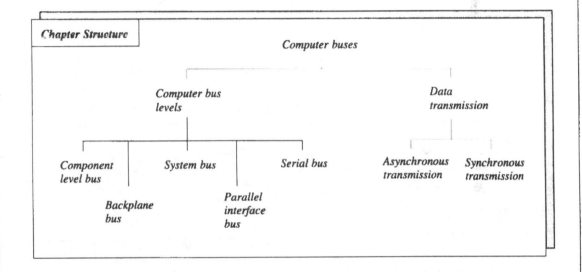

Learning Objectives

In this chapter you will learn about:

1. The many different bus levels in a computer system.
2. The features of the VME and Multibus backplane buses.
3. The properties of the RS–232 and other serial bus standards.
4. The difference between asynchronous and synchronous data transmission.

'In order for something to catch on it must be standardized.'

Anon

Chapter 6

Computer Buses

A **bus** in a computer system is a digital highway or transmission path which allows the memory, microprocessor and I/O devices within a typical computer to communicate with one another to exchange data, addresses and control information. The bus is usually an electrical connection consisting of one or more conductors wherein all attached devices receive all transmissions at the same time. It should be noted that the term bus implies the possibility of communication among more than two devices (connections limited to two devices are called wires, traces or signals). Furthermore, in the context of microprocessors, a bus usually implies parallel connections, where several related signals travel together along approximately the same route at approximately the same time. A parallel bus is one in which the separate address, data and control signals each occupy an individual track on the bus. Most of the buses used within a computer system are made of electrical conductors, for example copper line patterns etched on printed circuit boards, or fine aluminium line patterns within the integrated circuits. Access to use the bus means that some sort of contention control mechanisms must be incorporated into the system.

6.1 COMPUTER BUS LEVELS

The devices which can be interconnected by a bus range from processors on a circuit card to instruments in an instrument rack. This tends to mean that most computers, even small ones, contain several different buses, each optimized for a particular kind of communication.

Computer buses can be described in levels according to their speed, length, width and protocol. Generally speaking the larger the system the more levels it will support. These levels are illustrated in Figure 6.1 with a brief description of each indicated in Table 6.1.

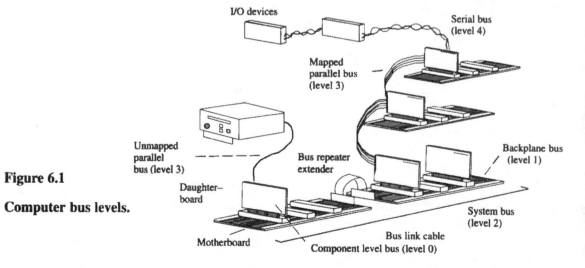

Figure 6.1

Computer bus levels.

Bus Level	Type	Description
LEVEL 0	Component bus	Connect individual integrated circuits to each other.
LEVEL 1	Backplane bus	Connect printed circuit cards to each other.
LEVEL 2	System bus	Connect racks of equipment to each other.
LEVEL 3	Parallel interface bus	Connect peripherals units to processing systems.
LEVEL 4	Serial interface bus	Connect processing systems to other processing systems in networks.

Table 6.1

Computer bus levels.

6.1.1   COMPONENT LEVEL BUS

The component level bus within a computer system links together the individual integrated circuit (IC) components of the computer on a printed circuit board (PCB). The bus is usually made up from a number of parallel tracks on the PCBs with all devices using the component bus tending to be electrically connected to each of the bus signal lines. The parallel component bus has the following class of lines or signals: address lines, data lines, control lines, interrupt lines and bus exchange lines. Component level buses can be distinguished from the other bus levels by the following attributes.

1. They are entirely confined to one printed circuit board. This sets a limit on the length of the signal lines and makes proper termination impractical, thus presenting the devices with mainly capacitive loads.
2. Large scale integrated (LSI) devices may have sufficient output drive capability to make buffering unnecessary, except on the interface to the backplane level.
3. All signals are specific to the processor or other LSI device in use, making the component level bus the most dedicated and inflexible of all the bus levels.
4. A number of basic housekeeping signals appear at this level, such as dynamic memory refresh.

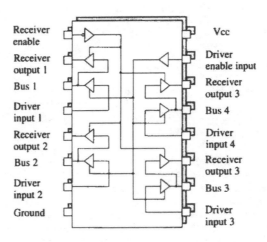

Figure 6.2

Bidirectional bus driver IC.

A computer bus acts as an electrical conductor such that when a transmitting device puts a voltage on the bus lines it subsequently appears everywhere along the bus and can be sensed by all the receivers. In this way the signal transmission within the bus is straight–

TRISTATE BUFFER

Article 6.1

The parallel bus systems within a computer are such that there may be many transmitters and many receivers connected onto the same bus but only one transmitter can use the bus at any one time. The selection of which unit is to transmit is controlled by the microprocessor, normally through a process called address decoding. However, since bus signal lines have to be shared, the bus must be designed such that two devices must never try to drive a line at the same time, otherwise interference would occur resulting in high current spikes, ill determined voltage levels on the bus, electrical noise and possible premature component failures. By making use of tristate (three–state) logic on the bus drivers, the digital output is disconnected when the device using the driver is not using the bus.

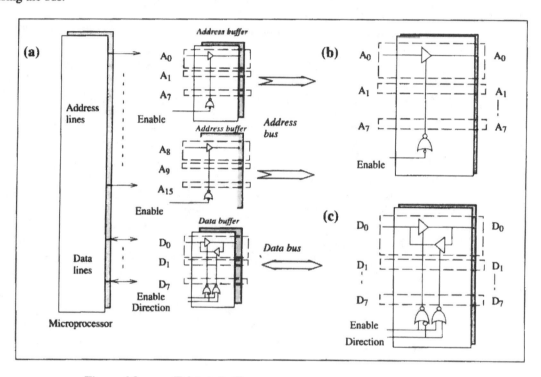

Figure 6.3   Tristate buffers
(a) Buffers used in a microprocessor system
(b) Internal logic of the address buffer
(c) Internal logic of the data buffer.

A tristate buffer is connected onto a bus line such that the output of the buffer can either be in one of three states: a Logic 1, a Logic 0 or an off (or high impedance) state. In the high impedance state the bus line is available for other devices to use without adversely affecting the other sources that can drive the line. A special input to the driver, usually called the output enable, turns it off when the device connected to the buffer is not using the bus. The output enable is controlled through address decoding from the microprocessor. In this way, by placing all but one transmitter in the tristate off mode, no interference occurs. Figure 6.3(a) shows some tristate buffers being used within a microprocessor system. The internal logic of a typical bidirectional address buffer is shown in Figure 6.3(b). Figure 6.3(c) shows the internal logic of a data buffer. Tristate buffers are constructed using a Schmitt trigger arrangement which has the effect of squaring any distorted pulses passing through the buffer, thereby restoring the original digital data after it may have been distorted by the bus.

forward since to transmit information over the bus the transmitting device simply changes the voltage in accordance with the desired digital information.

Component level bus

The circuitry that changes the voltage on each of the signal lines is called a **bus driver** or bus buffer. Bus drivers are available in IC form. The pin description of a 16–pin IC containing four bidirectional bus buffers is shown in Figure 6.2.

Within microprocessor systems the bus drivers are all tristate, Article 6.1.

6.1.2   BACKPLANE BUS

The backplane bus, or **motherboard**, serves a computer system by providing a communication pathway between the different printed circuit boards in the unit along a common set of backplane wires.

The devices which plug into this form of bus are no longer the individual ICs as in the component level bus but are now the PCBs, often called modules or **daughterboards**. Figure 6.4 shows a number of different system boards being connected by a backplane bus.

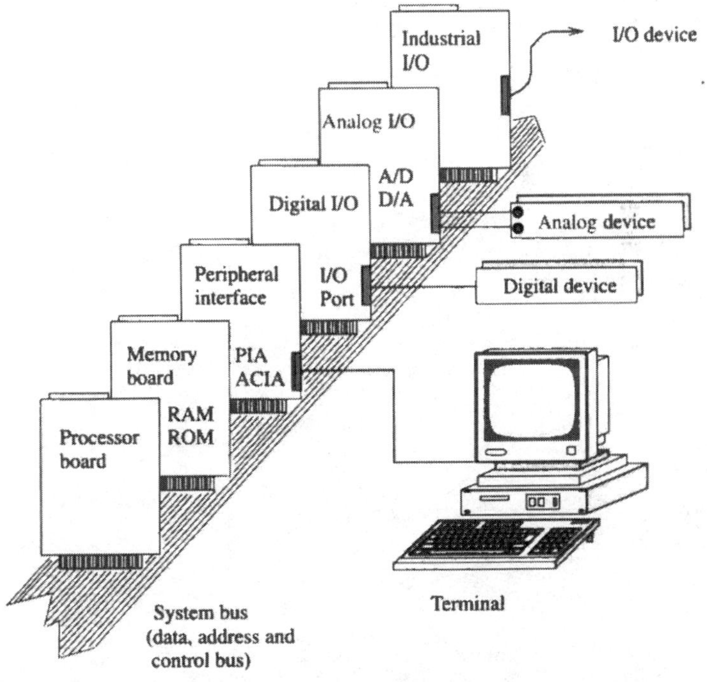

Figure 6.4

Computer system boards interconnected through a backplane bus.

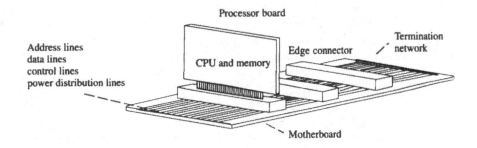

Figure 6.5

Backplane bus.

In addition to signal lines the backplane usually has heavy conductors providing the electrical power pathways needed by the PCBs. Figure 6.5 shows the backplane bus in more detail.

Backplane bus

The **connector** which connects the printed circuit boards to the backplane bus is the physical interface between the bus and board. It comes in two parts normally referred to as male and female components. Two types of connector are shown in Figure 6.6, the edge connector and the plug and socket connector. The edge connector uses metal contacts on the host board to connect to the socket. The plug and socket connector type has a separate connector, usually a plug, attached to the host.

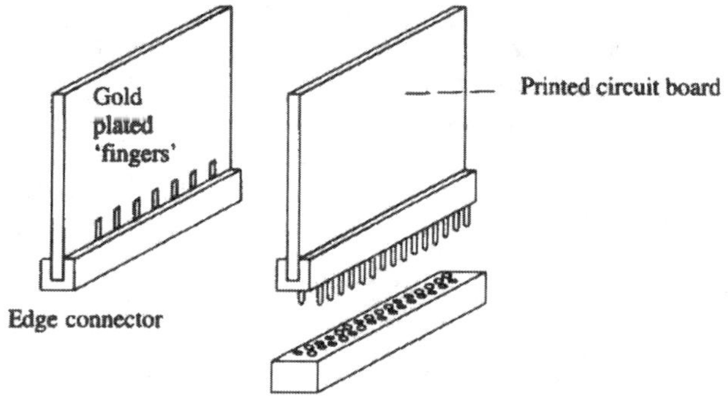

Figure 6.6

Two examples of computer connectors.

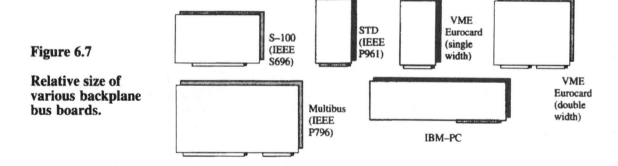

Figure 6.7 compares the relative sizes of boards for the different bus systems. Almost all new backplane buses now have module boards which conform to a double Eurocard standard size.

Figure 6.7

Relative size of various backplane bus boards.

6.1.3   SYSTEM BUS

The system bus within a computer comprises all the separate sections which make up the main parallel backplane bus within a computer. Certain buses at the system level have been designed to avoid a one–to–one correspondence with component level signals, since different manufacturer's cards and devices are more likely to be intermixed at the backplane level. Industry–accepted system buses, such as the S100, Multibus and the VME bus, all provide standard motherboards.

These buses have subsequently been extended to support higher system performance by adding more features. By using a standard bus system the design of the overall computer is made very much easier as it is then possible to buy populated PCBs with the correct connectors and pin connections. On the other hand, if the designers are working at the board level, they can buy unpopulated PCBs again with the correct connectors to match the backplane. Table 6.2 notes the properties of four of the more popular system buses.

	VME bus	Multibus II	Futurebus	Nubus
Originator	Motorola, Mostek, Signetics	Intel	IEEE	MIT
Data paths	Nonmultiplexed 16	Multiplexed 32	Multiplexed 32	Multiplexed 32
Primary address range	2^{32}	2^{32}	2^{32}	2^{32}
Communication type	Asynchronous	Synchronous	Asynchronous	Synchronous
Data rate (MB s^{-1})	Max 57	40	Max 117.7	37.5
Max no. of processors	1+4+chain	20	21	16
IEEE Standard	P1014	P1296	P896.1	P1196
Message passing	No	Yes	Yes	No
Board dimensions (mm)	233.4 x 160	233.4 x 220	366.7 x 280	366.7 x 280
Board area (cm^2)	373	514	1027	1027
Connectors (pins)	Indirect 96 + 96	Indirect 96	Indirect 96	Indirect 96
Bus drivers	TTL	TTL	Special (BTL)	TTL
Separate serial bus	Yes	Yes	Yes	No
Error detection	No	parity	Optional	Optional
Power supply (volts)	5 +/-12	5	5	5

Table 6.2

Properties of four standard system buses.

1. **S100 Bus** – The S100 bus, which derives its name from the fact that 100 pins are used on the bus, was originally developed for use on the Altair microcomputer, this was the first personal microcomputer and was based on the Intel 8080 microprocessor. There are a number of different S100 implementations. The S100 bus can transmit data over a distance of about 25 in at 6 MHz with a maximum of 21 printed circuit boards being able to be connected onto the bus. With its 24 address lines it is also suitable for 16–bit microprocessors.

2. **IBM Bus** – The bus used in the original IBM personal computer has become a type of standard often referred to as the ISA (industry standard architecture) bus. The ISA bus has been used in many personal computers built with the Intel 80286 and 80386 microprocessors. The bus has a 16–bit wide data bus and 24–bit address bus which is limiting for the newer 80386 and 80486 microprocessors. As an alternative the extended industrial standard architecture (EISA) bus system has a 32–bit data bus and bus master capabilities. EISA can achieve burst transfers at approximately 33 MBs^{-1} but 25 MBs^{-1} is a more reasonable sustained transfer rate.

   In 1987 IBM introduced the microchannel architecture (MCA) bus for its PS/2 computer models. The 32–bit MCA uses a 132–pin bus and can work at 20 MBs^{-1}, 160 MBs^{-1} in burst transfer mode. EISA and MCA are not compatible.

3. **Multibus Family** – The Intel Multibus system is a family of bus systems for microprocessor applications. It was initially targeted at the 8086 microprocessor but can be used with many other devices. The complete Multibus family is shown in Figure 6.8 where we have the central system bus and three other expansion bus structures.

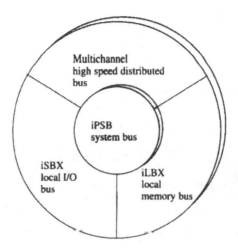

Figure 6.8

The Multibus family.

With the three extensions to the central system bus, a designer is provided with the ability to arrange the bus traffic in their microprocessor system to optimize the cost and performance of the design.

Multibus family

Figure 6.9 shows the relationship between the different bus systems. The characteristics of the Multibus system are also shown in Figure 6.9. We will look at each of the buses in a little more detail.

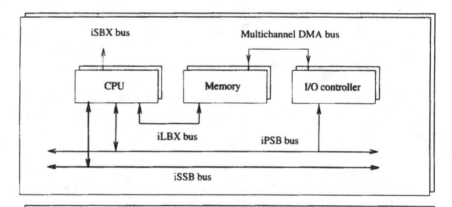

Figure 6.9

The different bus families in the Multibus system.

Bus	Description	Transfer rate MBs⁻¹	Bus width	Comms protocol
iPSB	Parallel system bus	40 (at 10 MHz)	32–bit multiplexed	Synchronous
iLBX	Local high speed memory extension	48 (at 12 MHz)	32–bit non– multiplexed	Synchronous
iSBX	Local I/O expansion	10	16–bit non– multiplexed	Asynchronous
Multichannel I/O	Remote DMA	8	16–bit non– multiplexed	Asynchronous
iSSB	Low cost serial system bus	2	1–bit	CSMA/CD
Bitbus	Local industrial I/O	2.4	Two signal pair	SDLC sync/ self–clocked

(a) **The System Bus***:* The system bus (iPSB) is the major data highway in the microprocessor based system. It is an asynchronous parallel backplane bus which carries five separate signal categories: a 24–bit address bus, a 16–line bidirectional data bus, multilevel interrupt lines, control and timing lines, and power distribution lines. Two connectors are required to interface Multibus printed circuit boards to the system bus. They plug into the backplane and are usually referred to as P1 (primary) with 86 pins and P2 (auxiliary) with 60 pins.

The system bus operates on a master/slave principle in that only one unit within the microprocessor based system can be controlling the operation of the bus at any one time. In this way multiple processing and multiple computing are supported with two to sixteen microprocessors, called bus masters, able to be supported on the same System bus. In order to allow correct control of the bus, the System bus provides an arbitration and bus control exchange method that guarantees that a bus master can access the system without another master obtaining it, and provides several reliable communication methods between bus masters through common resources such as memory and I/O.

(b) **The iLBX Bus***:* The iLBX bus provides a private high speed microprocessor to memory data path to memory expansion modules. The bus makes it possible to expand the local memory of a microprocessor or a small board computer (SBC) by using multiple buses. It is a standardized execution bus which when used with the system bus provides a further architectural extension of the Multibus system. Since the iLBX bus is local to the SBC, no contention for the bus occurs, thereby ensuring a guaranteed microprocessor to memory bandwidth, i.e. it helps prevent saturation of the system bus by removing most execution requirements from it. By providing this high speed tightly coupled connection between the microprocessor and its memory on another board, the iLBX bus permits the expansion of a board's local memory in a modular manner without using the system bus beyond what could fit on an SBC. The iLBX bus allows connection of up to four memory boards, yielding a total local expansion address space of up to 16 MB. The maximum transfer rate for the bus is 9.5 MBs^{-1} for 8–bit data transfers and 19 MBs^{-1} for 16–bit data transfers.

(c) **The iSBX Bus***:* The iSBX bus is a local on–board input/output expansion bus. It is implemented with a baseboard and iSBX multimodule boards. A **baseboard** provides an electrical and mechanical interface for the iSBX multimodule boards to connect them onto the iSBX bus. The multimodule boards then simply plug into the baseboard and are interfaced electrically to the bus, thereby providing the communication link between the multimodule boards and a main central processing unit (CPU) board. The baseboard is the master of that link in that it controls the address, chip select and command signals.

The iSBX **multimodule Board** is a small I/O expansion board that provides a processing board with application specific I/O. The iSBX boards are connected to the baseboard's local bus via the iSBX bus interface, which converts the iSBX bus signals to a defined specialized I/O interface. These boards therefore enable the user to configure the capabilities required for the system, in order to keep both system size and cost at a minimum level. Since the I/O expansion is local to a processing board, no system bus bandwidth is required.

By using a modular system where each board supports the iSBX bus interface, the designer can then apply new VLSI technology to individual SBCs without the need for redesigning the complete system.

Multibus family

4. The Multichannel Bus: The multichannel bus is a high speed path for block data transfers between a Multibus based system and peripherals or other remote computer systems. It is a block–oriented direct memory access bus which provides a standard high speed, 8 MBs^{-1}, block orientated gateway into and out of a Multibus based system.

By utilizing a standard interface, multiple heterogeneous devices, such as different high speed I/O devices and memory modules, can be easily connected together. The multichannel bus has the ability to link together up to sixteen devices that are distributed over a distance of up to 50 ft via a twisted pair flat ribbon cable. It has the addressing capability of up to 1 MB of memory and 16 MB of I/O space on each bus device. The data is transferred via an asynchronous handshake between devices, thereby allowing communication among devices that vary in speed and distance from one another. The multichannel bus can therefore remove heavy bus traffic from the Multibus system bus which would otherwise saturate this bus.

Multibus tended to dominate the 16–bit backplane market in the early 1980s and has been further enhanced with the introduction in 1983 of Multibus II which was developed for 32–bit microprocessors. This uses multiplexed address and data buses and synchronous transmission and is ideally targeted at multiple processing applications. The characteristics of the Multibus II system are shown in Table 6.3.

Table 6.3

Characteristics of the Multibus II system.

Characteristics	Multibus II
Originator	Intel
Data paths	Multiplexed
(primary bits)	32
Primary address range	2^{32}
Communication type	Synchronous
Data rate (MBs^{-1})	40
Max no. of processors	20
IEEE Standard	P1296 (in 1987)
Message passing	Yes
Board dimensions (mm)	233.4 x 220
Board area (cm^2)	514
Connectors (pins)	Indirect 96
Bus drivers	TTL
Separate serial bus	Yes
Error detection	Parity
Power supply (V)	5

4. <u>Futurebus</u> – Futurebus is a high performance system bus standard. The characteristics of the bus are given in Table 6.4.

Table 6.4

Properties of the Futurebus system.

Properties	Futurebus
Originator	IEEE
Data paths	Multiplexed 32–bit
Primary address range	2^{32}
Communication type	Asynchronous
Data rate (MBs^{-1})	Max 117.7
Max no of processors	21
IEEE Standard	P896.1
Message passing	Yes
Board dimensions (mm)	366.7 x 280
Board area (cm^2)	1027
Connectors (pins)	Indirect 96
Bus drivers	Special (BTL)
Separate serial bus	Yes
Error detection	Optional
Power supply (V)	5

Article 6.2

VME BUS

The versa module European (VME) bus (IEEE P1014) was developed by Motorola and was an extension of the older 16–bit Versabus (IEEE P970). The VME bus family was originally introduced to support the 68000 series of microprocessors from Motorola, although other processors have been packaged into this standard. The VME bus now offers the ability to handle 16– and 32–bit microprocessors.

The VME bus has two 96–pin connectors on a double Eurocard printed circuit board. The characteristics of the bus and the dimensions of the board are given in Figure 6.10.

Originator	Motorola, Mostek, Signetics
Standardization	IEEE P1014 (in 1987)
Data paths	Nonmultiplexed
Primary bits	16
Secondary bits	32, 24, 16, 8
Primary address range	2^{24}
Secondary address range	2^{32}
Communications type	Asynchronous
Data rate (MBs^{-1})	Max 57, Typical 24
Broadcast capability	No
Message passing	No
Arbitration	Parallel, daisy chain, 4 levels
Connectors (pins)	Indirect 96 + 96
Bus drivers	TTL
Separate serial bus	Yes
Error detection	No

(a)

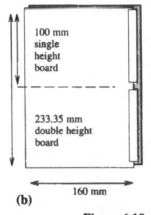

100 mm single height board

233.35 mm double height board

160 mm

(b)

P1   16– bit data and 24–bit address
Asynchronous control and address modifiers
Priority interrupt lines and bus arbitration lines
VMS bus, power and utilities

P2   32–bit extension
(16 data and 8 address lines)
user I/O or VMX bus
power

Figure 6.10

(a) VME bus characteristics
(b) VME board dimensions and connector descriptions.

Bus	Function	Transfer rate (MBs^{-1})	Data Path	Communication type
VME	Parallel bus system	40	32–bit non– multiplexed	Asynchronous
VMX	Local high speed memory and I/O extension	80	32–bit non– multiplexed	Asynchronous
I/O channel	Local I/O extension	2	8–bit	Asynchronous
VMS	Low cost serial system bus	3	1–bit	Token passing

Table 6.5   **VME bus types.**

(Continued)

(continued)

Article 6.2

VME BUS

Table 6.5 indicates the different bus types within the VME system. The VME bus system contains
1. **The VME** – the parallel systems bus (32–bit);
2. **The VMX** – a local high speed memory and I/O extension bus (32–bit);
3. **The I/O channel** – this is a local I/O extension (8–bit);
4. **The VMS** – a low cost serial system bus (1–bit).
It does not use any multiplexing of its data and address lines and works with an asynchronous form of transmission.

Article 6.2 gives details on the VME bus system.

6.1.4   PARALLEL INTERFACE BUS

Whenever designers connect peripherals such as keyboards or disk storage units onto a computer system, they usually make use of a structure that links the peripherals by a parallel bus, generally through a ribbon cable. There are two distinct classes of parallel interface buses which can suit this purpose.

1. <u>**The Mapped Parallel Interface Bus**</u> – This is a partial extension of the main system bus in the computer, in that some or all of the data, address and control lines signals may be carried from the System bus through latches to the outside world.

2. <u>**The Unmapped Parallel Interface Bus**</u> – This is a data highway system divorced from the normal system bus signals. Two examples of this type are the IEEE 488 general purpose interface bus (GPIB) and the small computer systems interface (SCSI) bus.

(a) *General Purpose Interface Bus (GPIB)/IEEE 488 Standard*: The IEEE 488 is a parallel transmission standard used for connecting laboratory instruments to a computer. It is synonymous with the general purpose interface bus (GPIB) which is an industry standard byte–serial, bit–parallel bus handling 8–bit words. The GPIB computer interface has been popularized by Hewlett–Packard for the transmission of parallel data across a limited range network usually to a series of laboratory based instruments. The standard was published in 1975 and was originally based on the Hewlett–Packard interface bus (HPIB).

The bus consists of eight data lines, eight control lines and eight ground lines.The data is transferred in an 8–bit parallel form with the bus able to support up to 15 devices. The specified maximum data rate is 1 or 2 MBs^{-1} but, typically, applications tend to operate at 250 KBs^{-1} or less. The maximum bus length is 20 m.

(b) *Small Computer Systems Interface (SCSI):* The small computer interface is a parallel, multimaster I/O bus that provides a standard interface between computers and peripheral devices. It supports a total of eight devices including the host computer. Communication is allowed between two devices at any one time. It is often used as a local I/O bus in order to interface mass storage devices such as floppy disks, hard disks, tape drives, CD ROMs and WORM drives to a host computer. SCSI has been an official ANSI standard since 1986 and comes in two forms. **Single ended SCSI** is such that each signal's logic level is determined by the voltage of a single wire relative to a common ground. The bus can be up to 6 m long.

Differential SCSI is such that the level is determined by the potential difference between two wires. The bus can be up to 25 m.

SCSI   The SCSI interface uses a 50–way connector arranged in two rows each of 25 ways. It can support communication up to 25 m. Current standards use an 8–bit data path, newer standards include 32–bit data width.

6.1.5   SERIAL BUS

A serial bus is used for communicating data and information over a long distance, it is comparatively slow compared to parallel bus systems. It relies on **serial transmission,** i.e. the transmission of the bits within a data word over a data circuit in which the individual bits of the data word are dealt with in succession, one bit at a time, rather than simultaneously as in a parallel transmission. The functional units required for a serial transmission are shown in Figure 6.11.

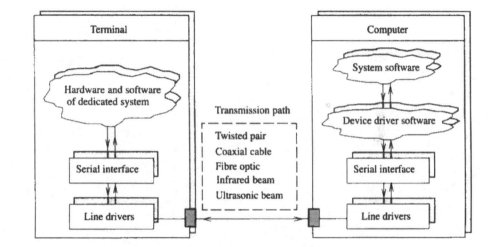

Figure 6.11

Functional units of a serial data link.

An example of a serial transmission between two devices is shown in Figure 6.12.

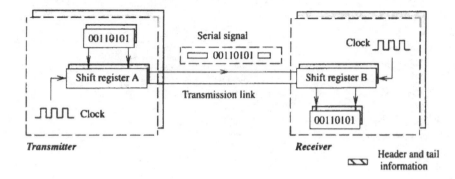

Figure 6.12

An example of a serial transmission.

For distances up to 1000 m the RS–422 standard can be used to transmit data without any form of coding or modulation. Over longer distances various transmission methods such as a modulated carrier, optical fibre or a radio system must be used. There are a number of standard serial buses.

1.   RS–232C – In order to achieve compatibility between different input and output devices and computers there is a need to standardize the I/O interface. By adopting a particular standard a computer manufacturer can then provide an interface which will permit communication with many different types and makes of I/O devices, terminals, peripherals instruments and even other computer systems. There is a large set of such standards, with perhaps the most popular being the RS–232C.

The RS–232C is an Electronics Industries Association (EIA) specified physical interface, with associated electrical signalling that is normally used between serially transmitting and receiving data circuit terminating equipment (DCE) and data terminal equipment (DTE). It is almost universally used for the interconnection of terminal equipment that receives or transmits serial asynchronous data, for example the interface between computers and modems. Because of its wide adoption, devices and microcomputers with built in RS–232 capability may be easily connected to other computer systems. The specifications for RS–232C is given in Figure 6.13.

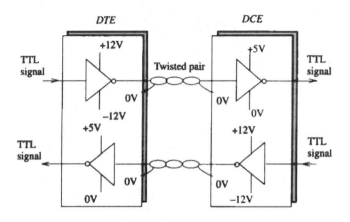

Figure 6.13

RS–232C interface characteristics.

Characteristic	Value
Operating mode	Single ended
Maximum cable length	15 m
Maximum data rate	20 kbaud
Driver maximum output voltage (open circuit)	–25 V < voltage < +25 V
Driver minimum output voltage (loaded output)	–25 V < voltage < –5 V or +5 V < voltage < +25 V
Driver minimum output resistance (power off)	300 Ω
Driver maximum output current (short circuit)	500 mA
Maximum driver output slew rate	30 V μs^{-1}
Receiver input resistance	3 to 7 kΩ
Receiver input voltage	–25 V < voltage < +25 V
Receiver output state when input open–circuit	Mark (high)
Receiver maximum input threshold	–3 V to + 3 V

RS–232 transmitters are specified to an output voltage more negative than –5 V for a logic 1 and more positive than +5V for logic 0, while a RS–232 receiver will interpret a voltage more negative than –3V as a logic 1 and a voltage more positive than +3V as a logic 0. This gives two volts of noise immunity for the transmission. Circuits are required to convert between transistor to transistor logic (TTL) and RS–232 logic and, while it is possible to generate these RS–232 voltage levels with discrete parts, special line drivers and receivers have been developed that meet all of the requirements of the EIA specifications. A universal asynchronous receiver/transmitter (UART) or similar device is also required to change the parallel data from the computer to serial data before the line drivers can transform the TTL logic signals to RS–232 levels.

RS–232C

At its maximum transmission rate of 20 kbaud the cable length for RS–232 is only 50 ft; however, it is common to find interfaces running at 1200 to 4800 baud with 1000–2000 ft cables. The baud is one of the units chosen to indicate the transmission speed of a serial data communications device. It is the maximum number of signalling elements or symbols that are generated per second. RS–232 are typically categorized by the transmission of digital data at 300 to 9600 baud. The most popular transmission rate is probably the 9600 baud used in computer visual display units.

The RS–232C interface approaches the problem of interfacing two modules by providing a separate wire between the two modules for each type of control signal that might be required. The wires for the modules are joined in a standard 25–pin connector, although usually many fewer wires are required. The transmission line type normally employed is a twisted pair of shielded wire.

2. RS–422 – The RS–232 details were first published in 1969. Since then it has been superseded by the higher performance RS–422 serial interface published in 1975 which permits higher data rates to be sent over longer distances (Figure 6.14). It has a maximum length of 4000 ft at 100 kbaud, although for distances up to 40 ft it can deliver signals at 10 Mbaud. It has a differential input and output.

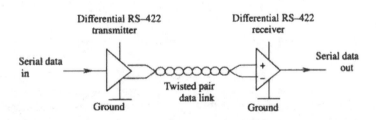

Figure 6.14

RS–422 interface characteristics.

Characteristic	Value
Operating mode	Differential
Maximum cable length	1300 m
Maximum data rate	10 Mbaud
Driver maximum output voltage (open circuit)	6 V between outputs
Driver minimum output voltage (loaded output)	2 V between outputs
Driver minimum output resistance (power off)	100 μA between 6 V and –0.25 V
Driver maximum output current (short circuit)	150 mA
Maximum driver output slew rate	No limit on slew rate necessary
Receiver input resistance	> 4 kΩ
Receiver maximum input voltage	–12 V to +12 V
Receiver maximum input threshold	–0.2 V to + 0.2 V

The greater capability of RS–422 is supported by a much smaller range of devices, generally for quite different purposes from those served by RS–232. The RS–422A standards are actually the electrical characteristic for yet another standard, the RS–449, which was introduced in 1977 as a replacement for RS–232C. This new standard uses a 37–pin connector to carry the main signals and a nine–pin connector to carry the secondary signals.

3. **RS–423** – Figure 6.15 gives the specifications for the RS–423 interface.

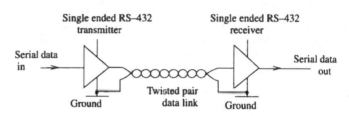

Characteristic	Value
Operating mode	Single ended
Maximum cable length	700 m
Maximum data rate	300 kbaud
Driver maximum output voltage (open circuit)	–6 V < voltage < 6 V
Driver minimum output voltage (loaded output)	–3.6 V < voltage < 3.6 V
Driver minimum output resistance (power off)	100 mA between –6 V and 6 V
Driver maximum output current (short circuit)	150 mA
Maximum driver output slew rate	Determined by cable length and modulation rate
Receiver input resistance	> 4 kΩ
Receiver maximum input voltage	–25 V to +25 V
Receiver maximum input threshold	–0.2 V to + 0.2 V

Figure 6.15

RS–432 interface characteristics.

6.2 DATA TRANSMISSION

All the buses so far described are involved in data communications or transmission. Data communications is the transfer, reception and validation of data between a data source and a data sink via one or more data links according to appropriate protocols (Glossary 6.1). In a computer system this is usually done through one of two ways, asynchronous and synchronous transmission methods.

6.2.1   ASYNCHRONOUS DATA TRANSMISSION

In asynchronous, or start–stop, communications we have a transmission method that is distinguished by individual characters, or bytes, being sent through a network encapsulated with start and stop bits. The start and stop bits are used by a receiver to derive the necessary timing for sampling the bits such that the start signal starts the receiver's clock and the stop signal indicates the end of transmission.

Using asynchronous transmission the transmitter then sends one character at a time, either immediately following one another with no time interval, e.g. during an automatic file transfer between a computer and a disk drive, or with some kind of variable time interval between the characters as in manual entry from a computer keyboard. The bits of each character are clocked into and out of the serial ports at each end of the link by locally generated clock signals.

Glossary 6.1

Communication

ACKNOWLEDGEMENT – An acknowledgement is a reply to messages or signals within a computer system.

ASYNCHRONOUS – The term asynchronous when used to refer to two computer systems implies that there is no inherent form of synchronization between the two systems, with each system operating independently, without reference to an overall central timing source. The speed or frequency of operation of an individual system is then not related to the frequency of the other parts of a system to which it is connected

BANDWIDTH – The bandwidth of a system is a measure of the maximum rate of data transfer. In communications the bandwidth is the difference expressed in hertz (Hz) between the highest and lowest frequencies of a transmission channel. The bandwidths for several typical types of signals are shown in Table 6.6.

Signals	Frequency
Telephone	0.3 – 3.4 kHz
Music	15 kHz
AM radio receiver, audio bandwidth	4.5 kHz
FM broadcast receiver radio bandwidth	180 kHz
audio bandwidth	15 kHz
TV signal	5.5 MHz
Radar radio frequency signal	10 MHz
Computer backplane bus	40 MHz

Table 6.6

Typical bandwidths of several types of signals.

CHANNEL – In its broadest sense a channel is a communications pathway or a circuit down which information can flow. It can be defined as a physical or logical path which allows the transmission of information between a data source (or transmitter) and a data sink (or receiver).

In computer systems it can also refer to a special limited capacity computer with its own logic for handling I/O operations. System peripherals such as tapes, disks and printers in a large computer can then be connected through special controllers to this form of channel. The channel takes over the task of input and output in order to free the processing unit to handle internal processing operations.

HANDSHAKING – Handshaking is a method of acknowledging signals passed between two devices, such that the data to be transmitted is not sent until the sending device receives a ready to receive message from the device that is to accept the data. Handshaking is also used for the verification that a proper data transfer has taken place such that the sender receives a data–received signal from the receiver. In this way both parties receive information of the communication action, thereby assuring an orderly data transfer.

PARITY BIT – A parity bit provides a elementary mechanism for detecting an erroneous transmission of data. It is an additional noninformation bit appended to a group of bits (typically with 7 or 8 bits) which indicates whether the number of ones in the message group is an odd or even number.

PARITY CHECK – A parity check is a process of error checking using a parity bit in order to determine the validity of a transmitted piece of data. The test determines whether the number of zeros or ones represented by the binary digits of the data word is odd or even, this is then compared with the parity bit giving an indication if an error has occurred.

START BIT – The start bit is a one–bit signal, used in asynchronous transmission as the first element in each character to indicate the start of a data transmission. It prepares the receiving device to recognize the incoming information elements.

STOP BIT – The stop bit (or stop bits) is a signal used in asynchronous data transmission as the last transmitted element in each character. This type of signal permits the receiver to come to an idle condition, before accepting another character.

TRANSCEIVER – Transceiver is the generic term describing any device that can both transmit and receive data.

TRANSMISSION – Transmission is the dispatching of a signal, message or other form of information across a communication medium.

Asynchronous data transmission

In order to synchronize the two communicating devices it is important that the clock used to input the data at the receiver must be switched on by the sending device. The clocks used by the transmitter and receiver are typically generated from crystal oscillators and are therefore on the whole highly accurate. However, even if they do differ fractionally from one another, by as small as one percent, it will only be necessary to run two of them for a relatively short period before they become completely out of step. In this way when using asynchronous transmission, based on the idea of restarting the clocks at the beginning of each character, only short messages can be sent. If the message length is kept small then the likelihood of the clocks at the transmitter and the receiver getting out of step during such a period of transmission is slight.

Asynchronous data transmission   The most common method transfer of data is accomplished by means of an **ASCII** (American Standard Code for Information Interchange) format. Here the 0s and 1s of a computer output are coded into a seven bit pattern giving a total of 128 discrete combinations. An eighth bit is used for error detection and correction, the parity bit. Finally, a stop bit or a series of stop bits will be transmitted to give the receiving device time to get rid of the incoming character and to enable the input clock to be reset. This allows the next character to follow immediately. An example of the transmission format of an ASCII character between a transmitter and receiver is shown in Figure 6.16.

Figure 6.16

Asynchronous data transmission.

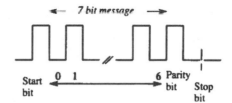

Compared to synchronous transmissions, asynchronous communication is more flexible in that it allows communicating systems to run at different speeds and to automatically adapt to the different speeds required to achieve intercommunication. However it is potentially slower than synchronous transmissions because two control signals are needed to accomplish each data transfer, the transmitter request and the receiver response, requiring two bus traverses plus additional logic delays before completion of the transfer.

6.2.2   SYNCHRONOUS DATA TRANSMISSION

For transmissions between computers, synchronous data communication occurs when characters, or bits, are sent at a fixed rate through a communications link, with the transmitting and receiving devices synchronized to an overall timing source. This eliminates the need for the start and stop bits encountered in asynchronous transmission, thereby increasing the rate of data throughput. With this form of transmission the transmitting station has to send a clock signal to the receiving system to synchronize the transmission and reception of the data. It is to be noted, however, that even when the clocks have been synchronized, they are unlikely to stay in step for too long. To avoid the danger of the clocks drifting during the transmission of a long block of data, the hardware line adapters may inject a short stream of synchronizing bits. This is usually handled by the periodic transmission of a stream of bits to the receiving device prior to the sending of a block of data to ensure that the reception clock is in step with that of the transmitting device. This operation introduces the concept of the SYNC (synchronizing) message which has a prescribed pattern of bits according to the line code being used. The ASCII SYNC character for example has a bit pattern of 0010110, plus parity bit. In some systems this can be generated automatically by the hardware after a predetermined period without synchronism. Hence, within synchronous transmission, character synchronism and the identification of the beginning of a transmission block is usually achieved in one operation.

6.3 KEY TERMS

Acknowledgement	Parallel interface bus
ASCII	Parity bit
Asynchronous	Parity check
Asynchronous data transmission	RS-232C
Backplane bus	RS-422
Bandwidth	RS-423
Bus (computer)	SCSI
Bus levels	Serial bus
Bus driver	Serial transmission
Channel	Start bit
Component level bus	Stop bit
Connector	Synchronous data transmission
Data transmission	System bus
Daughterboard	S100 bus
Futurebus	Transceiver
General purpose interface bus	Transmission
Handshaking	Tristate buffer
IBM bus	Unmapped parallel interface
IEEE 488 standard	bus
Motherboard	VME bus
Multibus family	

6.4 REVIEW QUESTIONS

1. How are a computer's internal components linked physically?
2. Define what the term 'bus' means in a microprocessor system.
3. List the five bus levels within a computer system.
4. What features distinguish the component level bus?
5. What does the term 'tristate' denote?
6. Why is it necessary to have tristate buffers to interconnect a microprocessor and its components onto the system bus?
7. In a microprocessor based system what is a motherboard and where is it located?
8. Name the different bus structures within Mulitbus.
9. Give a list of the benefits that having extra bus structures within a Multibus system has.
10. What is the SCSI bus most often used for?
11. What advantages does the RS-422 bus have over a RS-232C bus
12. When is asynchronous data transmission used in a computer system?
13. What do the following terms mean: channel, start bit, handshaking.
14. What are 'parity bits' used for in data transmission?
15. In synchronous data transmission why is it necessary to synchronize the transmission and reception clocks before communicating?

6.5 PROJECT QUESTIONS

1. Obtain the data sheets for a tristate buffer and note its operating characteristics.
2. On most computers all internal components are designed around a common word size. Why?
3. Explain how a computer's word size affects its processing speed, main memory capacity, precision and instruction set size.
4. What is 'bus contention'?
5. Look inside a personal computer and note the different connectors used.
6. By noting the specification for the VME bus compare it with the Multibus system.
7. Contrast the GPIB and SCSI bus structures.
8. Obtain more information on the SCSI interface standard.
9. On a personal computer system where would you find the serial and parallel port?
10. Why are the speed and distance of transmission inversely related in the RS–232 bus?

6.6 FURTHER READING

This chapter has provided an insight into the different types of bus systems in a computer system. A short list of articles provides more reading on the topic.

Bus (Articles)

1. Microprocessor bus structures and standards, P. L. Borrill, *IEEE Micro*, February 1981, pp. 84–95.
2. Computer buses–a tutorial, D. B. Gustavson, *IEEE Computer*, August 1984, pp. 7–22.
3. The SCSI bus, L. B. Glass, *Byte*, February 1990, pp. 267–274
4. High speed 32–bit buses for forward–looking computers, P. L. Borrill, *IEEE Spectrum*, July 1989, pp. 34–37.
5. Open buses broaden foothold at all levels, W. Andrews, *Computer Design*, May 1 1990, pp. 55–64.
6. Performance drives high–end bus standards, W. Andrews, *Computer Design*, May 1993, pp. 59–72.
7. A framework for computer design, W. Dawson and R. Dobinson, *IEEE Spectrum*, October 1986, pp. 65–75
8. Buses and bus standards, W. Dawson and R. Doninson, *Computer Standards and Interfaces*, June 1987, pp. 58–60.
9. Enhancing the performance of standard buses, W. Andrew, *Computer Design*, September 1991, pp. 61–71.

Bus Systems (Books)

1. *The Multibus Design Guidebook*, J. B. Johnson and S. Kassel, McGraw Hill, 1984.

Chapter 7

Computer Hardware Design

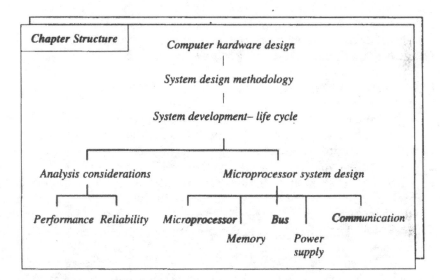

Chapter Structure

Computer hardware design
|
System design methodology
|
System development– life cycle

Analysis considerations   Microprocessor system design

Performance Reliability   Microprocessor   Bus   Communication

Memory   Power
supply

Learning Objectives

In this chapter you will learn about:

1. The systems life cycle and its various elements.
2. The top down development process.
3. The functional and nonfunctional requirements that have to be considered in the design of a computer system.
4. The purpose of benchmark tests.
5. The considerations required when selecting the microprocessor, memory and bus system for a computer.
6. The problem of noise in a computer system and the different sources of noise that can be identified.

> 'When you break the problem into parts you've determined its structure and the work that will be done around it forever.'
>
> Mark Miller's law of
> Irrevocable Subdivision.

Chapter 7

Computer Hardware Design

The design of a computer system can be broken down into the hardware and the software phases. However, the two must be integrated and tested to verify the overall design. In the study of system design it is important that a structured approach to the problem is undertaken, using a design methodology.

7.1 SYSTEM DESIGN METHODOLOGY

A system design methodology may be regarded as a study of the principles and procedures for creating designs. It is the role of a design methodology to deduce the structure and constituents of a system in order to satisfy the given specifications of the system. There are essentially two types of design methodologies.

1. Bottom Up – Bottom up is a design methodology in which parts, or modules, of a system are initially designed and tested separately, before being combined to form a larger system. In this approach the designer goes from a detailed level to a higher, more abstract level. It also extends the series of system components and modifies them to meet the requirement of a system.

2. Top Down – The most popular method used by engineers for designing all types of systems is top down design. This methodology uses the process of task decomposition and successive refinement to reach a final scheme, and is considered by most commentators to be more logical and systematic compared to a bottom up design.

Top down design tries to identify the total system function, then partition it into less complex subfunctions, each of which performs a specific task. These subfunctions are in turn, subdivided or partitioned into more subfunctions of even less complexity. The partitioning process continues until relatively low complexity subfunctions, which are amenable to easy implementation, are reached. In this way it is then possible to reduce the complexity of the hardware and software design within a computer system to manageable proportions.

It can be seen from such a description that each level of partitioning within the top down strategy involves adding increased detail to the design. Using the top down process, the initial analysis begins at a high, conceptual, level looking at the overall structure of the proposed system in a broad perspective before considering in more detail the design of the lower subfunctions. By proceeding from the general to the specific it is possible to have constant quality assurance checks throughout the design. It can also be appreciated that the way the design is partitioned at a particular level is constrained by previous decisions made at the more general higher levels and in turn constrains later partitioning decisions for the more detailed lower levels.

The bottom up approach is useful in hardware design since it consists of using already built very large scale integration (VLSI) chips or sub-assemblies. The top down approach on the other hand is more useful for the overall design of a system and is fundamental to software design in particular.

Glossary 7.1

Hardware Design

CAD – Computer aided design (CAD) is the use of a high performance computer to assist in the design process of an item. In its simplest form it is a computerized communication and design function within and between design engineering and manufacturing engineering. The first CAD system for the IBM PC style of computer was introduced in 1982 by Autodesk.

CAE – Computer Aided Engineering (CAE) functions include automated design, simulation analysis, process and tool design.

CAM – Computer Aided Manufacture (CAM) is a combination of many engineering and manufacturing disciplines. CAM functions include tool production, parts production, assembly automation, numerical control (NC) and inspection and testing; the disciplines are tied together with a common communications system and use a common data base.

OEM – An original equipment manufacturer (OEM) is the maker of equipment that is marketed by another vendor, usually under the name of the reseller. The OEM may either manufacture only certain components, or alternately complete computers which are then often configured with software and/or other hardware by the reseller.

OPTIMIZATION – Optimization in a computer environment refers to the design or modification of a system or program in order to achieve maximum efficiency. The particular type of efficiency sought after in the optimization depends on the requirement specified, typical examples are time, cost and storage capacity.

PLUG COMPATIBLE – Plug compatible implies being able to change between a manufacturer's original peripheral device and a similar peripheral device from another company.

PROTOTYPE – A prototype is an experimental system, usually developed at a research establishments or university. The ideas generated and the knowledge gained from the use of the prototype is usually applied to the production of commercial machines.

PROTOTYPING – Prototyping is the process of building a working model of a system, or part of a system, in order to clarify the characteristics and operation of the system. The prototype developers must consider what they intend to learn from building a prototype, whether it is a better understanding of user requirements in certain areas, the feasibility of a particular design approach or an understanding of system performance issues. When the objectives of the prototyping effort have been attained a basic decision must be made : either refine the prototype into the complete system or begin a new development process. This will depend on the functionality of the prototype.

Software prototyping is a development methodology based on building and using a model of a system for designing, implementing, testing and installing the software. The initial version of the prototype is not the full system software, but it does contain its designer's understanding of the database, screens, and reports required for the full system. As the user and information systems people begin to work with the prototype the prototype will usually require to be changed.

7.2 SYSTEM DEVELOPMENT – LIFE CYCLE

The development of a computer system is the process/activity of producing a computer to a given specification at a competitive cost and performance. It can be seen to involve the organization of various component parts into a coherent processing system, where the organization and implementation of the system involves the translation of the system specification into an architecture made up from functional units that are ultimately represented in an underlying electronic technology.

The design and development of a computer system is a complex process with many different stages in the development **life cycle**. The stages and the different tasks within each of the stages are shown in Figure 7.1.

The initial stage is the definition stage in which the functional requirements of the system are identified. After this, the requirements are evaluated and an analysis of the system's functions is performed. The design of the system can be broken down into two basic components, the major part of which involves the machine design stage in which the electronic circuits are designed. Hence it can be seen that in general in the production of a computer system there are three stages: analysis, development and implementation. We will consider each in more detail.

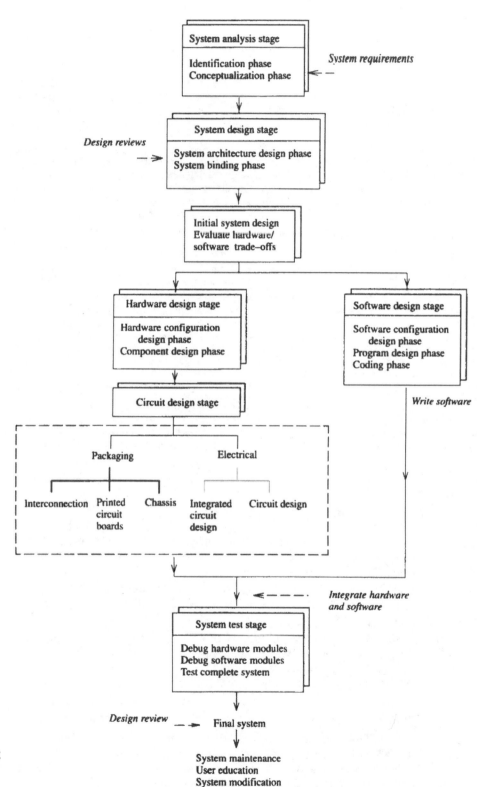

Figure 7.1

Computer development cycle.

7.2.1   SYSTEM ANALYSIS

It can be seen from Figure 7.1 that before any hardware or software design commences a detailed specification of the intended system must be decided by a team of **system analysts**. System analysis is concerned with the understanding of an existing, or proposed, system in terms of what it does, or has to do. It is a detailed step by step investigation of the system's organization for the purpose of determining what must be done in terms of updating an existing system or developing a new system and the best way to do it. In performing a system analysis the analyst will use **structured analysis** techniques. These are guidelines and techniques, that can be used to assist in stating the functional requirements of a system in logical terms. It tends to rely heavily on data flow diagrams.

In the design of a computer system the analyst will detail a complete description of the hardware and software requirements for the design. These details are documented in the functional requirements of the system. It may include such things as a statement of the inputs to be supplied by the user, the outputs desired by the user, the algorithms involved in any computations desired by the user and a description of such physical constraints as response time.

The **requirements specifications**, also known the **functional requirements** or **system specifications**, state the desired functional and performance characteristics required of the system, independent of any actual realization. They are derived from the requirements of the user which in turn result from some form of analysis of a user's computer requirement. In this case specification refers to the creation of a document from the customer that reflects the type of computer system that is required in a way that drives the design and that also serves to provide quantifiable validation of the results of the design.

This is a key point in any design in that the design must begin with, and be based on, a clear understanding and a concise specification of what must be designed and what constraints are to be placed on the design. In this way the system specifications may be seen as representing the interface between the user and the designer. They designate the overall purpose of the system, i.e. what it is to accomplish in terms of required, and often desirable, features and not how it will be accomplished or what will be used to accomplish it. The purpose of these requirements are to facilitate implementation decisions in the later stages of development. The best method to prevent errors in a system is to write careful specifications, create a good design, and conduct frequent reviews. Without such documentation the designer may solve the wrong problem.

Table 7.1

Topics to be considered in detailing the requirements specifications of a computer system.

1.	A complete description of what the system should do, identifying overall objectives in terms of the areas of work it is intended for.
2.	The performance requirements it must meet.
3.	The improvements that will result over present systems in terms of accuracy, availability and control.
4.	Estimate of the cost. This involves a cost evaluation of both the present system and the new projected system, to provide a comparison, taking into account direct costs and any indirect savings that may accrue.
5.	An estimate of the date by which the machine is expected to be installed and the system to become operational.
6.	The length of time that the machine can be expected to give reasonable service.
7.	A note on any technical developments which may just be round the corner that will make the processing techniques decided upon redundant or uneconomic in the near future.
8.	Specific details of the operator/system interaction.
9.	A specification of the system's interface with the external environment.
10.	Procedures for handling errors and diagnosing malfunctions.

Growing interest in requirements specification has been accompanied by the emergence of general guidelines regarding the properties of a good specification. A requirement specification should include the points noted in Table 7.1.

System analysis

The initial purpose of generating this form of requirements specification is to govern the design, development and testing of the hardware and software of the computer throughout its life cycle. Its continuing function is to serve as a requirements baseline against which the impact of proposed performance and design changes on the hardware and software may be assessed.

7.2.2   SYSTEM DESIGN

From the initial requirements specification the various hardware and software design processes can proceed. System design is the second stage within the development life cycle of a computer product. Within this stage, the designer attempts to find alternative solutions to the problems and tasks uncovered in the system analysis phase. The cost effectiveness of these alternatives is then determined and the final recommendations for the system design made.

There are a number of points that must be considered when undertaking the design of a computer system. These are given in Table 7.2. There are also a number of desirable properties that a computer should have, given in Article 7.1.

Hardware	1. What component technology should be used?
	 2. Which microprocessors should be selected?
	    (a) One already used in another company product.
	    (b) One which is compatible with software and or hardware which is to be used for another product.
	    (c) A new one:
	       are compatible components available?
	       are development tools available?
	       are the microprocessors components second sourced?
	 3. How is the computer to be packaged?
	    (a) Circuit board size.
	    (b) Connector types.
	    (c) Subrack.
	 4. Should off–the–shelf microprocessor boards be used rather than developing new ones? Which ones?
	 5. Does the computer have to be compatible with some other product?
	 6. What are the future needs?
	    (a) Should the computer system be expandable
	    (b) Is it going to be the basis for a product family.
	 7. What will be the architecture of the computer system ?

Operating System	1. Which operating system should be used?
	    (a) A commercially available one.
	    (b) One which has been used in another company product.
	    (c) Design a new one.

Application Software	1. What programming language should be used? 2. How should the software structuring be undertaken?

Table 7.2

System design considerations.

DESIRABLE PROPERTIES OF A COMPUTER SYSTEM

Article 7.1

There are a number of important features which a computer system is expected to have.

1. **Compatible** – Compatibility is the ability of different computers to run the same programs. Computer designers generally design for upward compatibility, which allows high performance members of a family to run the same programs as the lower performance members.

2. **Generality** – This is a measure of how wide is the range of applications a computer system can handle. Although the number of instructions in an instruction set is not a direct measure of a computer's generality, it does provide an indication

3. **Efficiency** – This is measure of the average amount of hardware in the computer that remains busy during its normal use. An efficient architecture permits, but does not ensure, an efficient implementation. Notice that a conflict exists between efficiency and generality. Also, because of the decreasing cost of computer components, efficiency is of much less concern now than it was early in the history of computer developments. An efficient computer design however does not tend to allow for both very high speed and very low cost implementation and in a large computer family both types of implementations are important.

4. **Modifiability** – The ability of a system to be changed or enhanced to meet the needs of user during its productive lifetime is known as modifiability.

5. **Response Time** – The response time is the time the system will take to react to a change in, or a stimulus from, its environment. For a computer it may be the time from the completion of data entry to a response from the computer. The system response time is the time within which it must detect an internal or external event and respond with an action. It is an important measure in a time–critical or real–time system.

6. **Ease Of Use** – In terms of the hardware ease of use is a measure of how simple it is for a systems programmer to develop software such as an operating systems or a compiler for the particular computer. The term user friendly, on the other hand, describes a terminal or computer which has input facilities specially designed to allow an uninformed user to use the computer. The term man–machine interface describes the dialogue between people and the computer system, often used in discussion of the ease of use of the system.

7. **Malleability** – This is a measure of how easy it is for a designer to implement a wide range of computers having a particular architecture. The more specific an architecture, the more difficult it is to produce machines that differ in size and performance from others.

8. **Expandability** – This is a measure of how easy it is for a designer to increase the capabilities of a computer system such as its maximum memory size or arithmetic capabilities.

7.2.3   SYSTEM IMPLEMENTATION

After the design stages have been completed the complete system must be assembled and tested. The system implementation phase is the third stage in the development procedure for a new, or revised, system. The goal of implementation is to ensure that the system is completely debugged, operational and is acceptable to the users.

Finally when the hardware elements of the computer have been assembled they will be integrated together to produce the complete hardware system.

7.2.4   TESTING

After the construction of any hardware system it must be tested. There are several practices in the hardware design of a computer which can improve testability, these are shown in Table 7.3. There are two approaches to the testing of the hardware of a computer; validation and verification.

Table 7.3

Design considerations to aid testability.

1.	Dividing complex functions into functionally complete smaller sections.
2.	Provide edge terminated test/control points on boards.
3.	Provide a means of isolating each logic section from a common bus.
4.	Provide a means of substituting an external clock.
5.	Use sockets for complex integrated circuits.
6.	Provide tester access to all inputs and outputs.
7.	Provide a means of initialising all storage elements.
8.	Provide a means of breaking all feedback loops.
9.	Avoid logical redundancy.
10.	Avoid the use of asynchronous logic.
11.	Avoid excessive gate fan out.

Testing

1.   Validation – In the design of a computer system the validation phase involves the design team producing convincing demonstrations of the soundness of the decisions made within the design process. Validation tends to rely more on empirical experiment (testing) than on mathematical proof. In this context acceptance tests are often used by manufacturers to demonstrate the correct functioning of their design. An **acceptance test** is a single test, or a series of tests, designed by manufacturers, or users of a computer system, to ensure that a new system works properly.

2.   Verification – In the design of any hardware or software system the term verification is used to denote that a particular hardware and software combination implements identically the requirements defined by the specification. This involves producing rigorous, formally justifiable proofs of the system design.

7.2.5   DOCUMENTING

Good system and program documentation is also a fundamental component of the structured design philosophy. In documenting a computer design, the designer goes through the process of describing in detail every phase of the design and implementation of the computer system from its hardware to any specific facts about the software produced for the machine. He may also include the instructions for a user to work with the system. Good documentation meets the following requirements:

1. Enhances the understandability of the system.
2. Is easy and inexpensive to produce an update.
3. Gives a high level view of the system, explaining its purpose and the relationship of the various data and procedural components.
4. Describes in detail the components of each data structure and procedural component.
5. Provides a blueprint for representing problem requirements in a system design and then for translating the design into program code.

7.2.6   MAINTENANCE

After the system has been shown to work and to have fulfilled the functional requirements noted for it, the computer may be released to the customer. The development phase may be complete but it is still normal for the system developers to continue to maintain and possibly upgrade the hardware and software for the particular design. The maintenance of the hardware in a computer system is aimed at reducing the downtime of a computer system and maximizing the serviceable time for running programs. There are various categories of such maintenance, these are preventative, routine and scheduled. Corrective maintenance on the other hand implies work that is required to correct a machine fault.

7.3 ANALYSIS CONSIDERATIONS

The most important considerations in the development of a computer are usually performance and cost. Secondary metrics which may vary in importance include fault tolerance, power and environmental factors such as size and cooling and noise. We will consider two of these properties, performance and reliability, in more detail.

7.3.1 PERFORMANCE

The performance of a computer system is a measure of its ability to perform its allotted tasks. Figure 7.2 shows an approximate comparison between the performance of different computer types. There are many factors which influence the performance of a computer, including the programming language used, the type of compiler, the run–time library functions, the memory architecture and the cache memory size. Typical units of performance are **MIPS** (million instructions per second) and **MFLOPS** (million floating point operations per second), other considerations are given in Glossary 7.2.

Figure 7.2

Relative performance of different computer classes.

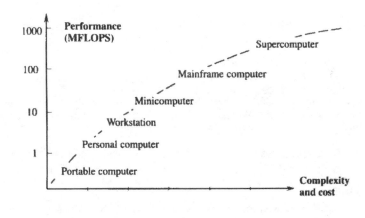

Figure 7.3

Performance range of various supercomputers.

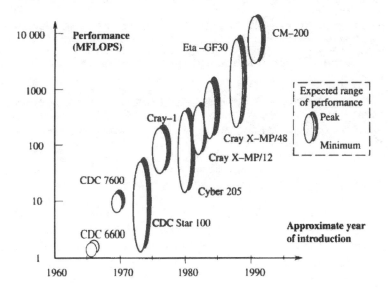

Glossary 7.2

Computer Performance

BALANCED SYSTEM – A balanced computer system is one in which the system is not input or output driven, i.e. the speed of the system is not solely dependent on the speed of the I/O operations. Instead the time taken to perform the input, output and computation operations within the system all take a comparable time.

BOTTLENECK – A bottleneck is a constraint on a system's performance that is usually a design error or oversight. A resource becomes a bottleneck when it cannot handle the work being routed to it. A resource operating near its capacity tends to become saturated and may become a bottleneck to a system's performance. Bottlenecks can usually be eliminated by design modifications.

FLOPS – In the area of scientific computing the most common unit of computer performance is the floating point operation per second (FLOPS). Millions of FLOPS are called megaflops, usually with the added specification of 'peak MFLOPS' or 'sustained MFLOPS'.

INSTRUCTION MIX – An instruction mix consists of a mixture of different instruction types that can be run on a computer to gauge its performance. One of the earliest instruction mixes was devised around 1960 by J.C. Gibson who analysed many programs written for the IBM 650 and 704 machines and came up with the Gibson Mix, which made it possible to compute the execution time of an average instruction. The inverse of this measure became the basis of the kilo instructions per second (KIPS) performance measure and later millions of instructions per second (MIPS). Today's ad hoc standard 'unit of computation' is the peak performance of a VAX 11/780 computer of about one million instructions executed per second (1 MIPS).

Major computer manufacturers however, rarely characterize their systems in terms of MIPS, since there is no standard procedure for computing MIPS. Vendors are therefore cautious about quoting a figure which may be too conservative and so implies less performance, or quoting too large a figure which may lead to law suits. Instead it is common for manufactures to compare their machine with their own previous models, where the relation may be decided by running a particular 'in–house' instruction mix, or some universally accepted benchmarks.

INSTRUCTION TRACE – In performing a dynamic instruction trace on a computer, either software or hardware instrumentation is used to record the type of each computer instruction as it is being executed. By performing such traces on a large, and presumably representative, body of programs, it is possible to determine how often a particular type of instruction is likely to be used in a typical program. Given the frequency, and the execution time for each instruction class, it is then possible to develop the execution time for an average instruction using a weighted average of the various instruction timings.

I/O BOUND – A central processing unit (CPU) that is I/O bound is slowed in its operation because of excessive I/O procedures which are extremely slow in comparison to the CPU's internal processing.

MIPS – The million instructions per second (MIPS) number is used as a general comparison gauge of a computer's raw processing speed.

OVERHEAD – An overhead is an amount of work that must be done to control a system resource which results in no useful computation on the task running on the computer system. Overheads at best can only be reduced by a more efficient implementation.

PERFORMANCE MONITORING – The performance monitoring of a computer system is the collection and analysis of information regarding the performance of the computer over a period of time. It is useful in that it can indicate any bottlenecks or excessive overheads in a system and provide information to improve the overall design of the computer. The information gathered may be, for example, the execution time of standard benchmark programs or may make use of special evaluation software.

PROCESS BOUND – A computer is said to be process bound when a program running on it monopolizes the processing facilities of the computer. This then makes it impossible for other programs to be executed. It may be contrasted with I/O bound computer systems.

THROUGHPUT – Throughput is the speed or productivity with which a computer either performs programs or processes data. It may also be regarded as the total amount of work the machine can accomplish in a given amount of time.

VON NEUMANN BOTTLENECK – The von Neumann bottleneck occurs in a computer system when the maximum speed of the system is limited by the processor to memory bandwidth.

Performance

Computer manufacturers have been continually increasing the performance of their products through the years. An example of this is shown in Figure 7.3 where the performance range is given for a number of supercomputers. It can be seen that the performance of different supercomputers has increased dramatically over the years. It can also be noted that for individual supercomputers there is a wide range of performance. This is dependent on the type of problem run on the supercomputer and illustrates that performance of any computer is highly susceptible not only to the architecture of the device but also the type of task being performed.

Benchmark Program – It has been realized by computer manufactures for a long time that instruction–level performance figures like MIPS would always be unreliable in comparing computer systems with dissimilar architectures. Furthermore, as more end–user's activity became directed to high level languages, the ability of the compiler to produce good, or optimized, code became an increasingly important performance factor. In a response to both these needs, synthetic benchmarks were developed, where the goal was to create a simple, relatively small program that approximated the behaviour of typical applications coded in high level languages. A benchmark program should be designed to be representative of the

Performance expected workload on a system, in order that it can be used to derive relevant information as to the system's performance. In particular, the synthetic benchmark is supposed to mimic both the relative frequencies of the various types of high level language statements and constructs, and the types of data structures that real programs have to deal with, in the framework of a small program. The program would then allow a comparison to be drawn on the speed and performance of different types of computer systems.

Benchmarks may also be real application programs, such as calculating a Fourier transform, sorting a series of numbers or performing a number of floating point multiplications. In this respect they may be regarded as workload simulation programs. Since the main purpose of any system is to run actual programs on actual data, benchmarks may be viewed as an important indication as to the ability of a computer to handle the real world. Benchmarks are useful in evaluating hardware as well as software and are particularly useful in comparing the performance of a system before and after certain changes have been made. There are also a number of widely used benchmarks (Article 7.2). Table 7.4 shows the benchmark performance of several computers using the Linpack benchmark.

Table 7.4

Linpack benchmark performance for a number of computers.

Computer class	Computer	Number of processors	Performance in Linpacks
Supercomputer	Cray Y–MP/832	8	275
		4	226
		2	144
		1	90
	Cray 1–S	1	27
	Alliant FX/2800	14	26
		8	22
		4	16
		2	9.9
		1	6.4
	CDC 7600	1	2.0
	CDC 6600	1	0.48
Mainframe	IBM 370/195	1	2.5
	IBM 370/165	1	0.77
Supermini	IBM RISC 6000–550	1	27
	IBM RISC 6000–320	1	9
Mini	DEC VAX 11/780	1	0.14
	DEC VAX 11/750	1	0.057
Workstation	Sun SPARC station	1	1.6
Personal computer	IBM PC AT w/80287	1	0.012

WHETSTONE BENCHMARK – The Whetstone benchmark was the first program in the literature explicitly designed for benchmarking. Its authors were H.J. Curnow and B.A. Wichmann from the National Physical Laboratory in Britain. It was published in 1976 with ALGOL 60 as the publication language although it is now mostly used in its FORTRAN version. The benchmark owes its name to the Whetstone ALGOL compiler system which was used by the authors to collect statistics about the distribution of instructions of the intermediate language used by this compiler (called Whetstone instructions) for a large number of numerical programs. A synthetic program was then designed consisting of several modules, each containing statements of some particular type, e.g. integer arithmetic, floating point arithmetic, 'if' statements, and ending with a statement printing the results. Weights were attached to the different modules, realized as loop counts around the individual module statements, such that the distribution of Whetstone instructions for the synthetic benchmark matched the distribution observed in the original program sample. Since the Whetstone has a high percentage of floating point operations it is often used to represent numerical programs.

DHRYSTONE BENCHMARK – The Dhrystone is a synthetic benchmark program which was published in 1984 by R.P. Weicker. The original language of publication was Ada, although a Pascal or C version can also be used. The basis for the Dhrystone is a literature survey on the distribution of source language features in system–type programming. The benchmark consists of 12 procedures included in one measurement loop with 94 statements. During one loop 101 statements are executed dynamically. The Dhrystone contains no floating point operations and a considerable percentage of execution time is spent in string functions, it also contains hardly any minor loops within the main measurement loop.

LINPACK BENCHMARK – The Linpack benchmark is a collection of linear algebra subroutines; the static code length for all the subprograms is 4537 bytes, often used to determine the performance of a computer system. It was constructed from a real purposeful program and was first published in 1976 by J. Dongarra in the USA. It measures the speed at which a computer solves a system of equations in a FORTRAN environment. The program operates on a large matrix of data, the standard matrix size is 100 x 100, and has a high percentage of floating point operations although it does not use any floating point division.

LIVERMORE LOOPS – These consist of twenty four FORTRAN loops that are assumed to be reflective of the type of computation common at Lawrence Livermore National Laboratories. The loops operate on data sets with 1001 or fewer elements.

SPEC – System Performance and Evaluation Cooperative (SPEC) is an international group of manufacturers, including Hewlett–Packard, MIPS and SUN Microsystems, of workstations, servers and other computer systems who have resolved to work together to provide performance summaries based on the results from a suite of programs run under specified conditions. The cooperative was formed in November 1988. The benchmark suite consists of ten real–world applications, taken from a variety of scientific and engineering applications, six of the programs use floating point operations.

Benchmark program

As has been stated, the benchmark is supposed to compare the relative performances of different computers; however, it is difficult to get a fair benchmark for all computers. Manufacturers often publish performance statistics such as MIPS, MFLOPS, Whetstones, Dhrystones, Linpack rates and Livermore Loops performance. Comparing these statistics for any two computers will likely show one faster according to some of the measurements and the other faster according to some other measurements. Hence, more than one benchmark should be tried on a machine to give an indication of its performance level, and an average taken to decide on a system's worth.

Glossary 7.3

AVAILABILITY – The term availability has several interpretations.
1. It can be taken as a measure of the capability of a piece of equipment to operate without failure when put into service.
2. It is a measure of an equipment's ability to perform its required function under known conditions for a certain period of time or its probability of survival over some specified time interval.
3. It can refer to the proportion of time for which a system is expected to be fully operational.
4. In the documentation for computer systems the term availability is the average probability that the system is operational at the instant of time considered.

Availability is typically used as a basis for evaluating the reliability of a system in which service can be delayed or denied for short periods of time without serious consequences. It may be calculated from:

$$\text{Availability} = \frac{\text{MTBF}}{\text{MTBF} + \text{MTTR}}$$

where MTBF = mean time before failure, and MTTR = mean time to repair.

The MTBF is usually mentioned with reference to the reliability of a system and is the average length of time a system retains some operational utility without external maintenance. The term MTTR of a system is an indication of how long it would take to repair a system.

BUG – A bug is a defect or an error in the formulation and writing of a computer program or in the hardware design or construction of a unit, which results in a system malfunctioning.

COVERAGE – One way of indicating the level of fault detection and reconfiguration within processing systems is to refer to their coverage. Coverage is the conditional probability that if a fault occurs the proper recovery procedures will take place.

DOWNTIME – The downtime of a computer system is that period when the computer or its resources are unavailable to end-users, usually due to a system failure.

FAIL–SAFE – The term fail–safe refers to a computer system which comes to a halt with minimal data loss if a serious failure occurs. Fail–safe is less acceptable than fail–soft; however, in some circumstances the two terms are synonymous.

FAIL–SOFT – The term fail–soft refers to a computer system which comes to a halt with no loss of data if a serious failure occurs.

INTEGRITY – The integrity of a system refers to the fail safe operation of that system. Integrity is an important consideration within any computer system design. It is important that failure in a system for whatever cause is treated correctly and the computer system does not produce any catastrophic problems.

MAINTAINABILITY – Where the term reliability gives an indication of how long a system can be expected to perform before a failure is expected, the term maintainability gives some indication of how long it will take to fix the system once a failure has occurred. The system maintainability law states that for a given system in its operational environment, the cost and effort associated with maintaining the system is principally dependent on three factors.
1. The inherent complexity of the functions it must perform.
2. How well it has been defined, and how well its development has been controlled through documentation.
3. The extent to which architectural features that facilitate maintainability have been employed in building the system.

MTBF – The mean time between failure (MTBF) is a stated or published period of time for which a user may reasonably expect the system (or device) to operate before an incapacitating failure occurs.

MTTR – The mean time to repair (MTTR) is the average time required to perform corrective maintenance on a failed device.

ROBUST – A robust computing system is one which can continue to provide service despite the impact of adverse operational circumstances, such as maltreatment or invalid usage. Issues such as security, safety and performance are naturally assumed.

7.3.2 RELIABILITY

The requirement that a processing system operate for long periods of time without any failure is usually taken for granted in the design of the system. This however requires careful consideration of the nature of failures within a system, and an understanding of the different techniques which can be implemented to increase the reliability of any design.

Reliability is a measure of the expected failure rate of a system and in a sense indicates the quality of the system.

It may also be seen as a measure of the continuous delivery of proper service, or equivalently of the time to failure and is sometimes expressed as mean time between failures (MTBF).

If a plot is made of the reliability of a system with time then it tends to look like the bathtub curve shown in Figure 7.4. The early failures shown in the curve are caused by manufacturing and constructional faults while the intermediate region of the curve exhibits a constant failure rate. This constant failure rate may be used in reliability estimation such that in this part of the curve the MTBF is defined as the reciprocal of the failure rate. Finally, after a certain time the curve adopts an upward trend, due to the system components coming to the end of their expected life. In order to have a reliable system the designer should aim to have R_1 and R_2 as low as possible, T_1 as short as possible, and T_2 as long as possible.

Figure 7.4

Typical reliability curve.

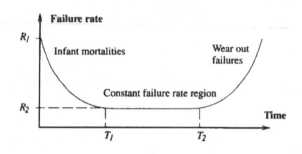

Reliability

Reliability should be considered at all stages in the design and manufacture of a system, from the initial system specification right through to the production and commissioning of the final article. At the system design stage, trade offs between reliability and other system parameters, such as size, weight and cost must be considered. Cost is always an important consideration where reliability is concerned in that the most reliable system design may not be the best choice in practice if the cost is too great. In all cases the cost of reliability should be contrasted with the full operational costs (including system failure) cost of subsequent maintenance, and the provision of spare parts.

Table 7.5

Reliability considerations for components in a system.

1. The components should have life test data available in the actual system environment. This data is needed for reliability prediction.
2. Basic component manufacture must be investigated to ensure that the manufacturing processes are well controlled.
3. A number of alternative sources of component suppliers should be obtained.
4. The specification of a computer system must be sufficiently detailed to allow if necessary, the replacement of one component by another.
5. The design of a computer should endeavour to use a standardization of circuits, where possible, to achieve a small number of basic components or subsystems. In this way it reduces the number of spare parts and testing procedures required for the system.
6. Components should be derated, ie. not run at the limit of their maximum specifications, or used conservatively with large safety factors.
7. The use of worst case and statistical circuit design methods must be considered.
8. Marginal testing facilities must be evaluated, bearing in mind the cost of the extra equipment required for the checking process.

In order to ensure the reliability of a system it is necessary for the designer to make the correct choice of components or subsystems, since these components determine the reliability of the ultimate system. When selecting components or circuits for a system the points noted in Table 7.5 must be borne in mind.

Reliability

It is also worth noting that an improvement in hardware reliability on a computer tends to makes software failures more sensitive. Similarly achieving an increase in hardware reliability is naturally accompanied, due to an increase in complexity, with an increase in the relative importance of hardware design faults.

Any device or system may at some stage in its lifetime break down or fail. We may define failure as the termination of a system's ability to perform its required function. A **system failure** occurs when the delivered service deviates from conditions stated in the service specification, the latter being an agreed description of the expected service. The process which causes the failure is known as the **failure mechanism**. The development of a system failure is shown in Figure 7.5.

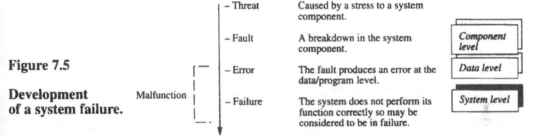

Figure 7.5

Development of a system failure.

The progression of a potential threat to a realizable failure within a system goes through various stage, such that from the initial threat, a fault (a mechanical or algorithmic defect) may go on to generate errors. An error in turn may be seen as an item of information that when processed by the normal algorithms operating within the system will produce a failure and a failure may be regarded as an event at which a system violates its specifications. We will consider each of the stages in more detail.

1. **Threat** – A threat within a computer system may be seen as some form of stress to a component in the system which may eventually lead to an anomalous condition, or a malfunction of the computer. Three categories of threat can be identified.

(a) Normal environmental stress that causes components to weaken or break down over a a long period of time.

(b) Abnormal environmental stress that can disrupt components in a short time.

(c) A design flaw in the system which over stresses the system components.

A system may be designed to stop threats from producing malfunctions, and/or it may be designed to cope in some acceptable manner with malfunctions. One possible solution to minimize the effect of threats within a computer system is by including backup, i.e. standby, components which can be used should parts of the system develop a fault. These components constitute redundant hardware and software items as they do no useful work until a malfunction occurs. It can be noted, however, that no number of redundant elements can eliminate every threat.

2.   Fault – The incorrect operation of a component or resource within a computer is called a fault. A fault in some senses behaves like a type of operator in that it moves a system from one existing processing state to another, where the latter is of less capability. Faults tend to manifest themselves through the presence of errors, such that when an error is perceived at the boundary of a system resource, the user must decide whether it represents a failure of the system to perform adequately or whether the failure simply leads to an acceptable, although less desired, level of system operation. A fault may be detected when either a failure of the resource is observed, or an error is discovered in the system. In most cases the fault can be identified; however, in some cases, particularly if the fault is transient, it may remain a hypothesis that cannot be adequately verified. Temporary faults, transients or intermittent faults constitute the majority of faults affecting a computing system, typically about 70–90 %. It is possible to partition faults into three classes and four different types. These are shown in Table 7.6.

Table 7.6

Fault classification and types.

Fault			Description
Classes	1.	**Limited**	For faults whose symptoms are limited in scope, redundancy is a possible counter measure.
	2.	**External**	Faults which are mechanically, thermally or electro-magnetically induced can usually be treated with some form of shielding.
	3.	**Incorrect design**	Neither redundancy nor shielding are of much use for a fault resulting from an incorrect design.
Types	1.	**Crash**	A component suffering a crash fault ceases to function without having previously violated its specification.
	2.	**Failstop**	This is similar to a crash fault except that the component suffering the crash notifies all components with which it communicates of its failure before ceasing to function. Fault-tolerant distributed algorithms are often designed to mask crash or failstop faults.
	3.	**Omission**	A component suffering this form of fault fails to perform some input/output required by its specification but otherwise continues to function correctly.
	4.	**Clock**	In a component containing a clock, a clock fault is any failure of the component to meet its specifications that result from an arbitrary failure of the clock.

A fault may be regarded as being latent as long as it has not caused any errors to occur, although it still exists as a potential cause of failure. Faults invariably are permanent malfunctions that do not fix themselves and therefore require the malfunctioning component to be replaced. Article 7.4 describes measures that can be applied in fault tolerant computer systems.

3.   Error – An error is a discrepancy between a computed, observed or measured quantity and the true, specified or theoretically correct value or condition. It is an item of incorrect information that, when processed by the algorithms within a computing system, may produce a system failure. An error may also be seen as that part of the system state with respect to the computational process which is liable to lead to failure. The hypothesized cause of an error is a fault. An error is thus the manifestation of a fault in a system and a failure is the effect of an error on the service; there are two types of error.

Article 7.3

FAULT TOLERANT COMPUTERS

In the design of a computer system care needs to be given to producing a reliable product. However in certain application areas it is important that the computer should keep going at all costs even when faults occur. This is referred to as fault tolerant computing. It may be regarded as the correct execution of a specified algorithm within a computer in the presence of hardware or software defects. For many computer systems there is now the requirement for a high level of fault tolerance, for example in a safety critical system, i.e. one whose failure may lead to loss or injury to life. Fault detection and reconfiguration is essential to some degree, in computers that require a high reliability without having to resort to the use of a large number of standby units. Achieving a dependable computer system calls for the combined utilization of a set of methods, classified in Table 7.7.

1.	Fault avoidance	How to prevent fault occurrence or introduction through the use of construction methods.
2.	Fault tolerance	How to provide a service complying with the specification in spite of faults through the use of redundancy techniques.
3.	Fault removal	How to minimize the presence of faults through the use of verification of the hardware and software design.
4.	Fault forecasting	How to estimate the presence, the creation and the consequence of faults in a computer system.

Table 7.7

Fault control techniques.

Failure control, as implemented within a computer system, usually involves many separate techniques.

1.   Fault–Avoidance: In the design of computer hardware it is important to try to eliminate any potential for faults. This may mean extensive simulation of the system before committing the design to production. In the design of software for a safety critical computer system one of the main techniques often used is to construct a number of program versions for a single specification using a number of different programming teams. In this way it is hoped that the failure domain of the resulting multi–version program is smaller than for any of the individual programs. This requires careful programming and coding. Fault tolerant programming techniques also usually start with the premise that the failure domains associated with program designs are never empty and then attempt to mask component program failures by relying on the use of design diversity.

2.   Fault Removal: Faults can be removed from computer systems prior to on–line operation by extensive hardware and software testing.

3.   Fault Testing : In a functioning computer system periodic fault testing can be used to identify when a faulty system element must be replaced repaired or reconfigured. Testing does not assure that equipment will work when needed, it does, however, note when a faulty system element must be replaced, repaired or reconfigured.

4.   Fault Detection: In certain complex computer systems diagnostic programs may continually operate to attempt to detect faults. This detection mechanism activates an appropriate recovery action which could be the restart of the program or if this fails and a permanent fault is assumed the failed module can be identified and replaced. The process of fault detection and location requires additional software to carry out the recovery procedures. After detection the software may reconfigure the system and/or analyse symptoms to infer the location of faults.

5.   System Reconfiguaration: If a computer system contains redundant stand–by computers then it is usually possible to substitute for any faulty module and then reconfigure the system to incorporate the new unit. Redundant systems are typically classified according to their most significant physical redundant structure such as triplex voting or duplex standby.

6.   System Recovery: If system reconfiguarion has been used then it is important to return the new system to a working error–free state. This may entail making a damage assessment followed by an appropriate initialization and remedial process.

(Continued)

(continued)

FAULT TOLERANT COMPUTERS

The designer of a fault tolerant computer system is faced with the task of coordinating distinct physical components so that they can cooperate to mask component failures and continue to deliver specified services, perhaps indefinitely. If components never fail, the task is trivial, whereas if components fail repeatedly and frequently and the mean time to replacement of a failed component exceeds the mean time to failure, then the task is hopeless. Between these two extremes, there are a a number of ways to decompose the task into parts with feasible algorithmic solutions. In practice, engineering judgement will be used to determine just what fault tolerance should be incorporated into which components within the computer and what exceptions should be defined and handled.

REDUNDANCY – Fault tolerant computers rely on redundancy to control the effects of faults. Redundancy within a computer system implies a multiple occurrence of certain hardware and software components, usually for the benefit of fault tolerance. Redundancy in a sense negates the effects of failures by simply deferring any form of fault location in preference to keeping the system going through the use of duplicate parts. We can identify two ways of using redundancy within a fault tolerant computer system.

(a) Static Redundancy: Static redundancy is achieved by employing redundant components in a hardware module so that the terminal activity of the module remains unaffected in the presence of hardware failures so long as the protection is effective. Special hardware is usually incorporated into each module to detect faults and switch in the appropriate parallel subunit. This sort of hardware which has been specifically designed for fault tolerance adds extra expense to a system without adding any extra computational features. Due to the cost and complexity of employing static redundancy in a processing system it is more common to find dynamic redundancy being employed.

(b) Dynamic Redundancy: Dynamic redundancy implies using a group of working processors such that if a fault occurs in one, the others simply take on the faulty processor's work without causing a major system failure. In a system using dynamic redundancy, faults are allowed to occur and propagate a number of steps until a detection mechanism is activated.

Dynamic redundancy techniques can also involve the reconfiguration of system components in response to failures, where the reconfiguration prevents failures from contributing their effects to the system operation. In many instances reconfiguration amounts to disconnecting the damaged units from the system. However, if fault masking is also used in the system as part of the fault tolerance design, the removal of failed components may be postponed until enough failures have accumulated to threaten an impending nonmaskable failure. Reconfiguration is triggered either by the internal detection of faults in a damaged subunit or by the detection of errors at the modules output.

In all cases of hardware redundancy the additional components must be continually checked to ensure they are still operational, unfortunately this is only really practical on a subsystem basis. It is more usual for fault tolerance to rely on the replication of complete functional units whether they be processing or I/O units rather than small submodules.

Error

(a) *A Systematic Error*: This is a constant error or one that varies in a systematic way, e.g. equipment misalignment

(b) *A Random Error:* This is an error that varies in a random fashion, e.g. an error resulting from electromagnetic interference. In a computer system, random or transient errors can be caused by any number of phenomena from alpha particle corruption of its dynamic random access memory cells to lightning noise being picked up in a computer's transmission medium. One of the major causes that particularly affects computer systems tends to be impulse noise generated by electromagnetic pick up from devices such as relays operating, or by surges on the power supply lines.

An error may be considered to be latent as long as it has not been detected by the different error detection algorithms that are used in computer systems and it has not caused a failure a failure in the system. However, no machine can be made entirely free from errors and faults, so failures in a system will always occur. There are two ways to deal with errors, these involve either using error detection codes or error correcting codes.

4.   Failure – A failure occurs when the user perceives that a resource within his system has ceased to deliver its expected service. The underlying cause of a malfunction is always some type of failure, such as a physical component functioning incorrectly. Failure itself may be further defined in a number of ways, depending on the cause, timing or degree of failure. A classification of failure is shown in Table 7.8.

Certain computer systems require extra methods to be incorporated into the system design in order to ensure a very high degree of reliability. These systems for example may be concerned with the control of processes where a failure on the part of the processing system would lead to commercially damaging results. The control of failures in a computer system can be an important part of the requirements specification for a new system (see Article 7.3).

Type	Temporary (or soft)	Due to either a transient or intermittent malfunction. 1. A transient failure is caused by some externally induced signal fluctuation. Transient failures can be minimized by careful design. 2. An intermittent failure often occurs when a component is in the process of developing a permanent failure.
	Permanent (or hard)	Caused by a component that breaks down due to a mechanical malfunction. The occurrence of permanent failures can be minimized by proper quality control.
Causes	Misuse	Attributed to the application of stresses beyond the system's stated capabilities. Such a failure could occur for example when too high a voltage is applied across a capacitor causing the dielectric to break down.
	Inherent weakness	Attributed to a weakness within a device or system, when the unit is subjected to a stress or stresses within its stated capability, Such a failure could be caused, for example if a resistor rated at 1 W breaks down when it only dissipates 0.5 W.
Magnitude	Partial	A failure which results in a deviation from the specified characteristic limits of a device or system, but which does not cause a complete breakdown of the system. A minor failure may only lower the quality of the service produced by the system.
	Severe	Causes complete breakdown of a device within a system.
	Catastrophic	Causes the complete system to have to stop, or crash.
Timing	Sudden	Impossible to anticipate by examination prior to use. A failure caused by inherent weakness may lead to a sudden failure.
	Gradual	Possible to anticipate by examination prior to use.
	Degradation	A failure which is both gradual and partial is known as a degradation failure.

Table 7.8

Failure classification.

7.4 MICROPROCESSOR SYSTEM DESIGN

We will now investigate the factors that must be considered when selecting the major components within a microprocessor system.

7.4.1   MICROPROCESSOR DEVICE SELECTION

In the design of a computer system there are a number of options for the central processing section. It is usual for a designer to start the development by choosing a microprocessor and then to attempt to produce a system design around it. The use of a conventional microprocessor provides a powerful combination of programmability, performance and low cost. These microprocessors have the data processing function and the control function, e.g. the decoding circuitry for the instructions are either microprogrammed or hardwired on the same chip, hence they are often termed fixed instruction set microprocessors. They therefore limit designers to a fixed instruction set and a fixed performance that cannot be changed to meet a particular application. These architectures, however, are familiar and easy to use. They are well defined and are supported by software, hardware tools and application notes, and supply sources are readily available.

Choosing a microprocessor to fulfil a given requirement is determined by many factors. The choice will be influenced by both hardware and software considerations and may change as new microprocessor designs become available. The designer, for example, may well base his judgement solely on performance or on previous experience of a microprocessor's instruction set. However, this restricted approach may not produce the best design since there are many more factors that must be considered.

1. **System Function** – The choice of a suitable microprocessor depends on the type of function the system must perform, for example number crunching, signal processing or general computing.

2. **Processor Speed** – Trying to decide on a microprocessor simply by its clock speed is a very restricted view of what is required in a system design. The speed of a processing unit in performing a given task depends on a number of factors and the raw performance can be dissipated in many ways such as:

(a)   utilizing task switching

(b)   responding to interrupts

(c)   handling memory management routines

(d)   waiting for I/O and disk transfers

(e)   polling I/O devices.

Clock speed is therefore not the most efficient way of telling the power of a device and the use of benchmarks and other performance evaluation techniques is important before committing a design to a particular choice of device.

3. **Interconnection** – The choice of parallel system bus to be used within a computer is likely to impact on the choice of processor in a number of ways. At the logical level the bus may be designed to take advantage of the characteristics of a specific microprocessor. This will give an optimal solution for that particular processing unit but will often make the use of another processor much more difficult. The Multibus system, for example, was specifically designed for the Intel 8086 series of microprocessors, while the VME bus has been tailored towards the Motorola range of devices.

MICROPROCESSOR DEVELOPMENT SYSTEM

A microprocessor development system is a special computer system designed specifically for developing programs and interfaces for other computer systems. A block layout of a microcomputer development system is shown in Figure 7.6. The development of the software and testing of the hardware is facilitated with the use of:

(a) Software tools: such as editors, assemblers, high level language compilers and debugging facilities.

(b) In circuit emulation (ICE) facilities: ICE allows the development system to be connected to the target hardware being developed so that the hardware and software can be tested and debugged together using the powerful facilities of the development system.

Emulators are used in microprocessor development systems to allow a wide range of processor types to be handled by a single computer without changing the system hardware. An emulator is a piece of software or hardware logic system, which allows the instruction set of one type of processor to be executed on a host computer system based upon a different type of processor.

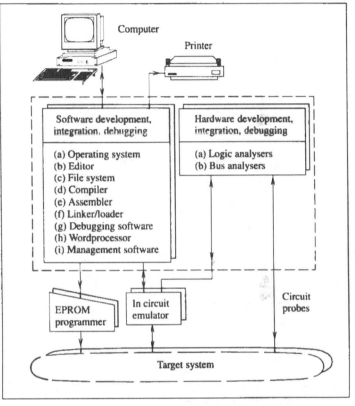

Figure 7.6   **Microcomputer development system.**

In operation the emulator makes use of microprogramming and or software to allow the host computer system to execute the programs for the different processor. The emulator may then be regarded as a type of virtual processor running on some host machine, such that the hardware microprograms and/or software within the emulator enable the host machine to execute programs written for another system. The main use of an emulator is in the computer design process.

With an emulator a designer can initially run the old programs of the virtual machine on a new or different processor in order to be able to verify that his or her hardware design operates correctly before they begin the software conversion of the programs for the old virtual machine to the software required for the new design. The ICE is connected from the host development system via an umbilical cable into the microprocessor socket of the target hardware. Pins on the ICE header are identical to the target microprocessor and the emulator generates the exact signals that would have been provided by the microprocessor, allowing the target system to be driven normally as if under the control of its own microprocessor. The ICE has facilities for memory substitution, emulation modes, debugging facilities through the use of break–points and single stepping. On very powerful development systems it will be possible to emulate a variety of different microprocessors simply by changing a personality card and the ICE header probe.

(c) Logic Analyser: A logic analyser is a piece of test equipment used to monitor and display a number of digital signals simultaneously. It is a digital version of an oscilloscope allowing the user to view and record logic signals from many different sources in the test hardware. They are ideal for testing and debugging hardware in a microprocessor based system.

4. Compatibility – The term compatible is used to describe two computer systems which can use the same program without any alteration. It can also be used to describes various computer units which can replace one another and still function correctly. If a processing system is to be flexible, then it may be necessary to continue to use the same microprocessor over a range of systems in order to achieve compatibility with existing designs. This in a sense has been why the Intel 80186 and 80286 microprocessors have been kept code compatible with the original 8086. The term upward compatible refers to a computer's capability to execute programs written for a previous computer, without a major alteration being required to the new computer.

5. Board Level Attributes – At the level of the circuit board there are many considerations that come into play when trying to chose a processing unit, such as the board level interconnections, the need for particular peripheral devices and the ability to handle error correction.

6. Availability – As manufacturers continue to improve their microprocessors by reducing the glue chips required for system integration, i.e. the extra logic chips required to produce certain system control signals, increasing the data width and adding features such as memory management to their devices. Designers must be aware of these new products, hence the date at which one of these new products is available is a critical factor in the choice of microprocessor.

*Micro–
processor
design
issues*

7. Software – Software generation is a major consideration in a computer design as in many situations the production of software may well exceed the cost of the hardware. It is therefore important that software generation tools including editors, compilers, linkers and diagnostic aids are available for the chosen microprocessor. One of the prime concerns then before adopting any particular processing device is to ascertain if a development system exists for the device and if there is much software already in existence for particular applications.

8. Cost – Cost is always a factor within any design.

9. Development System – The development of microprocessor systems is performed with the aid of a microprocessor development system (Article 7.4). it is therefore important that the chosen microprocessor should have a suitable development system.

Since single chip microprocessors have predefined and unchangeable word length architectures and instruction sets they are in general poorly structured for high speed input or output (I/O), due to the fact that such operations usually take several instructions and each instruction in turn takes several clock cycles. The use of special I/O processors such as the Intel 8089 coprocessor can improve the system throughput. This device has its own memory, direct memory access (DMA) facilities and special hardware and software to allow the device to operate independent of the main microprocessor. However, this tends to make the system much more costly and complex.

7.4.2   MEMORY DESIGN ISSUES

There are a number of fundamental issues that are associated with the design of a memory system for a computer. The designer must decide the various percentages of capacity which should be allocated to random access, direct access and sequential access devices. This affects the cost of the memory and the extent to which the speed of the computer will be bounded by the speed of the memory rather than by the speed of the I/O.

It must be remembered that throughout the time the computer is in use its central memory is continually in action, carrying the instructions which drive the program as well as the data being processed. The partial or final results of processing are also written into the central memory before being sent to the peripheral memories, such as disks and tapes. The memory must be large enough to handle these large amounts of data from the different devices in the system at the same time.

In deciding the amount of different memory for each type, the designer must consider a number of factors affecting the bandwidth of the memory in his processing system.

1. Processor Speed – A major consideration in the design of a memory architecture for a processing system is to minimize the impact of the processor to memory speed imbalance. The designer must attempt to find methods by which the overall processing system can run at the speed of the processor rather than at the slower speed of the memory. Hence a fast processor should have more memory than a slow processor since it will execute instructions and transform data at a rate that may require constant I/O operations to bring in more instructions or more data.

2. I/O Characteristics – In achieving a good memory design a designer must take note of the speed of the I/O devices and channels within the computer; the number of I/O devices in the system that can be read from or written to in parallel; and the spread of data and instructions across these I/O devices.

These factors all determine the rate at which data may flow into and out of the memory. The organization of the memory must also provide for parallel access from multiple devices. If the access rates are high, if the contention for the devices is low, and if many I/O devices can operate in parallel, then less memory may be required. This is because in such a system the processor will have the ability to refresh the contents of memory quickly.

Memory design issues

3. Bus Structure – The choice of bus structure within a computer can also affect the contention between different units accessing the same memory module. The designer must attempt to reduce this contention in an attempt to avoid having devices wait for memory references.

4. Program Size – In looking at the structure of a memory system it is important to consider the expected number and size of the programs to be run on the computer, as well as the expected number of instructions required by a processor to process a particular transaction.

5. Operating System Size – The size required for the underlying system software affects the amount of memory required for a design. This system software includes resident portions of the operating system and subsystems.

6. Cost – The cost of a memory system is the product of the capacity and the price per bit of the device. In general, random access memories prove more costly than direct or sequential access devices and (as a rule) the faster the device, the more expensive it is likely to be.

7. Memory Organization – In computer systems, the speed of the central processing unit (CPU) is usually higher than the speed with which large memories can deliver data to, or accept data from, the processing unit. Hence when a computer system adopts a simple memory design the memory tends to become the limiting factor in achievable performance for the system. For example where a CPU requires one time frame to process an instruction, the memory to drive the instruction pipeline may take on average six to ten time frames. This slows down any high speed instruction pipeline within the CPU. In this case the required demand ratio of the CPU is much higher than the available memory bandwidth, leading to a memory access, or **memory latency**, problem, the memory has become a bottleneck in the system and a bottleneck is a constraining influence on a computer's performance.

MAIN MEMORY ORGANIZATION

The physical locations of the main memory within a computer system may be organized in a number of ways to maximize the performance of the computer. The two most popular techniques are interleaving and banking. Both techniques divide the address space of the physical memory into a family of memory units with a distribution of addresses across each unit. This is usually done to try to redress the speed imbalance between the processor and the memory itself. The speed imbalance means that in most cases the processor must add certain wait states to its timing cycle to essentially wait for a memory device to respond to its request for data. By using interleaving or banking it is possible to reduce the number of wait states.

BANKING: The term banking, when applied to the main memory within a computer system, implies that the memory is divided into blocks in such a way that the upper bits of an address are used to select a particular memory bank. The memory **bank** itself contains addresses within a certain continuous range. Banking is useful in that it can allow more than one device to access the main memory at a time, thereby reducing contention among the different memory addressing units within a processor system. A memory bank is a directly addressable set of continuous registers or random access memory locations. Thus if a bank consists of 64 KB then addresses in the range 0 to 64 K and 64 K to 128 K would occur in different banks of random access memory (RAM) and instruction or data fetches or stores could proceed in parallel with for example one device accessing the lower bank while a second device uses the upper bank. Banking also provides better system reliability since a failed memory module affects only a localized area of the total address space. The register or other storage device that selects one particular bank from a series of them is called a bank switch register.

INTERLEAVING: Interleaving is a technique used in certain computers to enhance the memory accessing capabilities of the central processing unit (CPU). The primary design goal of interleaving is to address the speed imbalance between the CPU and the main memory by having a number of separate memory banks which can operate in parallel in the service of a single request. For example a single memory request to a series of four interleaved memory banks can produce simultaneously four memory locations in sequence, thereby allowing the processor to quickly read the data from the four memory locations in turn without having to put out a separate address for each of the four locations.

In a system which uses interleaving, the lower bits of the memory address from the CPU are used to select a particular block, or bank, of the main memory. For example, if just 1 bit was used, then two memory banks could be arranged such that one held all the even addresses and one held all the odd addresses, while if two bits were used four memory banks could be interleaved. The number of low order bits set aside for interleaving then determines the number of potential memory banks within a system that can be simultaneously accessed.

It is to be noted that interleaving may have the tendency to increase the memory contention in a system where more than one of the units can access the main memory, since it maximizes the number of memory banks that are involved in a single reference to memory. In such a system there is therefore a trade-off to be made in designing the processor/memory relationship, between minimizing memory contention and speeding up memory access operations.

There are two main techniques for reducing the memory access problem.

Memory design issues

(a) *Memory Partitioning* – Allocating, for example, half the memory for instructions and half for data allows both an instruction and a data item to be fetched simultaneously.

(b) *Memory Interleaving or Banking* – Increasing the number of memory accesses that are available simultaneously (Article 7.5).

7.4.3   BUS DESIGN CONSIDERATIONS

The bus is the backbone of any computer system architecture and must be designed, or selected, carefully if it is not to become a bottleneck to system communication. If the bus becomes a bottleneck then degradation will occur in the throughput of the system. A properly designed system bus must consider problems such as noise immunity, bus loading, reflection problems due to high speed logic pulses and crosstalk between parallel conductors. We will look at each of these four problems in more detail.

1. **Noise Immunity** – A microprocessor system can typically have a number of different types, or levels, of buses. They all work in the same way in that the transmitter places a logic 1 or a logic 0 on the bus, this signal propagates down the bus and is received as an input high or low voltage at the receiver. The transmitted data can be corrupted by the presence of noise on the bus.

Noise in an electrical circuit may be defined as unwanted signals caused by various electrical conditions, and which could give rise to erroneous operation of the circuit. In digital circuits, noise can interact with the data signals to distort the information being sent. This is a particularly troublesome problem when fast logic is used, or when logic signals are transmitted over long communication lines.

In circuits which encounter a large amount of noise there is always the danger that noise spikes entering the circuit may be interpreted by the receiver devices in the system as data pulses. The effect noise can have on digital data being transmitted between two devices is shown in Figure 7.7. It can be seen that the presence of a large amount of noise in the communication channel causes the receiver to interpret the corrupted data in such a way as to produce an erroneous transmission, i.e. the data received is not the same as the data transmitted.

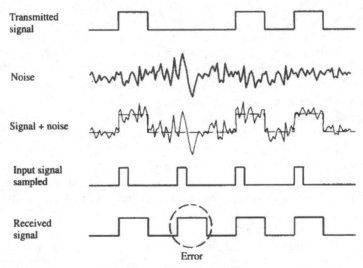

Figure 7.7

Effect of noise on a digital signal.

For a fixed noise level, the larger the voltage difference between the logic high and the logic low then the greater probability that the receiver will correctly interpret the transmitted data. The magnitude of the voltage between the logic high and logic low levels gives a measurement of the noise immunity of the circuit. It can be noted that since noise is an unpredictable characteristic of the circuits being used and the network of which it is a part, it can only be measured in statistical terms.

Internal noise	**Power noise**	This is noise originating from the power supply in the computer and can take a number of different forms such as power spikes or voltage ripple.
	Thermal noise	This a form of random noise in the conductors used in the computer due to the thermal motion of the electrons.
	Signal crosstalk	This form of noise results from signals on one wire being picked up on an adjacent wire.
External noise	**AC noise**	High frequency high voltage interference on an alternating current (AC) line, typically a power line.
	Power supply noise	Fluctuations in the AC power supply causing voltage spikes and dropouts. Voltage spikes are short voltage peaks in the supply, while dropouts are caused by short voltage reductions – a prolonged voltage reduction is known as a blackout.
	Radio frequency	Electrical noise stemming from devices such as televisions or radios.
	Electromagnetic interference	This comes from equipment such as motor drive devices, contact breakers, buzzers, solenoid operated devices, heating and cooling units.

Table 7.9

Noise types.

In a computer system there are two categories of noise, internal and external. The different noise types which fit under these two categories are shown in Table 7.9. Internal noise is generated by high frequency effects within the circuitry and backplane wiring of a digital system, while external noise tends to enter the computer by inductive or capacitance coupling from outside sources through, for example, radiated interference

Most externally generated noise enters a computer system either directly through its power lines or through electromagnetic pick up. It usually takes the form of short random bursts and can corrupt data blocks being transmitted at the time. This form of noise can create voltage spikes in the signal being transmitted, thereby leading to data errors at the receiver, and is especially noticeable in systems with high data rates. With low data rates the receiving device can be designed to ignore high frequency components and is therefore usually able to distinguish between a noise spike and the signal.

Noise immunity

Table 7.10 presents a number of different noise sources, their characteristics and techniques for their reduction. It can be seen that where for example the external noise is entering a circuit via the alternating current (AC) power lines, it is possible to make use of filters and power protection devices to reduce the noise. A filter is a piece of electronic circuitry that is able to remove the energy in unwanted frequencies from a transmission signal, and may be analog or digital in operation. The filters are attached to the regulated power supply, supplying the direct voltage for the circuit. Where the noise entering a computer system is being picked up electromagnetically it may be reduced by a series of measures, such as:

(a) attaching noise suppression circuitry to the appropriate part of the computer system, accomplished by using shielding on the transmission lines;

(b) by careful signal routing of any computer system bus;

(c) by grounding the computer cabinet.

Type of noise	Source	Method of introduction into the circuit	Typical symptons	Means of noise reduction
Supply voltage line transients	Voltage generated by current switching	Conduction through power lines	Intermittent operation, software malfunctions	Transient suppresssors, filters, good layout, good grounding
Electromagnetic interference	Spark discharges	Radiation	Intermittent operation, unrepeatable, incorrect operation	Circuit layout, shielding, coaxial cable
Electrostatic discharge	Static voltage discharge	Physical contact	Software malfunctions, damage or destruction of components	Transient suppressors, buffering circuits between CPU and I/O devices
Ground noise	Unwanted currents in ground line caused by potential difference among grounds in circuit	Conduction through ground lines	Intermittent operation, circuit damage software malfunction	Coaxial cable, circuit layout, ground plane, opto isolators

Table 7.10

Noise sources and means of radiation.

These last two methods are important since when digital computer systems are connected to external or remote equipment, as in the case of peripheral devices, the routing of power and logic signals, including earth returns, must be carefully organized if interference is to be avoided. For example, the power and ground wiring systems should never use the chassis or metal cabinet housing as a ground return, since the outer cabinet housing, which should be used as a noise shield, becomes ineffective if it is also used for carrying ground currents.

Noise immunity

If a computer system consists of several different subassemblies it must also be noted that where these subassemblies need to be connected together it is essential that the potential of their housings be stabilized in relation to each other. Thus, when interconnecting digital equipment via multicore cables, the common earth bus or cable shield must be grounded at the output connector on the cabinet housings. All other ground leads should, if possible, converge to the same common point with AC mains being earthed via a 50 mH choke.

2. Bus Loading – In bus design, the overloading of a transmitter device by receivers on the bus must be prevented. In general, each receiver added to a bus system will require a certain amount of source current from the transmitter in the high state and an amount of sink current into the transmitter in the low state. As microprocessors have a very limited drive capability, it is then normal to use bus buffers in the circuit to handle the sinking and sourcing of bus signals. As a rule, buffers should be used whenever the loading on the bus exceeds the drive capabilities of the microprocessor, or when it is necessary to drive receivers not on the main CPU printed circuit board. The requirement for buffers in this latter case is due to the capacitive loading associated with both the edge connector and the backplane wiring in a multicard system.

3. <u>Reflections</u> – Since the physical nature of a bus system usually entails the collecting together of a number of parallel traces on a printed circuit board, or a bundle of wires in a backplane harness, the designer must note that whenever these communication lines are long (more than about 30 cm for TTL buffers) they do not represent a lumped capacitive load but instead act as transmission lines with distributed inductance and are capacitance driven by nonlinear sources. One of the major problems with transmission lines is that if a transmitted signal encounters an open circuit then it is reflected from the open end back up the transmission line. If the input pulse is very short, the effect is a series of reflected pulses which, depending on their amplitude and duration, can cause clocked devices such as flip–flops and latches to produce erroneous results. The effect on longer pulses is to produce a series of ripples, sometimes called 'ringing', in the transmitted pulse. In some cases this may be so severe as to produce momentary invalid logic levels. Both these effects are undesirable in a bus.

In order to prevent reflections, a receiver placed on the end of a transmission line should absorb all the energy in the pulse. However, this can be shown to occur only if the input impedance of the receiver is matched to the impedance of the transmission line. Printed circuit board traces and wire wrap bundles usually present a characteristic impedance of between 100 and 150 Ω. Therefore one means of minimizing reflections in a circuit wire is to terminate each bus line with a 100–150 Ω resistor to ground. However, this can prove to be an excessive loading on the transmitter.

4. <u>Crosstalk</u> – Due to the effects of stray capacitance and inductance in electrical conductors it is possible for signals passing down one conductor to be coupled onto another. This is known as crosstalk. It is the unwanted transference of electrical energy from one transmission medium, the disturbing circuit, to another usually adjacent medium, the disturbed circuit. The absorbed electromagnetic radiation creates an inductance effect on the second circuit or line which induces currents to flow in the nearby circuit, thereby generating interference to the second set of signals.

Crosstalk can be a particular problem if wires or conductors on a printed circuit board (PCB) are situated close together and carry high frequency signals. Since many of the pulse waveforms that are transmitted though the backplane wiring of a computer system contain high frequency components, it is possible for the conductors in the backplane wiring to pick up these fast edges through crosstalk from any adjacent wires that carry the pulse. This problem is particularly accentuated if the wires or conductors in the backplane bus run close together in parallel, since the stray capacitance that exists between these parallel conductors will couple the noise capacitively between the lines. The magnitude of the crosstalk depends on the amplitude and rise time of the pulse and the impedance of the driven line. Crosstalk is therefore a particular problem in circuits using very fast chips with a very fast clock.

Crosstalk can be minimized in PCBs by increasing the conductor spacing, or shielding each signal line with alternate ground wires between them, i.e. running grounded wires between the conductors on the board. In wiring assemblies it can be reduced by using heavily insulated wire with good dielectric properties. This method of shielding, though helping to reduce capacitive crosstalk, which is proportional to frequency, does little for inductive crosstalk, which is proportional to signal current. The wiring layout is also important in preventing crosstalk and in general it is always advisable to use direct point to point wiring resulting in a random arrangement of wires rather than a neatly cabled system.

Furthermore, for capacitively coupled noise, lowering the impedance at the terminating end of a communication line will often cure this form of pickup. In addition, to handle transmission errors due to crosstalk, most computer buses with cycle times of less than 100 ns usually require error detection schemes to generate and check one or more parity bits for each bus cycle.

Crosstalk

A bus system must therefore be designed carefully in order to solve these four problems. Fast buses, for example, must limit the length of the wire traces on daughterboards, reduce as much as possible the capacitance of the transceivers and connectors, provide a good ground plane with plenty of ground pins, and specify transceivers that can handle the real problems of imperfect transmission lines.

7.4.4   POWER SUPPLY DESIGN CONSIDERATION

Most computers with the exception of the portable variety require a power supply to generate the DC (direct current) voltage that the various devices in the system require. A block layout of a power supply is shown in Figure 7.8.

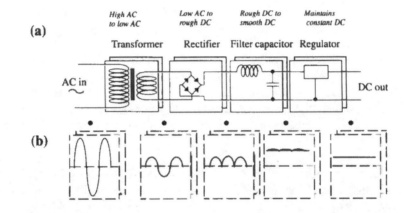

Figure 7.8

(a) Block diagram of a DC power supply.
(b) Waveforms at different parts of the system.

The DC voltage required to operate most computer systems is 5 V. It is generated from an AC mains supply, 120 V in the USA or 240 V in the UK. It can be seen from Figure 7.8 that initially the mains input voltage is lowered by a transformer. This voltage is then applied to a diode regulator to produce a regulated DC supply which is smoothed through a capacitor smoothing circuit. The smoothed DC voltage is applied to a regulator circuit.

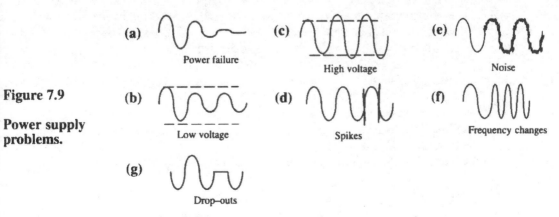

Figure 7.9

Power supply problems.

The regulator can either be based on linear circuitry or switched mode circuitry. Both types produce the smoothed DC supply required by the microprocessor and other circuits in the computer.

The AC input mains supply is subject to a number of problems which can cause difficulties with computer systems (Figure 7.9). The computer power supply should then be designed to minimize the effects these problems can cause.

Power supply design issues

Power supply lines can be a cause of electrical interference in a computer circuit, generated when large amounts of current are drawn along the narrow resistive power supply and ground lines on a PCB, thereby causing a significant voltage drop or noise spike. These noise spikes are usually caused by circuits changing logic levels and thus drawing large instantaneous currents for short periods of time. Voltage drops of this kind can be minimized by using ground planes and large conductive tracks for both the power distribution lines and the ground lines on the circuit boards. This has the effect of lowering the resistance of the power and ground supply lines.

The use of coupling and bypass capacitors can also minimize the transient noise pulses resulting from the ICs switching state. The capacitors are connected across the power supply leads on the PCB and are chosen to store sufficient charge in order to supply the ICs in the case of unwanted power drop spikes on the supply lines. The values of the capacitors chosen are typically in the range $0.001 - 0.1$ μF and must be distributed evenly over the circuit board to ensure a uniform protection for the board. Both ceramic and tantalum capacitors may be used since both have a low internal inductance; however, the tantalum capacitors have more capacitance per unit volume than the ceramics. It is to be noted that CMOS circuits are better able to cope with noise spikes than their bipolar counterparts since they do not draw such large currents and have a better noise threshold compared to, for example, TTL logic.

7.4.5   COMMUNICATION SYSTEM DESIGN ISSUES

In the development of circuitry to allow communication between computers over a large distance the designer must first be aware of a number of problems with such communication (Glossary 7.4). In a communications link (or system) noise, transmission echoes and crosstalk are the three most important problems to be dealt with.

1.   Noise – Communication noise is of two types, background or impulse.

(a) The Background White Noise: Information signals on a bus or wire can be regarded as a nonrandom movement of the electrons in the metal of the conductor whereas **thermal noise (Gaussian noise)** on a bus is caused by the random motion of the electrons in the conductors or wires. This random white noise forms a constant background hiss on communication buses against which the data signal must be sent. It is normally of a low level and, except for long connections inconvenient but mainly insignificant. The amount of thermal noise to be found in a bandwidth of 1 Hz in any device or conductor can be expressed as $N_0 = k\,T$ where N_0 = noise power (W Hz^{-1}), k = Boltzman's constant (1.3803×10^{-23} J K^{-1}) and T = temperature in degrees Kelvin.

From this formula it can be noted that the power of thermal white noise is proportional to the temperature. This means that unless the temperature of the conductor is at absolute zero the thermal noise in a circuit can only be minimized but never eliminated.

Glossary 7.4

Communication Design Problems

ATTENUATION – Attenuation is the difference in amplitude between a signal at transmission and reception. It is the reduction (or loss) of signal strength, usually measured in decibels and may be seen as the opposite of gain.

DISTORTION – In long distance data communication networks, the system elements not only have to contend with the fact that transmitted data can be corrupted by intrusion from different forms of noise, but also that the signal passing down the communication line can become distorted. Distortion is a completely different phenomenon from noise, in that it is a predictable property of any communications channel, while noise may be seen as a random corruption of the signal data. The distortion of a signal is then a predictable corruption of the information contained within that signal in which the frequency components within a signal are attenuated, or reduced, by different amounts as it travels through the transmission path. The amount of attenuation depends on the frequency of the signal, the transmission medium, and the length of the circuit. This nonuniform loss across the signal bandwidth can create amplitude distortion to the signal and is usually particularly noticeable for the high frequencies within a signal. Consequently those media that use high frequency bands require the signal to be conditioned during transmission.

Signal conditioners are circuits which may be fitted at various repeater stations along the communication network in order to restore the signal to its original form by essentially compensating for the distortion effects of the communication channel. Conditioning also involves the minimization of noise levels.

ELECTROMAGNETIC SPECTRUM – The electromagnetic spectrum presents the frequency range and wavelengths of the different elecromagentic waves. Figure 7.10 shows the electromagnetic spectrum.

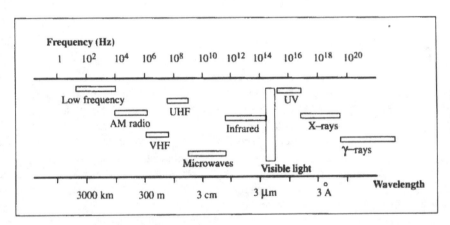

Figure 7.10

The electromagnetic spectrum.

EMI – Electromagnetic interference (EMI) is radiation leakage outside of a transmission medium that results mainly from the use of high frequency wave energy and signal modulation. It can be minimized with the use of metallic shielding.

EQUALIZATION – Equalization within communications devices, refers to the use of circuitry to compensate for the variation in attenuation at different frequencies by a transmission medium. Equalization in a telephone network uses amplifiers to provide the appropriate gain, per transmission frequency, thereby exactly compensating for the signal loss at the same frequency.

LOSS – Loss is the reduction in the strength of an analog communications signal. It is also called attenuation and may be regarded as the opposite of gain. Loss is usually expressed in decibels.

SIGNAL TO NOISE RATIO (S/N) – In a communication channel the S/N is the relative power of the signal as compared to the power of the noise in the channel. It is usually measured in decibels.

STATIC ELECTRICITY – Static electricity is produced when two nonconducting surfaces are rubbed together. Plastic surfaces in particular produce static. Although voltages of tens of thousands of volts can be produced, the actual charge is very small. A static charge of 500 V can produce erroneous data on a transmission line or within a memory circuit.

In practice thermal noise is not a problem in the logic circuits of a computer, however it is of concern when data has to be transferred over long distances. On long distance communication the signal to noise ratio must be maintained at level such that the reception equipment is able to separate the signal from the noise. As signal strength tends to drop with distance, while thermal noise remains constant, it is necessary to add repeater circuits at intervals along the communication medium in order that the signal strength does not fall below the threshold level at which the receiver can detect the signal.

(b) Impulse Noise: Impulse noise usually consists of short noise pulses and cracks. In telephone circuits this can arise from dialling pulses, switching exchanges, and crossover–induced noise pulses from adjacent circuits. The higher the ratio of the power of the signal to that of the noise, the signal to noise ratio, the easier it is for the receiver to separate the data pulses from the noise. On telephone lines, since there is a relatively low allowable limit on the signal power (to prevent overloading of the line equipment), a significant number of errors can occur from impulse noise.

The effect of impulse noise on a data stream depends on the power level of the transmitted signal and the power level of the noise spike. The signal to noise ratio (SNR) is usually measured at the receiver and is normally expressed in decibels (db):

$$\text{SNR db} = 10 \log_{10} \frac{\text{Signal power}}{\text{Noise power}}$$

2. Transmission Echoes – Care must be taken when signals are sent down wires in a digital system since when the wires are situated near a ground conductor they can behave very like a transmission line. If there are variations in the capacitance between the wire and the ground this will tend to resemble a transmission line in which there is a mismatch in the load impedance. On encountering this mismatch any signal sent down the transmission line will then cause an echo of that signal to be reflected back in the opposite direction to that of the original transmission. This is a particular problem when the wire is fairly long and is transmitting high frequency signals. Transmission line effects can be minimized by keeping the signal wire and the ground line at a fixed distance and attaching the correct termination between the end of the signal wire and the ground. Figure 7.11 indicates the proper terminations required for a coaxial cable and a twisted pair line.

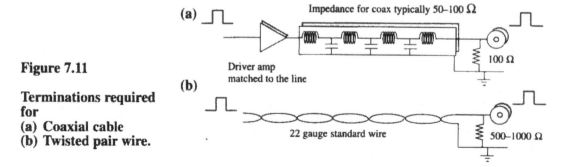

Figure 7.11

**Terminations required
for
(a) Coaxial cable
(b) Twisted pair wire.**

Echoes also occur where bus lines are connected at circuit junctions, again an example of an impedance change. These may be minimized by ensuring connections are perfectly matched in impedance, and by limiting the number of connections between circuits.

3. <u>Crosstalk</u> – Crosstalk is also typically encountered in long distance communications when logic signals are transmitted across unshielded twisted pair wires. While this type of noise is very difficult to completely eliminate, it is possible to reduce it by the use of filters on the output lines, by inserting a series inductor or by using shielded coaxial cables. In extremely noisy surroundings it may become necessary to use a balanced pair of signal lines with a common coaxial screen held at an earth potential or to use optic fibres which are inherently noise free.

From this description we can see that where computers are expected to use telephone lines these must be considered to be inherently unreliable. Hence where a computer system wishes to communicate over a long distance it has to be designed on the assumption that failures will occur, data will be lost and circuits will fail for short or long periods of time. This means designing into the communications hardware facilities for efficient error detection and possibly error correction.

(a) Error Detection: To implement error detection, extra data is included in the information transmitted within a system. This allows the receiver to use an error detecting code to decide if an error has in fact occurred during transmission. It can be noted that the receiver can only say whether an error has occurred and cannot decide where in the message the error actually is.

(b) Error Correction: Error correction is used in a data transmission system to allow the receiver to correct the majority of errors noted in the messages to which the code is attached. Error correction requires the transmitter of the data to add a large amount of extra coding information in the message being transmitted. Enough information must be included in the message packet to allow the receiver to decide the position within the transmission packet where the error has occurred, thereby allowing the receiver to be able to correct the incorrect bit(s). The error correcting information is usually included in the message header or trailer. The most widely used error correcting mechanism is the **cyclic redundancy check (CRC)**. A CRC is an error detecting code that is generated from a polynomial algorithm to check the integrity of a block of data. It is used in data communications, magnetic tape and disk devices to enable a receiving station or device to detect and possibly correct an error in a data transmission.

The CRC is a complex algorithm based on a generator polynomial which is applied to the data at the transmitting equipment. The result of this calculation is tagged onto the end of the data as additional check bits, usually 16 or 32 in length and often referred to as a CRC checksum. The receiving system performs an identical CRC calculation on the data it receives and compares its results with the received check bits. The two should be identical, and if not, an error involving one more bits has occurred during transmission. The CRC is usually good enough to allow the correction of any faulty bits in the message. It is normal in communication systems that after the receiver has checked the CRC on the packets it has received, it strips it off before giving the correct packet to the station.

7.5 KEY TERMS

Acceptance test
Analysis considerations
Attenuation
Availability
Balanced system
Bank
Banking
Benchmark program
Bottleneck
Bottom up design
Bug
Bus design considerations
Communication system design
 issues
Computer aided design (CAD)
Computer aided engineering
 (CAE)
Computer aided manufacture
 (CAM)
Coverage
Crosstalk
Cyclic redundancy check (CRC)
Design methodology
Development system
Dhrystone benchmark
Distortion
Documenting
Downtime
Echoes (transmission)
Electromagnetic spectrum
EMI
Emulator
Equalization
Error
Error correction
Error detection
Failure
Failure mechanism
Fault
Fault tolerant computers
Fail safe
Fail soft
Floating point operations per
 second (FLOPS)
Functional requirements
Gaussian noise
Impulse noise
In circuit emulation
Instruction mix

Instruction trace
Integrity
Interleaving
I/O bound
Life cycle
Linpack benchmark
Livermore loops
Logic analyser
Loss
Maintainability
Maintenance
Mean time between failure
 (MTBF)
Mean time to repair (MTTR)
Memory design issues
Memory (main) organization
Memory latency
Microprocessor: device selection
Microprocessor development
 system
Microprocessor system design
Million instructions per second
 (MIPS)
Million floating point operations
 per second (MFLOPS)
Noise
Noise (communication)
Original equipment
 manufacturer (OEM)
Optimization
Overhead
Performance (computer)
Performance monitoring
Plug compatible
Power supply design
 considerations
Process bound
Prototype
Prototyping
Redundancy
Reflections
Reliability
Requirements specification
Robust
Signal to noise ratio (S/N)
SPEC benchmark suite
Specifications

(Continued)

Key
terms

Static electricity	Testing (of hardware)
Structured analysis	Thermal noise
System analysis	Threat
Systems analyst	Throughput
System design	Top down design
System development–life cycle	Validation
System design methodology	Verification
System failure	Von Neumann bottleneck
System implementation	Whetstone benchmark
System specifications	White noise

7.6 REVIEW QUESTIONS

1. What is a system design methodology?
2. List and discuss the six stages in the system life cycle.
3. Which of the following is not a useful principle in a problem–solving process: a top down approach, a bottom up approach or a life cycle approach.
4. In the development of a computer system who is responsible for the modelling of a users requirements?
5. What do the terms CAD, CAM and CAE refer to?
6. Why do developers work with teams of users?
7. In the development of any product what will happen after the initial product specification has been drawn up?
8. Note four topics to be considered in specifying the requirements for a computer system.
9. Why is the maxim 'simple is beautiful' applicable to computer design?
10. List five of the desirable properties that a designer should ensure their computer has.
11. What is a prototype and why is one usually developed?
12. Contrast the approach of validation and verification in ensuring a computer system meets its specifications.
13. What is an acceptance test?
14. What is documentation? Describe its function in the system life cycle.
15. What is the purpose of system maintenance?
16. Why should maintenance be reduced if users are involved in the design of systems?
17. Give five performance measures that can be quoted for a computer system.
18. Why is a single performance figure not a fair assessment of a computer's true performance.
19. The typical reliability curve is known as a bathtub curve. Explain from what it derives its shape.
20. What do the following terms refer to when used to describe the reliability of a system: availability, maintainability and integrity?
21. What is meant by the term coverage?
22. Describe the mechanism of failure in a system.
23. Give several examples that could lead to a threat in a computer system.

24. Describe in a paragraph how the features added to fault tolerant computers increase their reliability.
25. Contrast misuse and inherent failure?
26. Give an example of a situation which could cause a random error in a computer system.
27. In a computer system why are temporary failures so difficult to trace?
28. · What are the advantages/disadvantages of choosing a microprocessor first in the design of a microprocessor system?
29. Note five factors to be considered in choosing a microprocessor for a computer system
30. Why are emulators important in a microprocessor development system?
31. What are the functions of a logic analyser?
32. What are the factors which dictate the type of memory system to be designed for a computer?
33. How does memory latency affect the performance of a computer system?
34. What is 'noise' and explain its effects on digital data?
35. List several 'noisy' environments.
36. What causes reflections in a computer bus?
37. What is 'crosstalk' and how can it be minimized in a computer bus?
38. Why are point–to–point wiring assemblies better in terms of noise reduction compared to a neatly cabled bus?
39. What is the difference between distortion and noise?
40. What effects would echoes in a transmission line have on the integrity of the data being transmitted across it?

7.7 PROJECT QUESTIONS

1. What would the functional requirements be for the development of a computer system for
   (a) an airline reservation system;
   (b) a centralized database network;
   (c) a stock quotation system?
2. Discuss the statement 'technology often drives design'.
3. Why is it important to develop the software for a computer system in parallel with the hardware?
4. Why is it said that design is a creative task?
5. List a number of reasons for poor design?
6. List several of the design techniques which can be used to reduce maintenance costs in a computer system.
7. Is it fair to say that the systems life cycle proceeds in an iterative fashion? Explain your answer.
8. What are trade–offs? What are some major trade–offs in designing a computer system?
9. How might you increase the performance of a computer through changes in its architecture?
10. Investigate the types of functions incorporated into the SPEC benchmark suite.

11. For a computer which is to be used to process images from a satellite describe the types of processes that should be incorporated into a benchmark program to allow different computer to be compared.
12. What factors affect the maintainability of a computer system?
13. How could a failure in a memory system be detected?
14. What factors would increase the reliability of a large computer system?

Project questions

15. In the design of a computer system explain why it is important to choose reliable system components.
16. What types of application software would a microprocessor development system have?
17. Why do you think it is important to minimize the number of ICs used in a design?
18. Why do systems working with very high clock speeds need more careful design compared to those using slower clocks?
19. Investigate the techniques used to achieve noise suppression within power supplies.
20. Note some of the properties of 'white' noise.

7.8 FURTHER READING

This chapter has considered the design of the hardware within a computer system, looking at the various considerations required in developing the different parts of the system. There are a number of books and articles on the topics covered in the chapter in particular performance and fault tolerant computing. Some of these are listed below.

Hardware Design (Books)

1. *Computer Appreciation*, 2nd Impression, T. F. Fry, 1972, Butterworths.
2. *Theory and Design of Digital Computer Systems*, D. Lewin, 1980, Nelson.
3. *The Principles of Computer Organization*, G. M. Schneider, 1985, J. Wiley.
4. *Structured Computer Organization*, A.S. Tanenbaum, 1984, Prentice Hall.

Hardware Design (Articles)

1. A framework for computer design, W. K. Dawson, *IEEE Spectrum*, October 1986, pp. 49–54.
2. A total system design framework, G. Roman et al, *IEEE Computer*, May 1984, pp. 15–26.
3. Protecting computer systems against power transients, F. Martzloff, *IEEE Spectrum*, April 1990, pp. 37–40.
4. A taxonomy of current issues in requirements engineering, G. Roman, *IEEE Computer*, April 1985, pp.14–21.
5. How to choose a high performance microprocessor, K. L. Ryder and B. W. Partridge, *Conf. Proc. MILCOMP 85*, London, October 1985., pp. 257–262.

Performance (Books)

1. *Computer System Performance*, H. Hellerman and T.F. Conroy, McGraw–Hill, 1975.

Benchmarks (Articles)

1. How fast is fast, B. Kindel, *Byte*, February 1989, pp. 251–254.
2. A benchmark tutorial, W. J. Price, *IEEE Micro*, October 1989, pp. 28–43.

3.   An overview of common benchmarks, R. P. Wiecker, *IEEE Computer*, December 1990, pp. 65–75.

Further reading

Fault tolerance (Articles)

1..   Fault tolerant system design: broad brush and fine paint, A. L. Hopkins, *IEEE Computer*, March 1980, pp. 39–45.
2.   Safety in numbers, V. P. Nelson, *Byte*, August 1991, pp. 175–184.

Software

Part Three

Software and Programming Languages

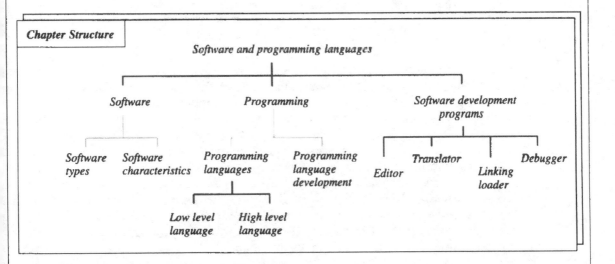

Chapter Structure

Software and programming languages

- Software
  - Software types
  - Software characteristics
- Programming
  - Programming languages
    - Low level language
    - High level language
  - Programming language development
- Software development programs
  - Editor
  - Translator
  - Linking loader
  - Debugger

Learning Objectives

In this chapter you will learn about:

1. The application packages most in demand for personal computers.
2. The reason for the development of high level languages.
3. The classification of programming languages.
4. The origins, capabilities and examples of various languages.
5. The tools required for developing application programs in assembly or high level languages.

> '*Most of the great computer languages represent the forceful vision, personal obsessions and unifying ideas of one or two individuals carried through with relentless consistency.*'
>
> *Ted Nelson*

Chapter 8

Software and Programming Languages

A processing system may be regarded as a physical construction, capable of carrying out computations by interpreting the control information within a program, where a computer program, often referred to as software, is in a sense a formal description of the computation.

8.1 SOFTWARE

The term **software** relates to any form of program that can be run on a processing system and although it can be represented in the form of listings, it cannot be seen when it is executing in the computer. In a more global form, software may be regarded as a set of computer programs, procedures, rules and associated documentation concerned with the operation of a computer. Software includes the programs, codes, sets of instructions, languages and commands that enable a computer to perform its functions. A program is a sequence of instructions describing how to perform a certain task.

In order for a program to run on a computer system, the electronic circuits of the computer must recognize and directly execute the instructions contained within the program. Programs can be held on punched cards, magnetic tape and other media but the essence of software is not the physical medium on which the program has been recorded but the set of instructions that makes up the program.

Software largely defines how a computer system looks to the user. As computers have evolved, it has become apparent that the software is as important as the hardware in getting the best out of any computer system.

8.1.1 SOFTWARE TYPES

The different **software types** are listed in Figure 8.1. The software that enables a computer to function can be broadly divided into system software and application software.

1. **System Software** – System software is a term often used to indicate any piece of software for a computer system which is not an application program. It includes the operating system plus general purpose programs provided by the manufacturer, such as assemblers, editors or compilers.

(a) *The operating system* is the interface software that controls the basic functions of the computer such as managing the system's hardware and software resources. The operating system may be viewed as a series of software routines that are executed by the computer to control the running of an application program by providing high level services from comparatively low level hardware. In a sense it is a set of software extensions to the primitive hardware that produces a processing system which can be used as a high level programming environment.

(b) *Utility software* is software that provides basic functions for use by a systems programmer. Utility programs include loading and saving programs, programs that edit files, send mail, compile programs or link them, programs to initiate program execution

Article 8.1

APPLICATION SOFTWARE

SPREADSHEET – A spreadsheet is a software application package usually used to display financial or statistical information. It takes its name from the way data is formatted on the screen in rows and columns, as in the traditional layout of figures in account books. The user can specify that the values displayed in particular positions are to be dependent on entries in other positions and are to be recalculated automatically when these entries are changed. A spreadsheet can then be seen as an expandable table of entries that are linked together through mathematically defined relations. Whenever one entry in the table is changed, the others linked to this entry are automatically updated. For example the effects of changes to selected data items on totals and subtotals can be explored. Other features include the ability to print reports from portions of the table, draw graphs of various kinds and perform simple database operations.

In 1978 Dan Bricklin and Bob Frankston teamed up to develop VisiCalc the first spreadsheet for the IBM PC. The introduction of VisiCalc saw an increased use of small computers in the business sector. Other examples are Lotus 1–2–3, Supercalc 5, Quatro, Microsoft Excel and VP–Planner.

WORDPROCESSOR – Wordprocessing is a particular application of computing dealing with the editing and production of typed letters and documents. Wordprocessing is a term coined by IBM in 1964 to describe electronic ways of handling a standard set of office activities. A computer system dedicated to this application is often referred to as a wordprocessor. A wordprocessing program is designed to produce professional documents more quickly and accurately than a typewriter. It contains features similar to an advanced text editor with features to aid in the creation, editing, printing, sorting and retrieving of text. It also incorporates features to facilitate the production of commercial documents including automatic page formatting and the ability to exchange blocks of text between documents

In 1976 Michael Shrayer wrote Electric Pencil, the first word processor for microcomputers, while in 1978 MicroPro International, formed by Seymour Rubenstein, announced WordMaster the precursor to WordStar. In 1979 the popular word processing package WordStar was released by MicroPro. Other examples are WordPerfect, Microsoft Word, and DisplayWrite. WordPerfect for PCs appeared in 1984.

DESKTOP PUBLISHING – Desktop publishing software is a form of advanced word processor which can deal with type styles of various sizes and shapes, in terms of fonts, picas and points. It also enables graphics to be fitted into a document. Desktop publishing software was introduced in 1984 by the Aldus Corporation with its software package Pagemaker which allowed personal computers to prepare and print a wide variety of typeset quality documents. Desktop publishing software allows text and graphics, possibly generated by other software packages, to be edited, changed in style and format and positioned before final printing. Desktop publishing software is often called page composition software. The features of a word processor are designed to let users control and manipulate a document's content, whereas a desktop publishing program is more concerned with changes in the style and presentation of text and graphics.

DATABASE – A database is a collection of digitally stored data records. It can also refer to a collection of data elements within records contained in files that have a relationship to records within other files with each group of records having a unique set of identifying characters. Database systems assist in the maintenance and recall of information that might otherwise be stored in a traditional filing cabinet.

Database software is available for almost all types of computers. In 1979 Wayne Ratcliff developed Vulcan, later developed into dBase II. The dBase series of software has become a standard for IBM PC–styled computers. In 1984 the software company Ashton Tate released dBase III and dBase IV in 1987. Other examples of this type of package are R:Base for DOS, PFS: Professional File and Q&A.

BUSINESS GRAPHICS SOFTWARE – Business graphics software packages provide facilities to allow users to produce charts and graphs to display, for example, sales figures and profits for the business market. Examples of this type of package are Harvard Graphics, DrawPerfect, Lotus Freelance, Microsoft Chart, Chart Master and Diagram Master.

COMMUNICATIONS SOFTWARE – This form of application package enables communications to be set up across a computer network. This allows one computer to access another device over very large distances. Examples are Crosstalk XVI and PC–Talk III.

(Continued)

APPLICATION SOFTWARE

INTEGRATED SOFTWARE – In the 1980s, a large collection of integrated business programs, which combined the capabilities of two or more general or special application programs, were developed, e.g. combining a wordprocessor with a spreadsheet, data base, graphic and communication capabilities. The first of the integrated software packages for the IBM PC was Lotus Symphony which appeared in 1984.

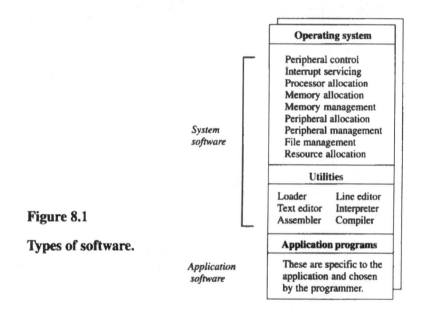

Figure 8.1

Types of software.

Software

and routines to observe and change the contents of memory locations. Utility software is usually written by, or for, the computer manufacturer to go with a particular operating system. While being part of the system executive, utility programs are not usually regarded as being part of the operating system and can be seen more as support programs that are invoked by users when and as required. Just as the operating system transforms real resources into virtual ones that are easier to use, the utility software transforms the virtual resources offered by the operating system into yet more user friendly facilities, thereby making the computer functionally richer and more meaningful to its human users.

It is the job of the system software to facilitate user access to the hardware resources of a computer system. Although the system programs do not solve end user problems, they do manage the resources of the computer, automate its operations and facilitate programming, testing and debugging of application programs. This software can be viewed as a buffer or interface between the hardware resources, such as the memory and disk drives and the high level operations that the typical user wants to perform, such as compile high level programs, execute application packages, or save information.

2. <u>**Application Software**</u> – An application package is a set of specialized programs and associated documentation usually supplied by an outside agency or software house to carry out a particular application. In this way it is a piece of software that is designed to solve a specific problem, or processing need, for a user. In terms of the layers of software within a computer system application programs may be viewed as being at the outside edge in that they require to use the operating system of the computer to run. The most widely used application programs are spreadsheets, wordprocessors and databases (Article 8.1).

A conceptualized view of the layers of software within a computer system is shown in Figure 8.2 where the application and utility software can be seen to be on the outer layer of the system. It can be seen that the central processing hardware is at the centre of the system with the operating system kernel surrounding it. The kernel is the heart of the operating system. The other parts of the operating system surround the kernel providing more services to the user or application program.

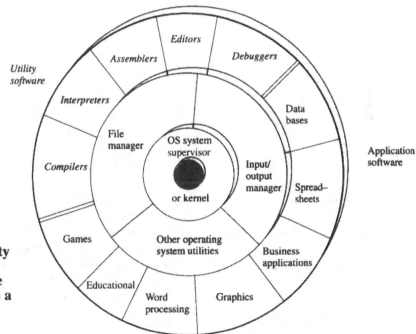

Figure 8.2

The position of utility and application programs within the layers of software in a computer system.

8.1.2   SOFTWARE CHARACTERISTICS

In terms of regarding software as a form of information it has certain characteristics.

1. <u>**Structured**</u> – The first characteristic that distinguishes software from other kinds of information is that it is arranged according to certain rules of logic in such a manner that it describes the function(s) to be performed, the input(s) to the function(s) and the output(s) generated by the function(s). Software is therefore structured with logical and functional properties and may be seen as embodying the function of a computer system.

2   Created And Maintained – Software usually has to go through a process of evolution before it emerges as the final product. This involves a transformation of human ideas into something that can operate the hardware within the computer. To reach the final operational stage the software will have to be processed in many different forms and representations. This in a sense distinguishes software from other kinds of information in that it is created and maintained in various forms and representations, during its life cycle. We can identify two basic forms in this evolution as nonexecutable and executable.

(a) Nonexecutable: Nonexecutable forms cannot be processed directly by the computer. Examples are the various forms of documentation that comprise the specifications of functions to be performed by the computer together with their inputs and outputs, listings of program language statements, and flowcharts that show the logic of the functions to be performed. This form is important in the software development process as without the proper documentation of a piece of software it is easy for the design process to be side tracked and fail to prepare software to do its prescribed task.

(b) Executable: The executable form of software consists of the sequences of information that may be processed by the computer to perform the function that the software was designed for.

Software characteristics

If software is to achieve its executable form it must first be translated from its initial non-executable forms into a form that is machine processable. Many languages may be used to describe the nonexecutable form of software, for example using English for a design specification, FORTRAN for computer programming, and hexadecimal for a derivative of the source code.

Software is also information that must be continually maintained. This means that in reality even after the software has been released for commercial use it may still require to be changed due to correction or improvements either called upon by the user or instigated by the programmer. Even if the software has been well designed it may still need to be changed as the hardware it interfaces to is altered.

3.   Machine Processible – The third characteristic that distinguishes software from other kinds of information is that it is intended to be processed by a machine. As most makes of computer are different from one another this invariably means that the executable form of the software must be matched to the particular computer chosen to run it. In this way when the software passes from its nonexecutable form to its nonexecutable form on the disks it must be transformed in such a way that it can be accepted by the hardware components of the machine. The hardware will then ultimately convert the non–executable form of software to its executable form.

8.2 PROGRAMMING

The language in which a program may be expressed is called a **programming language**. A computer program may then be seen as a series of instructions arranged in the proper sequence for directing a digital computer to perform a particular task.

Programming a computer is the process of determining the particular sequence of operations to implement a desired function. It may also be seen as the process of creating a sequence of instructions to implement the required operations. If properly programmed a computer system can implement any function as long as the time constraints on the function's execution are not too stringent.

The dominant requirement for a programming language is that it should facilitate in the production of reliable, maintainable software, within an acceptable time and at an acceptable cost. Thus programs written in the language should contribute to the documentation of the design and be understood both by other members of the design team and by those subsequently responsible for maintenance, so reducing the chances of obscure errors remaining undetected.

From a programmer's point of view there are two types of programming language, these being referred to as low level or high level.

8.2.1 LOW LEVEL (ASSEMBLY) LANGUAGE

An assembly language is a programming language in which the programmer can use mnemonic instruction codes, labels and symbolic names for data, to refer directly to their machine code equivalents. Such a language uses convenient abbreviations for low level instructions to assist in the definition of the instruction. In this way an assembly language program is easier to understand compared to a **machine language** and the programmer is saved the trouble of remembering the complex bit patterns in each machine level instruction and of having to keep track of data locations and instructions in the program as he or she would have to do if they chose to program in machine code. Assembly language is referred to as a low level language, since each assembly language instruction translates directly into a specific machine language instruction this allows the programmer a large degree of control over the microprocessor operations.

Low level language programming requires specifying in considerable detail the majority of program steps. In some cases, for example in non–numerical problems, this is important if maximum efficiency in terms of program run time and storage space is desired. However the organization of input and output routines, the general control of peripheral equipment, the tracking of store allocation, the updating of loop counts and the manipulation of subroutines can all give problems when programming a large software system in a low level language. Although an assembly language is more efficient in memory space and can run faster than a high level language it has a very restricted set of syntactic rules.

8.2.2 HIGH LEVEL LANGUAGE

In order to overcome many of the difficulties in using low level languages, and in an endeavour to make programming easier and available to a larger section of the technical and scientific community, high level computer languages have been developed. A high level language has the property that one instruction in the source program usually generates a number of low level instructions to directly manipulate the computer hardware. It is then usual for a single functional statement in a high level language to translate into a series of instructions or subroutines in the machine language.

These computer languages allow the program operations to be written in a more easily understood form and it is the computer itself, or rather the software compiler, which assembles together the necessary machine code instructions required to perform a particular operation. In this way mathematical problems, for example, may be expressed using standard mathematical formulae including both real and integer numbers and complete input/output routines may be initiated or called up by simply writing statements such as read or print into the program, with all the necessary housekeeping requirements being organized automatically by the compiler.

Glossary 8.1

ALGORITHM – An algorithm is a form of arithmetic computation which uses a set of rules to perform a task or solve a mathematical problem. An algorithm can also be described as defining some model of computation for execution on a computer. When an algorithm is evaluated on a computer it is expressed as a precise, fixed procedure, listed in a sequential operational form for solving a problem in a finite number of steps. The cost effectiveness of a computer for execution of an algorithm is heavily influenced by the mapping between the models of computation of the algorithm and the computer architecture.

CODE – In a computer language, code may be regarded as a representation of the text such that the various bit patterns of the code indicate the different letters of the text.

CODING – Coding is the phase in the design of a piece of software in which a programmer takes the flowcharts or data flow diagrams for the different software subsystems or modules and by choosing a suitable high or low level language, codes them into a form recognizable by the computer. If a high level language has been chosen this will further require that the program be compiled down to a machine coded program, while if a low level language is chosen this will require the use of an assembler. Coding is carried out after the specifications of the program have been developed by a systems analyst and after the structural design of the program has been carried out by a software designer. During the coding stage the software engineer makes use of various computational tools to aid in the software development. These tools include editors, assemblers, compilers, linkers, loaders and debuggers.

DECISION – A decision is an operation performed by a computer to choose between alternative courses of action. It is usually made by comparing the relative magnitude of two specified operands, a branch instruction being used to select the required path according to the result obtained.

DETERMINISTIC – Deterministic is a term used to describe a rule or procedure whose premises and conclusions are known with certainty.

FUNCTION – A function is a mathematical expression which associates an output value with a series of input values in a precise formulation.

HEURISTIC – A heuristic approach to problem solving is one which uses successive evaluations of trial and error to arrive at a final result. The term heuristic is often applied to rule of thumb procedures which might not be mathematically provable but which have proven experimentally to work in the past. The procedures may be based more on experience rather than mathematics. Heuristic rules and formulations are often used to solve complex problems in situations that do not yield to more conventional mathematic procedures, i.e. they are used when more mathematically robust methods cannot be applied. In software terms a heuristic program solves a problem by a method of trial and error in which the success of each attempt at a solution is assessed and used to improve the subsequent attempts until a result is obtained which is acceptable within the defined limits.

PROCEDURE – A procedure is a self contained set of codes in a program that affects some part of the larger program, much like a subroutine or a function.

PROCESS – A process is a software activity, such as a piece of a program, that is actually being executed by a processor. It is normally a self–contained transformation or a systematic sequence of operations that produces a specified result. A process contains its own code and private variables and coordinates its activities with other processes by sending and receiving messages. Furthermore a process is different from a program in that a process is a dynamic entity while a program is static. A process may also be seen as a block of sequential code executing some algorithm augmented by additional state information, the process control block, which allows it to be suspended and restarted and in some cases executed on different processors in the system.

ROUTINE – A routine is a software term which refers either to a set of computer instructions arranged in a correct sequence, or to a subprogram which is less complex than the main program. A program may contain many routines but in its most basic form a program can be considered to have an input, a main and an output routine.

STATEMENT – A statement is an instruction line in a high level language. Statements are then the commands within a language that perform actions and change the state of the computer. Typical statements are either
(a) **assignment,** which mean they change the values of variables;
(b) **conditional,** which means they have alternative causes of actions;
(c) **iterative,** which means they loop through many other statements until some condition is satisfied.

STRONG DATA TYPING – In a strongly typed language every variable and result has a type, such as an integer or a string, associated with it indicating the class of the object that it represents. This allows the compiler to check whether the destination object of a message is correct. Strong data typing automatically enables programmers to catch a significant number of errors that regularly creep into programs.

(Continued)

(continued)

Glossary 8.1

Programming

SUBROUTINE – A subroutine is a self contained part of a program that can be called from the main program to do a specific task. It normally has some well defined structure and may usually be reached from more than one place in a main program. The process of passing control from the main program to the subroutine is called a subroutine call. The main program can also modify the subroutine by passing data and addresses, both called parameters, by a process called parameter passing. Subroutines are important structures within a program since they allow preprogrammed and pretested subroutines to be incorporated into a main program. Also it is possible to save storage space by writing a section or program once and branching to it only when required in the main program.

SYNTAX – The syntax of a programming language describes the correct form in which programs must be written while the semantics of a programming language denotes the meaning that may be attached to the various syntactic constructs.

Some common high level languages are FORTRAN, COBOL, Pascal and BASIC. The primary benefit of a high level language lies in the possibility of defining abstract machines in a precise manner that is reasonably independent of the characteristics of the particular hardware within the processing system; thus the details of the microprocessor's architecture on which the object code is to be executed are not important during the writing of the high level language program. High level language programs require a complex piece of software called a compiler, to convert the source code into the machine instructions for the system.

The structural classification of high level computer languages can be based on the organizing principle of the language or on the procedures used within the language.

1. <u>**Organizing Principle**</u> – There are a number of ways in which the organizing principle of a language can be used for classification.

(a) *Imperative/Algorithmic*: With an imperative programming language the resulting program consists of instructions, or imperatives, on how to carry out procedures *High level* (algorithms). Imperative indicates that the language give instructions as to what values *language* should be placed in the computer's memory locations. Examples of this type of language are FORTRAN, BASIC, COBOL, ALGOL, PL/1, Pascal, Modula–2, Ada and C.

(b) *Functional/Applicative*: With a functional programming language the resulting program consists of functions that are applied to data. Example of this type of language are LISP and Logo for list processing, SNOBOL for string processing, and APL for mathematical algorithms.

(c) *Object Orientated*: Object orientated languages produce programs which consist of objects that are manipulated to obtain results. An object is defined as an area in computer memory that serves as a basic structural unit of analysis. In the process of writing programs users create new objects out of existing ones. Examples of this type of language are SIMULA and Smalltalk.

(d) *Forth*: The high level language Forth fits into a category by itself. Like object orientated languages it is made up of small units (words). Programs then consist of words that are built up from words previously defined.

(e) *Logic programming*: In logic programming a program consists of statements in first–order predicate calculus. An example of this type of language is PROLOG.

2. <u>**Procedures**</u> – There are two ways of categorizing high level languages according to procedures: A **procedural** language is such that the program tells how to reach a result while a **non procedural** language is such that the program tells what result is wanted.

Article 8.2 provides a longer description of procedural and nonprocedural languages. There are certain implementation attributes that are also expected from any high level language. These are shown in Table 8.1.

Table 8.1

Attributes expected of a high level language and its compiler.

1. The language must be easy to learn and easy to use with a standard and regular syntax as well as standard and well defined semantics.
2. The language should be simple, have high expressive power and have language constructs that are consistent – that is they should have a similar effect wherever they appear.
3. The notation used within the language must be convenient and compatible with widely used standards.
4. It must be safe from misinterpretation and misuse, i.e. free from insecurities.
5. It must be extensible without change of existing features. However, to achieve flexibility and power of expression in a programming language we must pay the price of greater complexity.
6. There must be a rigourous, mathematical definition of the language based on axioms.
7. It should support modern programming paradigms, such as
    (a) abstraction
    (b) top down design
    (c) modularity
    (d) reuse of code.
8. The definition must be machine independent, i.e. without reference to a particular mechanism, called portability of code.
9. It should have a good knowledge base of experienced programmers, with international adoption in industry and academia.
10. A language should also support the detection of as many errors as possible at compile time.
11. The compiler must make efficient use of the facilities of the available computer.
12. The compiler must be able to generate efficient code and economize storage.
13. The compiler must be fast and compact; it should be void of complex optimization routines that are rarely used.

8.3 PROGRAMMING LANGUAGE DEVELOPMENT

A number of high level computer languages have been developed over the years, to reflect the changing emphasis on different aspects of computer applications. The evolution of several of the different high level languages is shown in Figure 8.3. The diagram shows how the different languages have evolved from the regions of scientific, data processing, general purpose, artificial intelligence, text processing or systems programming.

Although there are many high level programming languages most have many more similarities than differences. For example, many tend to use ordinary English language type phrases and deal with abstract concepts such as constructs, data types, sequences and communication. This compares to the low level machine instruction format which deals with handling the hardware of the machine. High level languages also allow code to be developed that is easier to verify, thereby allowing a programmer to formulate an application precisely.

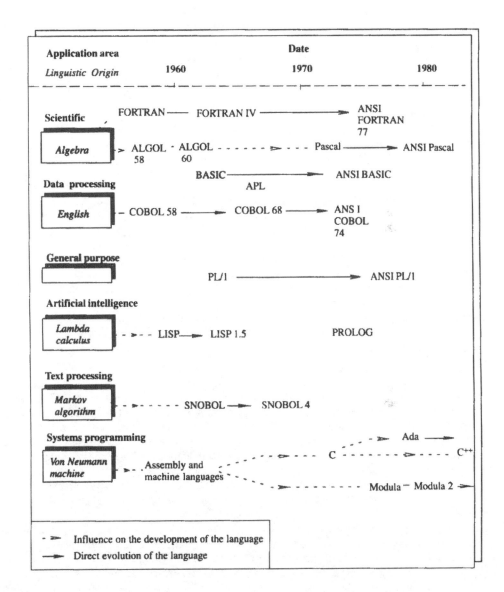

Figure 8.3

Historical perspective of several programming languages.

However they have been known to generate code anywhere from 10% to 200% less efficient than an equivalent low level language program. On the other hand for a large number of mathematical and business problems the program run time may not be as important as the program preparation time.

We will briefly note the features of a number of the languages shown in Figure 8.3. The languages are listed in chronological order.

Article 8.2

TYPES OF HIGH LEVEL LANGUAGES

PROCEDURAL LANGUAGE. – A procedural language is one in which the programmer must specify each of the major procedures or steps to indicate how the processing is to be accomplished. Most high level computing languages are procedural in nature, in that they describe in exhaustive detail the steps which a computer must carry out in order to perform a particular data processing task. They allow mathematical problems to be expressed using standard mathematical formulae, handling both real and integer numbers. Procedural high level languages also allow algorithms to be expressed in a level and style of writing which is easily understood by other programmers and are largely machine independent (as compared to assembly level languages which tend to be machine specific).

The control statements to be found in these languages are structured as jump actions in that they change the state of the computation, while the assignment statements imitate fetch and store operations between the CPU and memory. Central to this imperative model of computing is the concept of the present state of the program/computer, which in turn requires noting the state of the program counter, the values of all variables stored in registers and the state of the memory stack. Although iteration and recursion have been used in procedural languages as techniques for simulating parallel constructs, most procedural languages tend to be unsuitable for multiple processing systems where a large number of asynchronous processes are active.

APPLICATIVE LANGUAGE – An applicative language is the same as a functional language.

FUNCTIONAL LANGUAGE – In a functional (or applicative) language every executable action is a mathematical function, with each function in a program describing a data object, which may be an integer, a character string, a list or a more complex structure. In this way a program in a functional language may be regarded as a set of descriptions of data objects, rather than a set of recipes for obtaining data values. The resulting programs then consist entirely of functions, with the statements in the program not being interpreted sequentially but all being valid at the same time. Such a program is a function in the true mathematical sense in that it is applied to a set of input values and the resulting values are the program's output. The data dependencies within the program then only exist as a result of the function application, and the value of a function is completely determined by its arguments. One consequence of this is that there are no loops within a functional language, but instead a complete program may be seen as a function that calls on other functions to complete its operation.

One of the benefits of using functional languages is that programmers can focus on the structure of the solution to the problem, and the concurrency implied by that structure, without worrying about the details of how tasks are divided up or assigned to processors. They also do not need to express the details of the synchronization necessary to make the program work properly, since these responsibilities fall on the compiler and/or system architecture. One of the other benefits of this applicative approach include its rigorous approach to the logical structure of functions and data and the fact that it is amenable to formal proofs of correctness at all stages.

Examples of functional languages are LISP, Val, Id and Hope. Val and Id are single assignment languages in which an assignment statement implies a notational convenience for binding an expression to an identifier.

DECLARATIVE LANGUAGE – A declarative computing language describes computing tasks in terms of the intrinsic logical properties of the information or knowledge and the transformations required on this information. It is a command language dominated by statements which express what to do rather than how to do it. Declarative means the programmer supplies more 'whats' than 'hows', declaring what the relationship between objects are, and not so much how procedures are computed. The language is based on the general idea that computation can be viewed as controlled deduction, and an algorithm can be expressed as a combination of logic and control. It uses clauses to specify that certain facts follow logically from sets of other facts, where a fact generally states that certain elements, or expressions denoting elements, are in some relationship to one another. Declarative languages are typified by the artificial intelligence language PROLOG.

(Continued)

(continued)

TYPES OF HIGH LEVEL LANGUAGES

OBJECT ORIENTATED LANGUAGE – An object orientated programming language relies on a programmer breaking a problem into modules, called objects, which contain both data and instructions and can perform specific tasks. They can then organize their program around the collection of objects. It is characterized by the following concepts:

1. Objects: Objects are groups of data and their instructions, which are called methods, this is the opposite of the more conventional procedural programming in which data and instructions are always separate.

2. Class: Objects belong to a class or instances of a class. The class contains the characteristics that are common to all objects in the class and serves as a template from which specific objects are created. A collection of classes associated with a particular environment is called a library. Objects are also organized in a hierarchy of classes. The unit of modularity in an object orientated system is the class, so programmers often speak of developing classes rather than programs.

3. Inheritance: Each object knows its class membership and inherits, i.e. receives a copy (from its class and its ancestors) of its behaviour and structure. In other words a new object can be defined in terms of an old one. Inheritance also allows new classes to be built on top of older classes instead of being rewritten from scratch.

4. Messages: Messages are requests for an object to perform one of its tasks or methods. When an object receives a message, the method associated with the object is executed. Programming is done by sending messages from one object to another describing what is to be accomplished. Objects in turn decide, once they have received a message, how to process it.

5. Dynamic binding: Dynamic binding is the ability of a program to determine at run time which method to run in response to a message.

Object orientated programming techniques are claimed to result in a faster development time for producing computer programs. They are also claimed to offer lower maintenance, cost and improved flexibility for future revisions.

FORTRAN (Formula Translator)

John Backus of IBM was the chief architect of the high level programming language FORTRAN. It was initially proposed by Backus in December 1953 with the initial design of the language started in 1954. A FORTRAN compiler for the IBM 704 computer was produced in 1956 while the public version of the compiler was released in 1957. FORTRAN IV was released in 1962 while in 1966 ANSI (American National Standards Institute) FORTRAN was validated and internationally recognized. Languages which have the ANSI standard have been adopted as a standard and machines porting these languages have to conform to a certain set of rules. In 1977 FORTRAN 77 appeared.

FORTRAN is a universal high level language made up from statements which are designed to allow it to be used for programming scientific and mathematical problems. It is written in simple English type statements and permits mathematical expressions to be stated naturally in a form of algebraic notation. One of the most important constraints on the design of the language was to have a compiler which produced efficient programs rather than worry about the clarity or readability of the program. As a result it does not support many features for enhancing good structured programming, now considered important in any high level language. However, FORTRAN is an excellent language for performing calculations and supports vectors and multidimensional arrays. FORTRAN is known for its efficiency in computational speed as well as in memory usage, its large body of existing optimized code and its large and dedicated group of users.

The popularity of the language in the scientific and engineering community has resulted in numerous complex mathematical functions being preprogrammed as fixed subroutines and catalogued in various function libraries. So much is invested in already written and optimized FORTRAN code that it will necessarily continue to be used. FORTRAN 90 is the new standard.

COBOL (Common Business Orientated Language)

COBOL was developed in the late 1950s from the FLOW–MATIC language written by Grace Hopper at Remington Rand Univac in 1955. COBOL was designed by a committee in the USA, under the sponsorship of the Department of Defence, who initially met in early April 1959, in America. In April 1960 the short range committee of the Conference on Data Systems Languages (CODASYL) published the first version of COBOL (COBOL 60).

COBOL is a multipurpose international language designed to emphasize data formatting rather than data manipulation and is therefore more suited for commercial data processing applications. It is a problem–orientated imperative language using statements written in simple English that can be associated with normal terms used in business applications. COBOL was explicitly designed to be as English–like as possible in order to be readable by business managers. It is however not very easy to program. ANSI–74 COBOL was one of the authorized standard version of the language, COBOL 85 is the latest version language. It is estimated that there are over 70 billion lines of COBOL code in America.

ALGOL (Algorithmic Language)

The ALGOL language was designed in Europe in 1958 as a machine independent language to represent algorithms, an algorithm precisely describes a procedure for solving a particular problem, for processing scientific and mathematical applications. The initial 1958 proposal was incomplete and it was not until 1960 that ALGOL 60 was developed. The language emerged as a result of trying to standardize a large number of algebraic languages used on different machines. It is an early block–structured language providing many elegant features that were lacking in other early high level languages.

ALGOL is a problem–orientated language in which the source program provides a means of defining algorithms as a series of statements and declarations having a general resemblance to algebraic formulae and English statements. An ALGOL program consists of data items, simple statements and declarations organized to form compound statements and blocks. The block structure of the language allows different parts of a large program to be developed by different programmers. ALGOL 68 was the most recent refinement, although it is rarely used. Ideas from the language have been used in the generation of the programming languages Pascal, Modula–2 and Ada.

LISP (List Processing)

LISP was developed in 1958 by John McCarthy to process symbols in a way recognizable to speakers of human languages. When LISP was invented the major goal was to construct a language which could handle symbolic representation with the same ease that FORTRAN handled numbers. In the late 1950s McCarthy was a professor at Dartmouth College in the USA; he moved to Massachusetts Institute of Technology (MIT) in 1958. It was between 1958 and 1962 that McCarthy and his colleagues at MIT developed the first versions of LISP.

The language was initially intended for use by an artificial intelligence (AI) project. An AI language needs to be able to represent elements of knowledge, the properties of these elements and the relationships between the properties and the elements. LISP is also intended for problems which involve symbol manipulation and in part is based on the lambda calculus work of Alonzo Church. Some of the ways LISP handles programs, data etc. are given in Table 8.2.

LISP

Element	Comments
Programs	Programs in LISP are functions that define, manipulate and evaluate. A LISP program may then be seen as a data structure in which both data and program are expressed using the same syntactic structure, the list.
Lists	LISP is based on the creation of lists, made up of words, numbers or other lists, and on their manipulations, where the lists are used to represent information.
Functions	Everything in LISP is a pairing of functions and arguments; to run the program is to evaluate the arguments in terms of the functions. To run a program is then to produce a new result. The LISP language itself comes with about 200–300 functions already built in. It is up to the programmer to use these functions to create new ones.
Data	Data in LISP are expressed as functions that bind values to variables. The operation of the program is such that the data within the program are passed from function to function and the concurrent execution of the functions forms the basis for parallelism.
Recursion	Many LISP programs are built on the use of recursive operations, i.e. functions calling other copies of themselves.

Table 8.2

Description of the elements of the LISP language.

LISP has been widely used in the USA to the extent that Texas Instruments developed the Compact Lisp Machine (CLM) in 1987. This is a 32–bit CMOS microprocessor using 550 000 transistors which operates with a 40 MHz clock frequency and has been microprogrammed to support programming in LISP. The arithmetic logic unit within the microprocessor is capable of performing both arithmetic and logic operations on typed or untyped data. About half of the chip is used by the on–chip random access memory (over 114 Kbits). The microcode memory is 16 KB by 64–bits. The system also makes use of software 'garbage collection' routines to round up and recycle unused memory locations.

APL (A Programming Language)

The language APL was developed in 1962 by K. Iverson and A. Falkoff at IBM although the first actual implementation of the language was not running on the IBM360 mainframe computer until late 1966. It is a multipurpose functional (or applicative) language used primarily for data processing. The language began life as a functional notation to express mathematical algorithms and was subsequently adapted for use as a programming language.

APL is used interactively and is built on the manipulation of multidimensional arrays of basic data elements. There are operators within APL for restructuring arrays into new dimensions and sizes, for unravelling arrays into linear vectors, for combining arrays into new ones, for chopping parts of arrays, for rotating, reversing and transposing arrays and for testing membership of an array.

Although the language is very compact and can replace lengthy program segments, which might have required several nested loops, by single symbols, it does require a special characters set which means a special form of keyboard. As an interactive and interpreted language it is more suited to one off jobs and experimental tasks than to large program construction.

SNOBOL (String Orientated Symbolic Language)

SNOBOL was developed at Bell Labs in 1962 to implement pattern matching. It was also designed to process non–numeric data, including text and symbolic expressions, and can therefore provide a tool for applications involving symbolic manipulation of texts. Within the language the programmer defines a pattern that can be matched with some form of text. The basis of SNOBOL is its three data types, integers, reals and patterns, where the language supports a number of operations on patterns enabling the construction of complex patterns from simpler ones. SNOBOL 4 is the latest version (1967).

PL/1

PL/1 was written between 1963–64 by a committee sponsored by IBM. It was IBM's effort to produce a unified language with which to replace FORTRAN, ALGOL and COBOL. The language was intended to go with the IBM system/360 mainframe computer. IBM decided that as a service to all its customers it would create one machine, one operating system and one programming language that would be able to do anything. Work began on the language in October 1963 with the initial version of the language finished by late February 1964. It was called NPL (new programming language). The language went on the market in 1965 with its name changed to PL/1. However, PL/1 development had been rushed and it proved to be too large and complex to be acceptable to a wide number of programmers.

BASIC (Beginners All–purpose Symbolic Instruction Code)

BASIC is a high level computer language which was developed between 1963 and 1964 at Dartmouth College, Hanover, New Hampshire in the USA by Professor John G. Kemeny and Thomas E. Kurtz to enable computer novices to use computers as easily and naturally as possible.

It was the last in a series of experimental languages developed by the two writers. The genealogy of the language is given in Figure 8.4. It can be seen from the Figure that BASIC was influenced both by FORTRAN and ALGOL.

BASIC makes use of English words as instructions and symbolic names for data and although originally a compiled language most versions are now interpretive. With the interpreted version programmers can develop programs in a conversational mode in an on–line programming environment. In a conversational system each element of the program is input in sequence to the computer which carries out elementary checking and validation of each step before the next one is input. In this way it avoids any complications of structure or format and provides extensive diagnostic facilities as an aid to debugging and error correction.

In 1976 BASIC evolved onto microcomputers when Paul Allen and Bill Gates wrote an interpretive version of BASIC for the MTS Altair microcomputer. Their BASIC interpreter eventually developed into Microsoft BASIC. It became widely accepted as the high level language to allow programming of personal computers. Many varieties of the language have

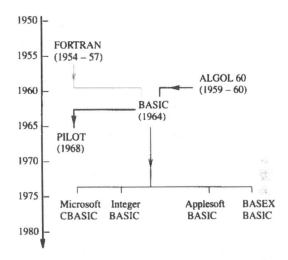

Figure 8.4

**Genealogy of the
BASIC language.**

evolved for different personal computers. In 1976 Stephen Garland, a professor of computer science at Dartmouth, developed a structured version of the language called SBASIC. It eliminated line numbers, GOTO statements, subroutines and primitive forms of conditional statements and looping statements and replaced them with more sophisticated structures. The first commercial structured BASIC interpreters began to appear between 1980 and 1981 with Hewlett Packard's HP BASIC.

BASIC

In the United Kingdom BBC BASIC for the Acorn BBC computer earned the language a wide following, particularly in schools. The BBC BASIC interpreter occupied 16 KB on a ROM chip in the machine.

LOGO

The language LOGO was designed in the mid 1960s by a team of researchers and consultants at Bolt, Beranek and Newman (BBN) a research firm in Cambridge Massachusetts (USA). When the Swiss educationalist Seymour Papert joined the group as a consultant a new design effort was undertaken to create a graphics orientated language aimed at providing children with hands on experience of programming. What resulted was a scaled down user friendly version of LISP which they called LOGO. The language was used in schools in 1967. However, it was not until 1981 that microcomputers were supplied with the language, the Texas Instruments 99/4 was the first.

SIMULA (Simulation Language)

The object orientated programming language SIMULA was created in the mid 1960s at the Norwegian Computing Centre in Oslo. The language is a descendant of the high level language ALGOL 60 and was designed to simulate the operation of systems composed of discrete events. At the core of SIMULA is the entire contents of ALGOL 60, an imperative or algorithmic language with a block structure and a process of describing data and control structures. SIMULA 1 was developed in 1965 while SIMULA 67 was released in 1967. The language uses a model based in communications in which the universe of discourse to be modelled is decomposed into objects and the objects communicate with one another.

C

The language C can trace its origins back to CPL (combined programming language) developed in 1963 by a group of programmers from the Universities of Cambridge and London in England. Compared to the high level languages of its day, CPL was different in that it also allowed access at a low level to the hardware of a computer. CPL was a large language; however, in 1967 Martin Richards at Cambridge developed a smaller version of the language specifically aimed at systems programming, i.e. for writing operating systems, compilers and interpreters, data base packages and editors. This language was called BCPL (basic combined programming language).

In 1970 Ken Thompson of Bell Laboratories in the USA developed a smaller version of BCPL called B and, in 1972, Dennis Ritchie reworked B to produce the C language, primarily for minicomputers using the UNIX operating system, initially the DEC PDP–11.

C was developed as an alternative to assembly language for the development of system software as it has a fast operating speed and efficient memory requirements. It offers high level abstraction when required and portability of programs without sacrificing control over individual bits of information. C manages to remain low level yet machine independent, by placing all of its machine specific instructions in a run–time library that is distinct from the language itself. For example, I/O is not directly defined in the language but is implemented by means of a standard library. This has the advantage that it makes the code written highly flexible and portable.

The source text of a C program can be distributed among several files, each of which can be compiled independently. C is also a structured language which breaks a program down into components, called functions, and makes use of strong data typing. However, the C compiler does not check to make sure you have indeed used variables the way you said you were going to in the writing and running of the program.

As a final comment on the language it is claimed by many that 'C is a programmer's language, written by a programmer for programmers, which should really only be used by people who already know how to program well'.

SMALLTALK

Smalltalk was designed during the 1970s by Alan Kay at Rank Xerox's Palo Alto research centre in the USA. It is a descendant of the programming language SIMULA and the language FLEX (flexible extensible language) which Kay had developed as a simulation language. Smalltalk is a modular and extensible language which enables users to easily create and manipulate information. To implement Smalltalk Xerox needed to create a new set of hardware so a Smalltalk system was developed. The system encompassed not only the formal language but a hardware and software environment including high resolution graphics, a mouse, the use of overlapping windows and an operating system. There are a number of versions of the language such as Smalltalk 72, 74, 76 and 80.

PASCAL

Pascal was developed in 1971 by Nicolaus Wirth at Zurich, Switzerland. The language was named after the French mathematician Blaise Pascal. It is intended to be used to teach students high level programming techniques and is at present widely used in the computing science community for this purpose.

Pascal

Pascal is a block structured language which allows formalized programming techniques to be used and as such it is seen as an easy language to use and debug. A Pascal program consists of subroutines nested inside one another with various rules governing the scope of variables. The language offers a wide range of data structure facilities including constructs for defining data types. Pascal was also one of the first programming languages to support the concept of strongly typed data.

FORTH

Forth was developed over a decade by Charles H. Moore up until 1971 when it was first used. Moore used the language as a means of controlling an astronomical telescope at the Kitt Peak Observatory in the USA. Moore left the Kitt Peak Observatory in 1973 to form Forth Incorporated, a company dedicated to producing application programming in Forth.

The language follows the principles of structured programming and has been designed primarily for use on mini– and microcomputers. Forth makes extensive use of a resident dictionary in which the user is able to define new words and add them to the dictionary, thus making Forth extendible. In this way it not only provides the user with all the essential tools necessary to build a program, but also to create his or her own language. Each program can be said to constitute a new language since in the process of writing a program, the user creates new words out of older ones. Outside of a small shared core, the set of elements used in one program can look radically different from the elements in another. Syntax is also extendible in Forth, allowing programmers to construct new data types, data structures and classes of operators.

Like C and Modula–2, Forth is both a high level and a low level language and it is claimed that with this flexibility Forth programs can run up to ten times faster than equivalent programs written in BASIC.

PROLOG (Programming Logic)

In 1981 Japan's newly formed Institute for New Generation Technology known as ICOT announced a daring 10 year project intended to change the face of computing. The goal was to produce a new 5th generation of computer hardware that would be able to process knowledge. The high level language chosen was PROLOG, which had been developed by Alain Colmerauer and Philippe Roussel at the University of Marseilles (France) in 1972. It was based on the logic–programming work of Robert Kowalski. Logic is a formal system for stating relationships between assumptions and conclusions and for proving these relationships necessarily hold.

PROLOG is a declarative language designed for expressing a problem as a set of rules. Programming in PROLOG consists of telling the computer what is true and then asking it to draw conclusions about the initial statements of truth. The user does not have to specify how the conclusions are to be drawn. Programming then consists of three steps:

1. specify the facts you know about objects and relationships between them;
2. specify the rules that apply to objects and their relationships;
3. ask questions about these same objects and their relationships.

A program then consists of a set of known facts, a set of clauses and a question. The question asks whether a certain fact is true and the program attempts to answer it by figuring out whether the fact in the question follows logically from the known facts.

PROLOG

Built into the language is an inference mechanism which searches for possible ways to answer the question and if it fails it backtracks to take another approach until it runs out of options and reports back failures. Programming in PROLOG may then be seen as constructing a data base and asking questions about its contents. The answers are all there before the program is run. However, the purpose of running the program is to confirm whether the question to be answered is already part of the database.

PROLOG is a highly extendible language in that, although it comes with a number of built in basic elements, most programming requires that the user create basic elements in order to solve the problems in which he or she is interested.

MODULA-2

The Modula-2 language was developed by N. Wirth, the designer of Pascal, and as such its syntax and structure rely strongly on Pascal. The language was designed to offer multiprogramming facilities to allow the expression of the concurrent execution of several activities and to offer facilities for operating a computer's peripheral devices.

ADA

Ada is a high level programming language developed for the USA Department of Defense (DOD) by a French team of software engineers led by Jean Ichbiah of Cii Honeywell-Bull over a period of about 7 years from 1975. It is a structured programming language developed from Pascal with many improvements in control structures and subprograms. Some of Ada's important features are noted in Table 8.3.

Table 8.3

Features of Ada.

Feature	Comments
1. Strong typing	Ada is a strongly typed language which means that all data elements must be declared explicitly before they can be used in programs. This makes more errors detectable at compilation.
2. Programming in the large	Ada provides the mechanics for encapsulation and separate compilation.
3. Exception handling	Exception handling gives the user appropriate control of error recovery.
4. Tasking	Ada is one of only a few languages which contain embedded facilities for multiple tasking. Ada tasks can provide a set of resources to other tasks by means of a task specification which is independent of the implementation.
5. Data abstraction	Ada provides private data types and representation specifications which separate the absolute properties of data from its physical realization.
6. Concurrency description	The overall logical structure of the language is such that it provides a powerful set of concepts for the description of concurrency in a computer system. This structure not only allows the user to package logically related collections of resources but it also facilitates the definitions of common data pools, collections of related subprograms and abstract data types. A program can then be easily split into various tasks in such a way that the tasks can subsequently be executed concurrently.
7. Intertask communication	To assist with implementation on different multiple architectures Ada allows two forms of communications between tasks : (a) Communications by shared nonlocal variables. (b) Communications by message passing. The concurrent execution of tasks has been included to support shared memory architectures and distributed processing architectures.

The language was developed for use in computer control and communication systems, where instruments or systems are monitored or governed by a program. A major emphasis during Ada's design was to incorporate features for programming real time computer systems used as a part of large machines such as missile guidance systems. The large multipurpose language is then primarily intended for real time embedded systems such as those found in missile guidance computers. The Ada reference was issued in July 1980 with a revised version appearing in 1982, while in 1983 ANSI finally approved the language. The Ada standard is defined in MIL–STD–1815 (10th December 1980).

Ada

Ada is intentionally more complex than many procedural languages; for example, the Modula–2 compiler is about 5000 lines, whereas the Ada compiler is several hundred thousand lines long.

C⁺⁺

C⁺⁺ was developed by Bjarne Stroustrup at AT&T's Bell Labs in the early 1980s. It is an extension of the C language with object orientated features. It is claimed to combine conventional imperative and object orientated features, thereby giving programmers the best of both worlds.

OCCAM

Occam was specifically developed as a high level language for the 32–bit transputer microprocessor manufactured by the integrated circuit manufacturer Inmos. The language was primarily developed in joint collaboration between Inmos and Professor A. Hoare of Oxford University, in England. It is intended for use in a parallel processing environment so that although basically a procedural language based on the idea of implementing sequential processes, it also contains constructs which allow the creation of a large network of communicating processes. Since the transputer's hardware is specially tailored to the structure of the language it is possible to design a program in the form of a data flow diagram and then map the resulting program onto an appropriate array of transputers with minimal alteration to the structure of the program. Table 8.4 notes several of the features of the language.

Table 8.4

Features of Occam 2.

1.	It is a high–level programming language with parallel processing primitives based on the rules of communicating sequential processes (CSP).
2.	It is a block structured language that supports both sequential and parallel programming.
3.	It is a small language, 54 keywords and 30 symbols, with an elegant structure, concise command vocabulary and simple direct syntax.
4.	It is a strongly typed language which contains features which facilitate the development of reliable as well as efficient concurrent programs.
5.	It can directly express concurrent programs.
6.	It can be directly implemented by a network of transputer elements.
7.	It can describe and program transputer systems as a collection of processes which operate concurrently and communicate through channels.
8.	It provides easy access to hardware address locations and simple implementation of multitasking.
9.	It can also be used as a hardware description language.

8.4 SOFTWARE DEVELOPMENT PROGRAMS

In the development of software for a system the design team will make use of a number of application programs to assist in the process. These pieces of software are typically an editor, a compiler or assembler, a linker/loader, a debugger and a software simulator.

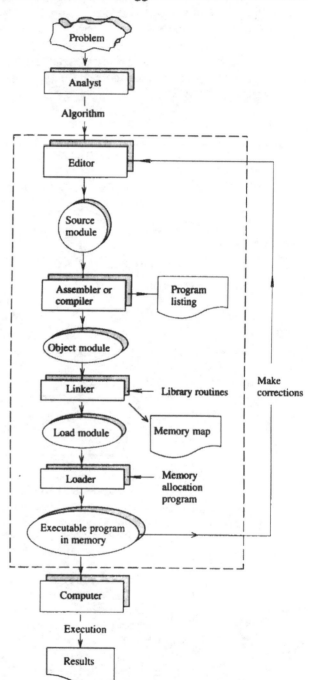

Figure 8.5

The software development process.

The software team may also use more sophisticated computer assisted software engineering (CASE) tools which are able to take the functional specifications of a program and produce the source code for the task.

The software development process is shown in Figure 8.5. The figure indicates the different software packages used in the development and the different input and output files used at each stage. We will note the different software utility and application packages in more detail.

8.4.1   EDITOR

The initial creation of a piece of software consists of producing text in a suitable programming language for each procedure and module required by the program and storing this text in suitably named files An editor is used to write the text (source module) of the program by allowing the creation and modification of a source program file. It allows the listing and correction of the text as well as providing other sophisticated functions that can enable an operator, for example, to move sections of code throughout the program, or delete large sections of text. These operations are all keyboard or mouse controlled. The user commands within the editor are therefore orientated towards creating and modifying a file, making changes in the text, correcting design or programming errors, re–positioning lines of text and adding them to or deleting them from the program file. The program file produced by the editor is usually saved on a computer floppy or hard disk and can then be recalled for processing whenever necessary.

After the program has been edited by the programmer into a source file this must then be translated into machine code. This requires the use of a translator program.

8.4.2   TRANSLATOR

A translator is a piece of software which is able to take a source program and translate its instructions into machine code, or object code. There are three different types of translators: assemblers, compilers and interpreters. In making the conversion between source code and object code, the translator program is stored in the central memory of the computer, and the source program is read in instruction by instruction and converted into machine code, object module. The machine code instructions are then written out to some form of storage, usually magnetic disk, so giving a permanent object program that can in turn be run on the computer. During this process the translator indicates any violations of the grammatical rules of the particular language the source code represents.

1. **Assembler** – An assembler is a piece of software that translates an assembly language program into the machine code which can be executed by a processor.

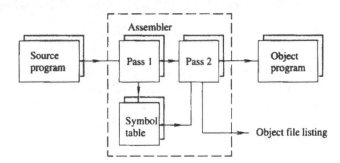

Figure 8.6

**Operation of
an assembler.**

Glossary 8.2

CROSS ASSEMBLER – A cross assembler is an assembler program for a microprocessor which runs on a computer using a different type of processor. A cross assembler is often used in universal development systems to allow software to be developed for any processor without the need to change the hardware system.

HIGH LEVEL TRANSLATOR – High level programming languages must be translated into machine code before they can be run on a computer. There are two forms of high level translator; the compiler and the interpreter. The difference between the two is that with compilation a separate program is first constructed from the high level language source code before it runs on the computer, while with interpretation the high level program is read in line at a time and changed to the low level language.

LOW LEVEL TRANSLATOR – A low level translator is another name for an assembler. It converts a low level language into machine code. Each statement in the low level language directly translates into a few machine code instructions, typically one or two.

MACHINE LANGUAGE – A machine language is a programming language which a computer can directly execute. The actual machine code is a set of bit patterns representing the computer instructions and data, normally in octal or hexadecimal.

MACROINSTRUCTION (MACRO) – A macro is a user–developed sequence of instructions which can be referenced by name within a program. It is generally used when writing assembly language programs and usually consists of two or more ordinary assembly language instructions. The macro can then be effectively copied in–line during translation as a result of being invoked by name, with the assembler or compiler replacing every reference to the macro with a copy of the instruction sequence. It is chiefly used to relieve the programmer from having to write out frequently occurring sequences within a program, thereby save programming time.

OBJECT CODE – Object code is the executable machine code programs that have been compiled or assembled by a translator. An object code program may however need to be linked and loaded before it can be executed by the central processing unit.

RELOCATABLE – The word relocatable pertains to programs where the machine coded instructions can be loaded into any stated area of memory, without changes being required to the code. The program can then occupy any set of consecutive addresses and be executed successfully irrespective of its position in main storage.

SIMULATOR – A simulator is a piece of software that runs on one computer to emulate the operation of another target machine. While the host computer runs the simulation program it can accept any code for the target machine performing the necessary manipulations of data between the simulated registers and memory locations in exactly the same way that the target machine would do. The big advantage of the use of a simulator during the software development process is that it allows the software to be tested in a host system before being loaded onto the target system. In this way it is possible to develop the software independent of the target machine's hardware. The simulator software is also able to mimic the actions of the target computer in detail and hence can be used for the debugging or testing of a machine language program on a machine that is not the final target system of the program.

For a detailed analysis, the simulator usually allows a step by step approach of the program operation, to let the programmer to see the effects the program would have on the different parts of the target system. As the whole system does not run in real time it therefore allows the programmer more of an insight into the effect that the software under test would have on the target machine. In addition, the simulator provides the user with a number of execution monitoring commands to allow logical errors in a program to be detected, i.e. mistakes in the use of instructions to implement the required algorithm. Furthermore it is possible from a simulation to evaluate the performance of the target machine running the proposed software, although for proper testing it is still important to run the piece of code in a real time environment.

In general, however, simulators are effective for testing small segments of code. Problems do arise when attempting to test complete programs of several segments. They are slow and inappropriate, especially if time is at a premium.

SOURCE CODE – Source code is a term referring to a program written in a programming language which needs translation before it can be executed. Hence a source program is one written by a programmer using a computer language, while an object program on the other hand refers to the program after it has been compiled or assembled into machine code.

Assembler

The term assembler was introduced by Wilkes, Wheeler and Gill in the 1950s to denote a program which assembles a master program with several subroutines into a single run time program. In operation the assembler translates the mnemonic instruction codes of the assembly language into an ordered sequence of binary machine instructions and data, the binary 1s and 0s of machine code.

Assemblers tend to operate on a one for one basis in that each phrase of an assembly language program translates directly into a specific machine language word.

Assembler

The process of converting a source program into an object program is usually performed by the assembler making two complete passes over the source code (Figure 8.6). After pass 1 the assembler creates a symbol table noting the relationship between different program labels, data values and addresses. It makes use of this table in the second pass to convert the labels and data values to their equivalent numbers. Typical features supported by assemblers are shown in Table 8.5.

Table 8.5

Features supported by most assemblers.

1. **Mnemonic instruction opcodes** – These produce more meaningful and readable programs.

2. **Symbolic referencing** – At the level of assembly language programming, all the addresses used in memory references for data transfers or transfers of control must be directly specified by the programmer. Many assemblers, however, allow the programmer to use symbolic names for various addresses, for program lines, instruction sequences, memory locations, I/O ports etc. This not only makes it easier to write a program but also facilitates the insertion or deletion of instructions. Using these labels the assembler automatically computes the correct addresses of all memory references and branch instructions and substitutes them for their symbolic representations.

3. **Assembler directives** – These control and simplify the production of source code.

4. **Numbers systems** – Assemblers allow a choice of binary, octal, hexadecimal or decimal.

5. **Comments** –The use of comments in the source code to improve program presentation and documentation.

6. **Syntax error detection** – This includes error messages explaining the errors and their locations.

7. **Conditional assembly** – This allows assembly of certain sections of source code to be made conditional on parameters set at assembly time.

There are several types of assemblers. An assembler that runs on a microprocessor and creates an object code for that same microprocessor is called a self assemble, whereas an assembler which runs on one processor but assembles programs written for another processor is called a cross assembler. Cross assemblers are usually written in a high level language and are more powerful than self assemblers because of the additional features that can be supported.

2. Compiler – A compiler is a special piece of software used within computer systems to automatically translate a particular high level language application program into its low level machine language equivalent. Thus by running the source, problem orientated, program with the compiler program, an object machine instruction program is produced which can then be used on the computer hardware to perform the task its source program was designed for. This conversion process is illustrated in Figure 8.8. The conversion, or compilation, is a precondition for using the high level language as a programming tool for the machine.

INTERPRETER

When using an assembler or a compiler within the software development process, the complete source code entered by the programmer is assembled or compiled into machine–executable object code prior to execution. The alternative to this form of translation is to pretranslate the source code into an intermediate code and then to execute that code, with the aid of a different type of translator called an interpreter. In this way both the source program and the interpreter program are read into the main memory and as each source program instruction is accessed, reference is made to the interpreter program which converts the instruction to machine–coded instruction(s) and executes.

An interpreter may be seen as a piece of software which directly translates instructions from a high level language to machine code, as the high level source code is being executed. The interpreter fetches the instructions from memory, examines the instructions in the source code as they occur, determines what they mean, and then takes the actions that the instruction indicate. The machine then proceeds to the next source program instruction, converts and executes this, and so on until the program has been completed. The use of an interpreter program in this way is known as **interpreting**. This is in contrast to the one step process of compilation.

The technique of interpretation was pioneered for interactive languages, particularly BASIC, but interpreters for COBOL and other high level languages are also in use. An interpreter has the advantage that the result of each operation can be seen individually, in contrast to a compiled program where the result of the program operation can only be seen after the complete program has been compiled and run. Interpreters may then be regarded as an aid to programming a computer, because they enable the writing of programs to be an interactive dialogue between the computer and the programmer. Interpretation also has advantages in timesharing systems when one copy of the interpreter is shared among multiple users. Compiled programs however generally run much faster than interpreted ones since the translation into machine language has already been done.

Figure 8.7 shows the operation of three types of interpreters. As an intermediate step it can be noted that the program is pretranslated before the interpreter takes the intermediate language instructions and produces the machine code. Interpreters may be stored in firmware, e.g. a read–only memory (ROM), rather than on a disk in order that the run time, or execution time, is reduced and the main memory otherwise occupied by the interpreter is freed.

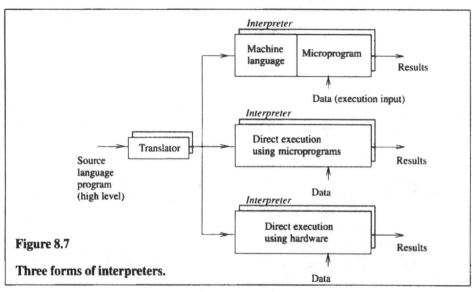

Figure 8.7

Three forms of interpreters.

(Continued)

Article 8.3

INTERPRETER

(continued)

An interpreter can also be a program written for one computer, called a host machine, causing it to imitate the behaviour of another, called the target machine. The task of the interpreter in this case is to translate the incoming code which is aimed at the target system into a format suitable to control the host computer. The interpreter must then manipulate the different states, which the target machine can assume, i.e. the numerical values found in all recognized stores, registers, control flags, condition codes, modes and status words, into a form that the host machine recognizes. It must also interpret the transition rules within the target software which determine for any given state of the target system, what the next state should be when the machine is working correctly and then convert it into similar transition rules for the host system. These transition rules are essentially embodied in the instruction set of the target machine. Interpreters are often used in this way when target software already exists and is valued enough to make its imitation worthwhile, since the use of an interpreter can allow this software to be ported to other systems. The interpretation program in this case is frequently held in firmware as a series of microprograms and is collectively known as an **Emulator**.

Figure 8.8

Compilation process.

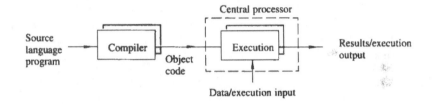

In essence the compiler is simply a program containing a list of the statements used in the high level language and for each statement the compiler has a sequence of machine instructions necessary to perform the statement.

It is usual for a compiler program to reside on an input/output device, such as a disk or tape and be summoned into the main memory of a system when a programmer wishes to compile a program. During the compilation process the compiler is run on the computer system taking as its input the high level language program and producing at its output a representation of the program in a machine executable form, It can be noted that since many machine instructions are usually generated for each high level instruction the resulting compiled program can be quite large compared to the source code.

Compiler

Most compilers consist of four separate stages.

(a) **Lexical analysis:** The first processing stage of a compiler removes redundant information, condenses statements by eliminating spaces and displays error messages for statements that cannot be processed. If any errors are found, the compiler will give instructions for them to be printed out for the attention of the programmer indicating the type of error detected. These error message are known as diagnostics.

(b) **Syntactical analysis:** This is the second stage of a compiler in which the syntax of each line within the source program is tested. As each source program instruction is read in, the compiler scans it for errors in construction of the statements. Error messages are displayed accordingly. It detects errors such as the incorrect order of statements or omissions from statements, misuse of names of variables, incorrect punctuation or conflict with other statements through using the same reference.

(c) **Code generation:** If no errors have been detected the third stage generates the appropriate machine code for each particular statement or group of statements.

(d) **Optimization process:** The optimization process increases the efficiency of the object program. It can be performed before or after the generation of machine code and on any part of the machine language program. Optimization for loops for example performs as many tasks as possible outside the loop to reduce the number of cycles within the loop. Optimization greatly increases the complexity of the compiler.

Compiler

Several implementations of high level programming languages claim to have reasonable optimizing compilers. However, the optimization of computer programs in this way has several disadvantages. An optimizing compiler is usually large, slow and unreliable. Secondly even with a reliable compiler there is no guarantee that an optimized program will give the same results as a normally compiled one. Lastly a more subtle danger with optimization is that it tends to remove from the programmer the fundamental control over, and the responsibility for, the quality of their program.

It will be appreciated that since different types of machine have different machine code formats, each must have its own compiler program to convert the appropriate programming language into the correct machine code. The process of compilation enables syntax errors involving incorrect coding to be detected. An error in the logic of the program cannot be detected by a compiler as any correctly formatted statement will be translated, even if the instruction will cause the program to perform a calculation incorrectly. On observing that a program produces incorrect results, logic errors, usually the result of incorrect appreciation of the problem, the program must be suspended.

3. **Interpreter**   Interpreters are described in Article 8.3.

8.4.3   LINKING LOADER

A linking loader is a special piece of software used to tie or link together independently assembled or compiled subprograms to form an executable program. The linking loader links the object program from a system residence device (e.g. disk) to primary memory storage and does this by assigning the appropriate primary storage addresses to each byte of the object program. Figure 8.5 shows the position of the linking loader in the development of a computer program. It consists of two separate pieces of software.

1. **Linker** – The linker is responsible for connecting together all the different programming modules that have been written, including any library routines. It provides the required interconnections between the subprograms allowing programs to be developed in small modules and then linked together to form a whole.

2. **Loader** – The loader has the function of transferring the object code from an external medium, such as a floppy disk, to the microprocessor's main memory. In operation the loader runs on the central processing unit and loads the object code into either the semiconductor random access memory for execution or, if the object code is to be loaded into a programmable read only memory (PROM), it will interface to a PROM programmer.

The loader is therefore responsible for the address–binding of the program. There are essentially two types of loaders which allow address–binding at four different stages in the life of a program.

(a) *Absolute loader:* The earliest point at which binding can occur is programming time. In this case all the actual physical addresses are directly specified by the programmer within the program. The next time binding can occur is at assembly time or compile time. The loader for these first two options is usually referred to as an absolute loader. If the machine code is said to be absolute then it means that it must be loaded into specific memory locations.

(b) *Relocatable loader:* The third time that the program addresses can be bound is at load time. In this case the assembler or compiler does not translate the program symbols into absolute addresses but rather into relative addresses. The program is initially assembled or compiled with an assumed base address of zero for the first instruction and embedded addresses are then corrected by the loader at the time of loading. This type of loader is called a relocatable loader. It is used for loading programs generated by relocatable assemblers or compilers to any part of the allowable main memory space, since the memory locations are relative rather than absolute.

Linking loader

The fourth time binding can occur is at execution time where the address field that is loaded into memory is not yet bound to a physical address. It is not until a particular statement has to be executed that the address of the physical location that will be referenced is determined. In this case the code is said to be position independent and all address references are program counter relative. The program can then be loaded anywhere in memory without address modification, so providing the system with maximum flexibility. Systems that use this form of binding use virtual memory addressing techniques.

8.4.4 DEBUGGER

The final software package used in program development is the debugger. To **debug** a program means to locate any software mistakes. The two main types of error that can occur are logic errors and syntax errors. The former are a result of incorrect appreciation of a problem while the latter are the result of incorrect coding. Errors are detected by observing that programs do not produce the results expected from them, or by the failure of a program to compile or assemble properly. The process of compilation enables syntax errors involving incorrect handling of the symbolic language to be detected. However, errors in logic cannot usually be detected by compilers. Errors of syntax detected by compilers are usually recorded by the compiler and the type of error identified. Programs which have caused compilation errors cannot be run and the errors must be corrected before the programs can be tested.

Debugging is the process of detecting, diagnosing and correcting errors, also known as bugs, which have occurred in programs or systems, both hardware and software. A debugger is a piece of software that aids in the detection, location, isolation and elimination of errors or bugs in a program. It allows the execution of one instruction at a time on the computer for analysis, or alternatively it can be used by a programmer to run his or her program on the system and to stop the program executing at the occurrence of specified events, i.e. at a breakpoint. A **breakpoint** may be seen as a pointer to a memory location which is used for debugging. When a program under test addresses the location to which the breakpoint is pointing, processing is interrupted and control is returned to a monitor program.

By so doing, it allows the examination and modification of the contents of registers and memory locations. A debugger may be part of a simulator or, for example in a microcomputer development system, debugging capabilities are often included in the operating system of the device.

8.5 KEY TERMS

Ada
ALGOL
Algorithm
APL
Application software
Applicative language
Assembler
Assembly language
BASIC
Breakpoint
Business graphics software
C
C++
COBOL
Code
Coding
Communications software
Compiler
Cross assembler
Database
Debug
Debugger
Decision
Declarative language
Desk top publishing
Deterministic
Editor
Emulator
Forth
FORTRAN
Function
Functional language
Heuristic
High level language
High level translator
Integrated software
Interpreter
Linking loader
LISP
LOGO

Low level language
Low level translator
Machine language
Macroinstruction
Modula–2
Object code
Object orientated language
Occam
Operating system
Pascal
Pl/1
Procedural language
Procedure
Process
Program (and programming)
Programming language
Programming language
  development
PROLOG
Relocatable
Routine
SIMULA
Simulator
Smalltalk
SNOBOL
Software
Software characteristics
Software development program
Software types
Source code
Spreadsheet
System software
Statement
Strong data Typing
Subroutine
Syntax
Translator
Utility software
Wordprocessor

8.6 REVIEW QUESTIONS

1. Write a short paragraph explaining what is meant by the term software.
2. What are computer instructions?
3. List several kinds of application software packages that make today's computer easy to use.
4. Identify the features of spreadsheet software.

5. What is word processing and why has it become so important?
6. List some of the tasks for which people use wordprocessors.
7. What are some of the advantages of using a wordprocessor over a standard type-writer?
8. Describe the features and functions supported by desktop publishing application software.
9. What type of printer would be most suitable for desktop publishing software?
10. Describe the differences between system software, application software and utility software.
11. List three characteristics of software.
12. What is meant by nonexecutable software?
13. Why are programming languages necessary?
14. Outline the main features of low and high level languages.
15. Contrast the use of machine, assembly and high level programming languages.
16. Why is machine code regarded as a low level form of software?
17. Why do we say that there is a one to one correspondence between an assembly instruction and the machine code of a given computer?
18. What does the syntax of a programming language represent?
19. What do the following terms mean: algorithm, heuristic and coding?
20. Note the five categories that result when high level languages are divided according to their organizing principle.
21. Describe the features of a good high level programming language.
22. What does the word procedure convey in a procedure orientated language?
23. What are the differences between a procedural language and a functional language?
24. For what kind of problems is FORTRAN intended to be used for?
25. Why are new high level programming languages being developed?
26. Which of the following high level languages is mainly used for commercial programming: BASIC, COBOL, FORTRAN or ALGOL?
27. What is reputedly the easiest high level language to learn?
28. BASIC is a popular language for personal computers. Why?
29. What is meant when we say that C is an efficient language?
30. Match the language to the description:

    (a) Reinforces the principles of structured   FORTRAN
        programming for teaching students.
    (b) A popular high level language which has   Ada
        never been standardized.
    (c) Ideally suited to evaluating complex   Pascal
        mathematical expressions.
    (d) Advanced programming language with   BASIC
        sophisticated control structures.

31. What are the functions of the editor in software development?
32. Explain briefly the meaning of the terms object program and source program.
33. What is the function of a compiler in the software development process?
34. Explain what is meant by the phrase 'a one to many' correspondence between a high level language statement and the machine code into which it is translated by the compiler.

35. What is the input to a compiler and what is the output?
36. Which software tool is responsible for translating each programming statement interactively into an immediately usable machine language instruction?
37. How do interpreters differ from assemblers and compilers. When is it preferable to use an interpreter rather than a compiler?
38. What is the difference between an emulator and a simulator?
39. Which software tool is used to combine software modules into an executable program?
40. What is the major function of a debugger?

8.7 PROJECT QUESTIONS

1. Which application software packages would be used in
   (a) businesses
   (b) education
   (c) medicine
   (d) police
   (e) publishing companies.
2. What does the phrase 'user friendly software' mean?
3. Describe how a spreadsheet can be used to form a business plan for a new business starting up.
4. Describe the differences between a menu driven and a command driven word processor.
5. What is the word wrap facility found in word processing application software?
6. Find out what 'information managers' are in database application software.
7. Investigate the use of computerized information systems to hold data records.
8. Database software can only be used for computerized record keeping. Is this a fair statement?
9. Describe a data base and discuss data base design concepts.
10. Categorize the following application software: Lotus 1–2–3, dBase IV, Word, Excel, Quattro Pro.
11. What functions would communication software perform in interfacing a modem to a PC?
12. What do 'paint programs' do?
13. Discuss the features found in an integrated application package.
14. Give three examples of the use of integrated application software.
15. In terms of application software what do you think the purpose of a 'tutorial' is?
16. What is the user's role in making a decision on a software application package?
17. What effect does software piracy have on software companies ?
18. How do you think the ability to program a computer would influence the way the computer would be viewed by the user?
19. Are computer literacy and the ability to program a computer interlinked?
20. What would you envisage were the prerequisites for leaning to program a computer?
21. What are the distinguishing features which separate hardware, software and firmware?

22. Choose one of the high level programming languages noted in the text and produce a detailed account of its features and uses.

23. Devise a suitable program of studies for a computer company that wishes to introduce BASIC to its employees.

Project questions

24. Find out as much as you can about fourth generation languages.

25. Why is Ada expected to have a major impact on programming?

26. Why is PROLOG the language for the fifth generation computers?

27. List some of the factors to be considered in the selection of a programming language for a particular task.

28. Why do you think standardization of a computer language is important?

29. Discover what the term 'two pass' refers to in a 'two pass assembler' and note the functions performed with each pass.

30. Investigate the features supported by an optimizing compiler.

8.8 FURTHER READING

Each of the programming languages mentioned in the text has many books written on it. A few articles and reference texts for the more popular computing languages are listed below.

Programming Languages (Articles)

1. Classic languages: BASIC, D. Appleby, *Byte*, March 1992, pp. 155–158.
2. Classic languages: SNOBOL, D. Appleby, *Byte*, January 1992, pp. 149–1156.
3. Classic languages: COBOL, D. Appleby, *Byte*, October 1991, pp. 129–132.
4. Smalltalk yesterday, today and tomorrow, L. P. Deutsch and A. Golberg, *Byte*, August 1991, pp. 108–115.
5. Classic languages: FORTRAN, D. Appleby, *Byte*, September 1991, pp. 147–150.

Programming (Books)

1. *Programmer's Phrasebooks: C Programming*, I. Clark, Sigma Press, 1991 .
2. *Programmer's Phrasebooks: Basic Programming*, I. Clark, Sigma Press, 1991.
3. *Programmer's Phrasebooks: Pascal Programming*, I. Clark, Sigma Press, 1992.

Software Engineering

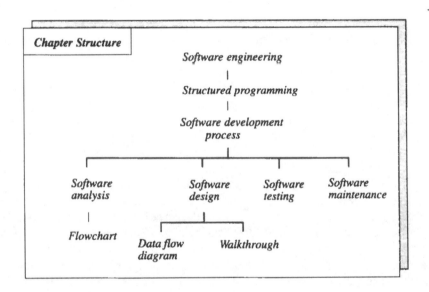

Chapter Structure

Software engineering
|
Structured programming
|
Software development
process

Software analysis — Software design — Software testing — Software maintenance

Flowchart

Data flow diagram — Walkthrough

Learning Objectives

In this chapter you will learn about:
1. The qualities of good programs.
2. Structured programming concepts.
3. The nature and process of software development.
4. Systems analysis and system design techniques and the importance of them in the development of software.

> '... programs must be written for people to read, and only incidentally for machines to execute.'
>
> *Abelson and Sussman.*

Chapter 9

Software Engineering

In the previous chapter the process of producing software for computers was discussed. In this chapter we will consider a more detailed approach to the process of software development.

Over the past ten years computer systems have been used to control more and more critical operations, this has led to the demands for software correctness and reliability to become more important. To cope with these demands, together with the growing complexity of computer applications, the **software engineering** approach to software development has emerged. Software engineering is the establishment and use of sound engineering principles in order to obtain economical software that is reliable and works efficiently on real machines. It is an outgrowth of hardware and system engineering and may be seen as a profession standing at the intersection of science and engineering which contains elements of formal mathematics and logic, computing science, economics and management as well as a tightly developed skill in programming.

In terms of the software development process, software engineering is the application of scientific principles to the orderly transformation of a problem into a working software solution and the subsequent maintenance of that software through to the end of its useful life. The software engineer attempts to combine the disciplines mentioned above such that in response to a software problem, he or she can formulate the task in a way that uncovers the structure of the problem. Using engineering principles, they can analyse solutions to the problem and attempt to develop a computer program in a managed fashion, subject to strict cost, time and performance constraints. It is normal for the software engineer to operate in a team environment. This means the group of software engineers assigned to a particular software project usually divide into small teams of people to handle the various identified tasks within the project. In this way we can see that software engineering encompasses methods, tools and procedures that enable the manager of the process to control the process of software development and provide the practitioner with a foundation for building high quality software in a productive manner. With software engineering structured programming has become an important practice.

9.1 STRUCTURED PROGRAMMING

In 1968 Edsger Dijkstra published the paper 'Go To Statement Considered Harmful' which suggested that the use of the 'go to' statement in programming was the source of a large number of programming errors. This introduced the concept of adopting good programming practices called structured programming. Dijkstra advocated structured programming as a way of writing computer programs

1. to minimize the number of errors that occur during the software development process;
2. to minimize the effort required to correct errors in sections of code found to be deficient and to upgrade sections when more reliable, functional, or efficient techniques are discovered;
3. to minimize the life cycle costs of the software.

Dijkstra in his dissertations essentially discussed a comprehensive programming process that anticipated top down development, programming stepwise refinement and program verification. This precipitated a decade of intense focus on programming techniques and altered radically the way software is developed. Pre–1970, programming was regarded as a black art practised by a few skilled individuals and highly dependent on experience; now, however, programming is regarded as a public, logically based activity of restructuring software specifications into working programs.

Structured programming can be understood as the application of a basic problem decomposition method to establishing a manageable hierarchical problem structure. It is a top–down modular approach to designing computer programs that emphasizes dividing a program into logical sections. Before Dijkstra, top down design programming was widely regarded as a synthesis process of assembling instructions into a program rather than as an analytic process of restructuring specifications into a program, called structured analysis.

Structured programming uses a set of guidelines and techniques to write programs as a nested set of single entry, single exit blocks of code, using a restricted number of constructs. By using this methodology it is claimed to increase the productivity of programmers, to bring clarity to their programs as well as reducing the testing time for the programs. With a structured approach to design, computer programs should feature modularity, security, understandability, ease of modification, error handling, testability and portability.

The use of structured programming techniques has reduced much of the unnecessary complexity of programming and has allowed many programs to be written, and with little or no debugging, to run first time. Other researchers following on from Dijkstra's initial postulates showed that any problem which permitted a flowchartable solution, could be expressed in a structured program form. The more popular of the structured programming methodologies are

1. Jackson structured programming, devised in 1975 by Jackson;
2. structured system analysis and design, as propounded by Yourdon and Constatine, De Marco and by Gane and Sarson in 1978;
3. Jackson structured design, advanced by Jackson in 1983.

Structured thinking can be used in all phases of program production right from the initial concept to final program coding. With structured programming a programmer can quite easily decompose a large problem into small manageable sections, such that with the proper documentation and adhering to strict interfacing guidelines it then becomes immaterial as to which section of the program is written first, or which of the programming team has to write it. It can be noted that for the sake of clarity most programs are written in an ordered chronological fashion since in this way every line of the program can be verified by reference only to lines already written, and not to lines yet to be written. With the advent of structured programming, syntactic and typographical aspects, such as indentation conventions, were introduced. These allowed programmers to easily read each other's programs, permitted them to conduct structured walk–through program inspections, and also permitted managers to understand the progress of software development as a process of stepwise refinement that allowed progressively more accurate estimates of project completion.

In writing structured programs other techniques should also be used, these include: adopting a team organization controlled by a chief programmer; ensuring highly visible program development; making use of library procedures; utilizing intensive program reading.

Structured programming

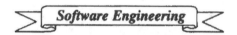

Glossary 9.1

INCREMENTAL IMPLEMENTATION – Incremental implementation is a software testing and implementation strategy for adding a new software module to a tested collection of existing modules and then testing the new combination.

SOFTWARE QUALITY ASSURANCE – In the writing of a piece of software it is important to consider its quality. Software quality assurance may be defined as 'whatever is necessary to make sure that the program does what the customer wants and needs, and doesn't do anything else'. It is ensuring conformance to explicitly stated functional and performance requirements, explicitly documented development standards, and implicit characteristics.

STEPWISE REFINEMENT – Stepwise refinement is an approach used in structured programming where small program modules are written and tested before being incorporated into a much larger software system. The term is synonymous with incremental implementation.

STRUCTURE CHART – A structure chart is a graphic representation of top–down programming which displays the different modules within a piece of software and the relationships between these modules.

STRUCTURED DESIGN – Structured design in a software environment is a top down approach to programming which uses a set of guidelines and techniques to assist a software designer in determining the appropriate software modules which will best solve a well stated problem. By using such a disciplined approach to both systems analysis and design, users and professional designers are assisted in the area of intercommunications.

These techniques are important within a structured programming environment particularly in the ability to defer and delegate design tasks through specifications of the small programming modules.

9.2 SOFTWARE DEVELOPMENT PROCESS

There are many stages through which a piece of software must pass before it is able to be run on a computer system (Figure 9.1).

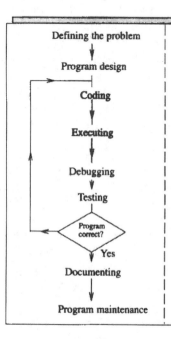

Figure 9.1

Software development cycle.

Defining the problem	The problem is studied and defined ;a method of solution is developed.
Program design	The solution is represented as a algorithm.
Coding	Each step of the solution is converted into computer instructions in the form of a computer language.
Executing	The set of instructions is placed into the computer and the computer is directed to execute the program.
Debugging	The program is checked to eliminate errors, bugs.
Testing	The program is tested to determine if it does what it is supposed to do.
Program correct? Yes	
Documenting	Write–ups, program listings, operating instructions etc., are assembled for future program modification or for individuals who may want to use the program.
Program maintenance	The program is kept functioning at an acceptable level.

In Figure 9.1 can be seen the process of software development. Software development is a process that starts with a particular design problem and culminates with the creation and verification of a pattern of zeros and ones which, when placed in the appropriate memory locations of a specific microprocessor system and executed, causes the system to implement its intended function.

In general, software development consists of the five steps analysis, design, coding, test and support.

1. System Analysis: Given the job of generating a piece of software for a computer system the first task that must be undertaken is system analysis. System analysis is concerned with the investigation of the present position in terms of hardware and software of the target computer system. The major part of this process involves fact finding and information recording. The details derived from this process must be organized in such a way that a clear and precise picture is built up of the present organization of the system under consideration. From this process, information is derived to help in the development of the new software system.

2. Software Design: On completion of the analysis phase, decisions can then be made regarding the allocation of suitable portions of processing to either software or hardware elements. This allocation is subsequently evaluated and validated in terms of performance characteristics and resource utilization. Within the design process we constrain the boundaries of the problem search for alternatives and construct a system on paper. Software design is concerned with the determination of the overall structure of the program and the data.

Mathematical models and functional flowcharts should appear at this stage. The approach taken in design should also define the relationship among the computer program components. During this phase, a draft of the computer program development specification is prepared. Finally the design is represented or specified to the level of detail necessary to support the programming step.

3. Coding and Checkout: The third stage of the software development process consists of the coding and debugging all program components, producing detailed test procedures, testing the components in groups until all components have been assembled into the program, demonstrating the validity of the program by means of a test in a simulated environment, and completing the documentation of the program.

The coding process is such that the high level source program, or assembly language program, to solve the software task is written by a programmer with the use of some type of editor program. The source program is subsequently input to the translator program, which produces the equivalent machine code along with a listing of the original source program. In the group of translator programs are such things as assemblers, compilers, emulators and interpreters which allow source programs to be changed into the object machine code. In the process of assembling or compiling the source program, the translator checks for any violations of the rules governing the structure of instructions, syntactic errors, and for consistency and completeness of the programs. The syntax errors are usually indicated on the program listing adjacent to each incorrect program statement.

On discovering errors in the source code the programmer must reuse the editor to make the required corrections in the source statements before the program can be re–assembled or re–compiled. In practice it may take several translations followed by use of the editor to locate and correct all the syntax errors in a program.

Article 9.1

COMPUTER AIDED SOFTWARE ENGINEERING (CASE)

CASE is the use of software tools to aid in the design of computer programs. Many of these tools use microprocessor based workstations with powerful graphic capabilities and user friendly interfaces. CASE tools redefine the software development environment allowing all phases of the software development cycle to be linked together. CASE software includes the following facilities:

1. Diagramming tools for drawing structure diagrams and creating pictorial system specifications.
2. Screen and report painters for creating system specifications and a form of simple prototyping.
3. Dictionaries, data base management systems and reporting facilities for storing, reporting and querying technical and project management system information.
4. Specification checking tools to automatically detect incomplete, syntactically incorrect and inconsistent system specifications.
5. Code generators to generate executable code automatically from pictorial system specifications. Some CASE systems also include code generators that when given the specifications for a part of a system produce high level language programs that implement the part of the system. Manual programming in some cases can be replaced by CASE code generators that automatically produce 80 to 100% of the code.
6. Documentation generators to produce technical and user system documentation required by structure analysts.

It is claimed by many software engineers that the use of software tools has a number of advantages. These are noted in the Table 9.1.

1. Makes structured techniques practical.
2. Enforces software/information engineering.
3. Improves software quality through automated checking. Error checking is one of the most important capabilities of CASE workbenches since getting errors out of the way is an excellent way of reducing software costs.
4. Makes prototyping practical.
5. Simplifies programming maintenance.
6. Speeds up the development process.
7. Frees developers to focus on the creative part of software development. With CASE, system developers can spend more of their effort on analysis and design and less on coding and testing.
8. Encourages evolutional and incremental development.
9. Enables reuse of software components.

Table 9.1

Advantages of using CASE software tools.

Software development process

The linker/loader program links the code produced from the compiler with other software units and loads the resulting module into memory for execution. The linked object code can then read by the central processing unit (CPU) causing the internal functions of the CPU to be manipulated through the enabling of the logic gates within the computer.

4. Integration and Test: After the software has been checked on its own it must be integrated with the hardware of the target system and formally tested against the requirements specified in the computer program performance specification.

(a) Testing – Testing is the determination of whether the object code, when executed, actually causes the system to implement its intended function.

(b) Debugging – Debugging is the process of determining the source of any failures found during testing and eliminating them.

At the completion of this phase, the software is demonstrated to the user or customer and the development is considered complete.

Software development process

5. Operation and Support: Once the development process is completed it is still important to monitor the behaviour of the system when it enters its operational phase. During this phase the customer will use the software to perform its intended function. In this way the operational suitability of the software and the program's capability to operate on the total set of input data in an operational environment can be assessed. It is to be noted that invariably the software will still need to be modified as changes are made to the design to adapt the software to new operating or hardware environments. Modification or enhancements to the program to improve performance or add capability are also a part of this phase.

Modern software development makes use of computer aided software engineering (CASE) tools to aid in the process (Article 9.1).

We will consider each of the stages in the software development process in more detail.

9.3 SOFTWARE ANALYSIS

The software analysis phase usually begins with the release of the specifications for the required software and terminates with the successful completion of the preliminary design review. During this phase the functional performance requirements allocated to the computer program are also defined and the computer program performance specification completed, normally as a document separate from the overall system design.

The **program specification** for a computer program may be seen as the precise description of the requirements for the program. It includes a statement of the inputs to be supplied to the program, the outputs desired from the program, the algorithms involved in any computations and a description of such physical constraints as execution speed and memory limitations. The input–output relationship determined by a program may be thought of as a what specification, or what the program does, while the program itself is a how specification.

A **software analyst** is an individual who is responsible for analysing the requirements for a particular piece of software and creating a specification of the needs of a user. This must be completed before a system design proceeds.

In the analysis stage flowcharts are often used to assist in software documentation.

9.3.1   FLOWCHART
There are two types of flowcharts :
1. Within the software development procedure it is usual for a system flowchart to be drawn up by a software analyst, on the basis of the particular program requirements.
2. A program flowchart is a graphical representation of a software procedure or computer program that is used by programmers to illustrate the different steps in a procedure or program. By showing the logical sequence of the major steps in a program, a programmer can easily determine what the program is doing and see the different interactions between the parts of a problem.

In drawing a system or program flowchart, an analyst or programmer often uses a prescribed set of symbols, directional marks and other representations to indicate the stepped procedures of the computer operation. The symbols for drawing program flowcharts are shown in Figure 9.2.

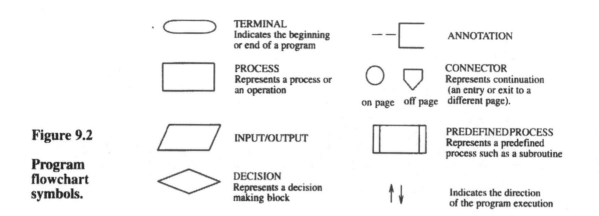

Figure 9.2

Program flowchart symbols.

9.4 SOFTWARE DESIGN

Software design is concerned with determining how the required specifications of the software can be achieved in the most effective way, and specifying, in detail all the processes involved. Furthermore, design determines precisely what is to be done and specifies the interfaces between the different parts of the software as well as between the target computer and the software. Software design differs from software analysis, in that design deals with solving the problems rather than stating the problems as in software analysis. The design process can be regarded as beginning when the software is initially decomposed into a number of different programming modules at the data flow diagram stage. The data flow charts will show the overall information flow for the proposed system and from that a structured design can be drawn which defines the top level structure and the functional relationship of the software structure diagram as devised from the data flow charts.

Table 9.2

Software characteristics.

1.	Correctness	The extent to which a program satisfies its specification and fulfils the customer's mission objectives.
2.	Reliability	The extent to which a program can be expected to perform its intended function with required precision.
3.	Efficiency	The amount of computing resources and code required by a program to perform its function.
4.	Integrity	The extent to which access to the software or data by unauthorized persons can be controlled.
5.	Usability	The effort required to learn, operate, prepare input and interpret output of a program.
6.	Maintainability	The effort required to locate and fix an error in a program.
7.	Flexibility	The effort required to modify an operational program.
8.	Testability	The effort required to test a program to ensure that it performs its function.
9.	Portability	The effort required to transfer a program from one hardware and/or software system environment to another.
10.	Reusability	The extent to which a program can be reused in other applications.

Software design is also concerned with the determination, from the functional specification, the overall structure of the program and the data as well as the determination of any algorithms required to implement the necessary functions.

The characteristics that are important for a piece of software are given in Table 9.2; they must be considered by the software designer in the development of the software.

Most software consists of many different small tasks, interlinked within the complete program structure. The tasks in turn usually consist of several small pieces of code which can be executed in sequence, i.e. short sequential routines. In this way a processing job can be decomposed or broken up into a number of small routines. There are also other intermediate levels which can be added to further categorize the functional decomposition of processing jobs.

Figure 9.3

Decomposition of a computing job.

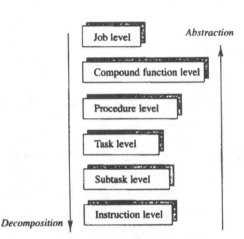

Software decomposition is a term used in computer programming to indicate the process by which a computing problem is broken down into smaller modules. Figure 9.3 indicates the different levels in software decomposition ranging from the all encompassing job level down to the fine detail of the individual instructions required in the programming of the job.

In terms of handling the programming of any large task the software engineer must then partition the problem into a number of different tasks or subtasks. When applied to a piece of software, **partitioning** is the identification and separation of the functions within the software into the different submodules or routines necessary to a program solution.

Software design

The term **granularity** is sometimes used to describe the number of operations in a processing task. It can vary between 'fine grain', with a small number of operations per task, and 'coarse grain', with a large number of operations per task. The level of granularity is highly dependent on the particular application and on factors such as maintaining load sharing between computer and I/O devices, the limited interconnection links, the cost of the hardware and the data flow patterns between tasks.

In partitioning a software job into its constituent parts the software designer/analyst often uses data flow diagrams.

Glossary 9.2

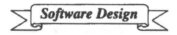

COHESION – Cohesion is a way of categorizing the amount of functional relatedness of processing elements within a single program module. It is a term used to describe the glue that holds a programming or software module together. It may also be seen as a measure of the strength of the internal structure of a software module and can be thought of as the type of association among the component elements of a software module.

There are different forms of cohesion. These are noted in Table 9.3 together with a brief description of their properties. Generally for elements within a programming module one wants the highest level of cohesion possible.

Type of Cohesion	Elements related by
High, strong and desirable	
Functional	Contribution to a single task.
Sequential	Performance of a sequence of tasks on the same data object.
Communicational	Performance of a collection of tasks on the same data object.
Procedural	Performance of a sequence of tasks selected by control flow.
Temporal	Performance of a collection of tasks related by being needed at the same time.
Logical	Performance of a collection of tasks of a similar nature.
CoincidentAl	Performance of an arbitrary collection of tasks.
Low and weak	

Table 9.3

Different forms of cohesion.

COUPLING – Coupling is a term used in software systems to describe the strength and complexity of the interconnection and interrelationship between one software module and another.

It is also often used as a measure of communication bandwidth between software modules. Task communication occurs when a producer task needs to pass information to a consumer task usually through the exchange of messages. The communication may be referred to as being one of the following.
1. **Tightly Coupled** in which the components require extensive synchronization to complete their execution.
2. **Closely Coupled** in which each time the producer sends a message, it waits for a response from the consumer.
3. **Loosely Coupled** in which the producer and consumer proceed at their own rates and a queue of messages builds up between the producer and consumer.

9.4.1   DATA FLOW DIAGRAM (DFD)

A data flow diagram shows graphically the movement of data. It is a directed graph, a type of flow chart, used to represent the flow of data streams through successive transform in a program. A data flow graph contains transform bubbles representing the function carried out by the system, arrows representing the data flows between transforms, and dat stores representing its data repositories. An example of a data flow diagram is given in Figur 9.4. It shows data being transferred from a data source to a data sink and on the way bein processed by three processes. Processes 1 and 3 make use of a data store to store and retriev values.

1. **Data Flow** – Data flows do not represent the flow of control through the system bu instead represent the flow of data. A data flow is a pipeline through which packets o information of known composition flow. In a DFD, a data flow is represented as a directec edge and is labeled with the name of the data flow. In Figure 9.4 there are seven data flows

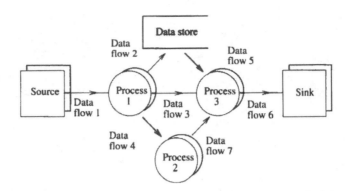

Figure 9.4

Example of a data flow diagram.

2.   Transform – A transform is a process that converts one or more incoming data flows into outgoing data flows. In a DFD, each transform represents a detailed action that is performed within the overall system and which contributes to the aims and objectives of the system. We can also regard such a process as transforming one or more inflows of data from other processes, files or sources into one or more outflows of data, to other processes, files or sinks. Each transform is labelled with a unique name and a reference number, and in drawing the actual DFD convention dictates that the transforms or processes are shown as circles.

3.   Data Store – A data store is a time delayed repository of information. Data stores can be thought of as manual or automated files, databases or any other accumulations of data. They also can be regarded as repositories of data, such as tapes and disks.

4.   Terminator – A terminator is a net originator or receiver of system data. Its purpose is to mark the boundary of a model, i.e. the edge of the area of interest. Terminators are also known as sources and sinks and can be viewed as entities which exist beyond the immediate scope of the system, but are originators or receivers of data.

Data flow diagram

Hence in a DFD, the nodes represent the program instructions, or processes, whose outputs pass along the links, or arcs, in the graph to subsequent processes. In Figure 9.4 the circles represent the processes, while the arrows indicate the direction of data flow. A node executes if all its inputs have been enabled with data. The main advantage of the diagram in the area of computer programming is that it encourages a clear representation of information processing tasks in a top down fashion allowing a software designer to move from a high conceptual level to lower ones in turn.

Data flow diagrams are part of the stepwise software refinement process expounded by researchers such as Jackson, Myers, Yourdon, Constantine and de Marco. They are useful in that they show how a program is made up of modules and the interfaces between the modules. As the data flow chart defines the transformation that the data undergos in the process of flowing through a software system it can be used to define structured diagrams and vice versa. Hence it is often used as a common interface between the user and analyst with the customer using it for recording changes in data, for documenting his business and for telling the analyst what he wants.

The graphic notation adopted within a DFD is useful for indicating the partitioning of a large software system into several smaller software systems and showing the information interfaces among the partitioned sections. The basic principles are similar to those encountered in Petri nets. An example of this decomposition is shown in Figure 9.5, where the diagram of Figure 9.4 has been expanded to show the data flows within each of the processes.

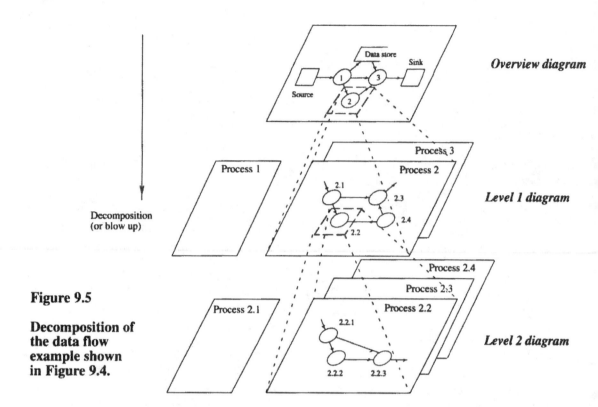

Figure 9.5

Decomposition of the data flow example shown in Figure 9.4.

The procedures within the processes shown in Figure 9.5 can be further decomposed until the only operations within the nodes consist of basic transforms such as additions and multiplications.

A **data dictionary** is a form of document holding a description of the terms used in a data flow graph. It typically stores data definitions representing data flows and data stores in terms of their components. In the data dictionary, data is kept principally about the following.

1. Each data flow item which appears on a data flow diagram. All data flows and data items must have an entry in the data dictionary. The entry is used to describe the decomposition of the data flows either in terms of other data flows, which appear in lower level data flow diagrams, or in terms of data items, which are individual packets of data generated and consumed by processes at either end of the data flow.
2. The structure of files and the data which appears within the data flow diagrams. This is known as the file specification.
3. How incoming data flows are transformed into outgoing data flows. This process specification notes the different processes in the data flow diagram and may be documented in various forms of notation.
4. The interfaces declared in the integrated set of data flow diagrams.

In software design, the bottom up method gathers groups of instruction sequences together into action clusters, initially starting at the atomic machine instruction level and working its way up to the complete solution. Bottom up design is not recommended for software design. Top down decomposition therefore implies that an initial program design is transformed into a series of smaller and smaller independent design problems which have their own interrelated specifications. It is simply a divide and conquer technique.

In using DFDs the software designer can partition a task into its different parts and indicate the flow of data between the parts. The level of decomposition usually stops at the **module** level. A software module is a segment of a program that accomplishes a single function. It is normally a contiguous sequence of program statements that has a minimal dependence on the other modules within a program. If possible the grouping within each module should be done so that the amount of interaction, or coupling, between the program modules is small, weak, and that each software module is dynamically well behaved. By using this approach to programming the final structured program will posses a natural hierarchy among its instructions. On viewing the program, the instructions can be seen to be repeatedly nested into larger and larger parts of the program by sequence, selection and iteration structures. The technique of **modular programming** emphasizes the organization and coding of logical program units, usually on the basis of function.

The early high level languages such as FORTRAN, ALGOL and COBOL do not provide powerful modularized facilities to ease the design of understandable, attractive and reusable software components. This has led to the development of newer high level languages that provide such facilities. Ada and Modula–2, for example, encourage the development and use of libraries of standard packages, collections of data and subprograms that act upon that data, and can allow the separate compilation mechanisms to facilitate easy reuse of programs. They are also able to efficiently handle parallel processing and to cope with a standardization of the interface between application programs and operating systems.

Table 9.4

Guide-lines in decomposing a software task into modules.

1.	Each module should represent one, logical self–contained task.
2.	Modules should be simple.
3.	Modules should be separately testable.
4.	Each module should be implemented as a single independent program function with information being passed via parameters and shared data.
5.	Each module should have a single entry point and a single exit point. Modules have control connections via their entry points and exit points.
6.	Each module should exit to a standard return point in the module from which it was executed.
7.	Modules should be able to be combined into larger modules without the knowledge of the internal construction of the modules.
8.	Modules should have well defined interfaces.

In deciding the different modules within a program the software designer can use several guidelines, these are shown in Table 9.4.

Often after a module has been written and tested it can be incorporated into a library. A **library** is a collection of standard and proven software routines, subroutines and programs usually stored on disk to provide a programmer with common functions and operations. The library routines may have been developed by the user or a software manufacturer. These routines can then be selected and inserted into a main program as required during the compilation process. Article 9.2 describes the difficulties in developing software for a real time system.

REAL TIME SOFTWARE DESIGN

The term real time can be used to describe a data processing mode of operation where the reception and processing of the input data and the return of the results usually occurs so quickly that it interacts instantaneously with a user. Real time may also be used to describe control systems where the results of the computations are used to influence the operation of a process. A real time control system must be able to disseminate data fast enough to respond within the time scale required by the process being controlled. The response time of such a system typically depends upon changes in the environment being controlled, the dynamics of the process and any time constants inherent in the system. The performance of a real time system is often characterized by its response time or its transfer rate.

The development of software for real time systems is far more exacting than software written for atemporal systems since real time computer systems have several features which make them difficult to deal with.

1.   Complexity of the hardware: Real time systems are typically large and complex, making coordination of the program task a problem in itself.

2.   Interaction with the environment: Real time computers systems are usually concerned with the monitoring and control of a complex system. This means the computer must interact with devices that sense the environment and must be able to interface with standard and non standard computer peripherals. A typical real time program will have an input and output process which is in a loop continually processing the input and output. The input and output values will be communicated to other processes in the system via communication channels.

3.   Time constraints: The correctness of the computing function within a real time system depends not only on the logical result of the computation but also on the time at which the results are produced. Thus the most important property of a real time system should be predictability ; that is its functional and timing behaviour should be deterministic as necessary to satisfy system specification. Real time computing must then meet the individual timing requirements of each task and be able to respond to external events within a guaranteed response time.

4.   Fault tolerance requirements: The critical feature of a real time embedded systems is its need for reliability. A real time computer is expected to run continuously in an automated fashion with an extremely high reliability, often in demanding environmental conditions. This is important from both a performance and a cost point of view. The requirement for high reliability is implemented with extensive error detection and recovery methods. Some computers are also designed with duplicate standby hardware components where the switch over to the standby component is performed when an error condition is detected. Other systems are designed with a fail–soft capability, where the control program performs a graceful degradation.

9.4.2   WALKTHROUGH

In software design it is important at all times to maintain clear lines of communication among all members of the design team and the customer. Hence periodic meetings of the software development team are important to review the progress of the project. Within these reviews walkthrough meetings will be arranged. A walkthrough is part of the software development process in which a programmer, or team of programmers, performs a step by step analysis of the software developed. Walkthroughs elicit and encourage peer review of the development process.

The walkthrough may also be seen as a process with these purposes.

1.   The review of a particular document such as a data flow diagram for correctness and completeness. Errors in DFDs are usually the result of a misunderstanding between a user and an analyst.

2.   By using the walkthrough process it may be possible to identify any logic errors and other difficulties that may have been overlooked during the design and/or coding of the program.

9.5 SOFTWARE TESTING

The final stage in the software development procedure involves testing the software on the target machine. Verification is the act of finding out whether or not a program does what it is supposed to do and nothing else. There are four approaches which can be used in program verification.

1. **Formal Proofs**: At its basic level any program can be regarded as a mechanism which converts an initial state of data into a final state. Since programs define logical functions, it is possible to abstract out all detail of execution, including even which language or which computer is used. It is then possible to discuss the correctness of a program with respect to its specification as a purely logical formulation. Formal proofs can subsequently be used in some circumstances to prove that the program is correct. It is worth noting that modern fourth generation software development tools include various forms of program generators, where a program is entered in a specification language and the code is generated automatically. These generators are often claimed to produce correct easily verified code.

2. **Exhaustive Testing**: Current methods of testing programs involve manual or automatic tests of program modules under as many conditions as possible, and comparing the outputs with those expected, according to the specifications of the modules. The problem is that no matter how exhaustive the tests are, they can never simulate all the possible permutations of data values and program states. Thus the status of almost every operational computer program is that it is conjectured to be correct until this is refuted by errors encountered in operations.

3. **Selective Testing**: Detecting errors by means of observing a restricted number of results from the operation of a program on a system, involves testing programs with selected samples of data which the program would be expected to be able to handle in normal operation. This process is designed to isolate any logic failure in the software by running it on the target computer system and monitoring its behaviour with test data. The test data would be of the type and volume that the software has been designed to process; it should also test the boundary conditions of use. The nature of this data would be predetermined and the results expected from the program calculated by the programmer or system analyst prior to running the program.

4. **Random Testing**: Random testing is a form of functional testing in which the input data are randomly chosen, then applied to the program and the results compared with what should be expected. The data may be chosen from the set of all possible inputs, or from the expected run time distribution of inputs. As well as the testing of sample data in programs it is also important to test all the conditions that the program would expected to meet.

Any failure of a program to produce the correct data or to respond to the different control conditions in the correct fashion requires that error **diagnosis** be carried out on the program. To perform this diagnosis the programmer may make use of a step by step analysis of the program to determine any logic errors.

The programmer may also be able to make use of a diagnostic program to provide a printed record of the operation of each of the program's instructions. On detecting an error within the logic of the program, the programmer will usually have to alter the source code with the use of some kind of editor facility. Any changes that have to be made to the system at this stage may well require that the whole sequence of analysis, design and coding be repeated.

The next stage before full scale production begins is the formal update of the software design **documentation**. The software patches developed during the verification tests are reintroduced into source code, and the erroneous software modules replaced by the correct ones. The rebuilt software is tested for a final time to ensure that no new errors have been introduced by the fixes that were made. Eventually all obvious failures are wrung out, and the system is deemed acceptable for release to the customer. Before release all documentation must be completed and finally updated. Documentation refers to written information on how the program works.

9.6 SOFTWARE MAINTENANCE

After a piece of software has been tested and proven to work correctly it is released to the customer. Thereafter we enter the maintenance stage of software development. Software maintenance covers two quite distinct activities.

1. The correction of errors that were missed at an earlier stage in the software development process but have been detected after the program has been in active service.
2. Modification and enhancements to the program to take account of additions or changes in a user's requirements.

9.7 KEY TERMS

Computer aided software engineering (CASE)	Real time software design
Cohesion	Software analyst
Coupling	Software decomposition
Data dictionary	Software design
Data flow diagram	Software development process
Diagnosis (software)	Software engineering
Documentation	Software maintenance
Flowchart	Software quality assurance
Granularity	Software testing
Incremental implementation	Stepwise refinement
Library	Structure chart
Modular programming	Structured design
Module	Structured programming
Partitioning	Software analysis
Program specification	Walkthrough

9.8 REVIEW QUESTIONS

1. What is the term given to the process of managing the environment to ensure that a program is correctly and efficiently produced.
2. Briefly describe what is meant by software engineering.
3. What was Edsger Dijkstra's contribution to software engineering?

4. What features does structured programming bring to the development of a piece of software?
5. Briefly describe each stage carried out by a programmer in the development of an application program.
6. What do the following terms mean when used in software engineering:
   (a) stepwise refinement,
   (b) incremental implementation?
7. What is system analysis and why is it desirable before computerization takes place?
8. Why is the systems analyst regarded as the interface between a customer and the design team?
9. Outline the nature and purpose of a program specification.

10. What is a flowchart? What does it illustrate?
11. Where are flowcharts used?
12. Explain the difference between a systems flowchart and a program flowchart.
13. What advantages do the use of CASE tools bring to the software development process?
14. What do the following terms mean in software design:
    (a) software decomposition,
    (b) partitioning,
    (c) granularity?
15. What do data flow diagrams illustrate?
16. Note the four constructs which can be used to draw a data flow diagram.
17. What do the terms 'cohesion' and 'coupling' mean when used in software design?
18. List some of the features of good data flow charts.
19. Why is a data dictionary important in the development of data flow diagrams?
20. What are programming modules?
21. What could be expected to happen at a walkthrough meeting?
22. List a number of reasons for using subroutines within a program.
23. What is the purpose of testing a program?
24. What is the most common way of testing a program?
25. In the testing phase of a complex software system why would exhaustive testing be inappropriate?

9.9 PROJECT QUESTIONS

1. Is it fair to say that the major objective of software engineering is to discover ways to make the creation of software easier and more productive? Explain your answer.
2. Define the nature and purpose of the following approaches to program design:
   (a) top down approach,
   (b) functional decomposition.
3. Why is a structured approach to software design similar to a form of hierarchical design?
4. What skills does a system analyst have?
5. What is the difference between an applications programmer and a systems programmer?

6. Distinguish between the roles of a programmer and a systems analyst.
7. What does the term 'compatible' refer to when using an application piece of software on a computer system?
8. Investigate why flowcharts are not regarded as being acceptable for structured design.

9. Contrast the two methods of detailing a program by either using data flow diagrams or flowcharts.
10. Software modules should be data transforms. Explain.
11. At the coding phase what dictates the choice of programming language?
12. Is a useful estimate of programmer productivity to count the number of lines coded per day?
13. Describe the function of a structure chart.
14. What skills are needed by a programmer?
15. Detail the use of algorithms within a computing environment.
16. What is pseudocode?
17. What are software bugs?
18. What does the term 'debugging' refer to in program development?
19. Explain the nature of a logical error in a program.
20. What types of procedures are undertaken at the testing stage.

9.10 FURTHER READING

Software engineering is a vast topic to cover in such a short space. It is hoped that the chapter has introduced a number of the terms and techniques used by software engineers. There are many books and journals devoted to expounding the practise of software engineering. A few are listed below.

Software Engineering (Books)

1. *An Introduction to Software Engineering*, G. Jones, Wiley, 1990.
2. *The Practice of Structured Analysis*, R. Keller, Yourdon Press, 1983.
3. *Systems Analysis and Development*, P. J. Layzell and P. Loucopoulos, Chartwell Bratt, 1986.
4. *Concise Notes on software Engineering*, T. De Marco, Yourdon Press, 1979.

Software Design (Articles)

1. Software design, R. W. Jensen and R.L. Van Tilburg, *IEEE Computer*, February 1981, pp. 61–83 .
2. A software design method for real–time systems, H. Gomma, *Comms of the ACM*, September 1984, pp. 938–949 .
3. Software quality assurance and CAD user interfaces, E. Grinthal, *IEEE Design and Test*, October 1986, pp. 39–47.
4. Structured programming, R. W. Jensen, *IEEE Computer*, March 1981, pp. 31–48
5. Structured programming retrospect and prospect, H. D. Mills, *IEEE Software*, November 1986, pp. 58–66.
6. No silver bullet essence and accidents of software engineering, F. P. Brooks, *IEEE Computer*, April 1987, pp. 10–19 .

Operating Systems

Part Four

The Operating System (Part 1)

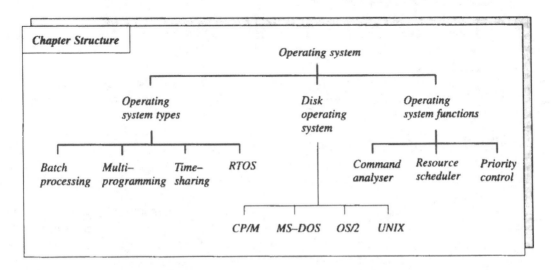

Chapter Structure

Operating system

Operating system types — Disk operating system — Operating system functions

Batch processing · Multi-programming · Time-sharing · RTOS

CP/M · MS–DOS · OS/2 · UNIX

Command analyser · Resource scheduler · Priority control

Learning Objectives

In this chapter you will learn about:

1. The purpose of operating system software.
2. The characteristics of batch and multi-programming operating systems.
3. The use of time sharing.
4. The different disk operating systems for workstations and small personal computers.
5. The job management functions of an operating system

'A program is a structure to planned events – not just a series of things to take place, but an interweaving of many different things that either will or may take place, depending on the data or the time of day or the choice of a user sitting at a keyboard or whatever.'

Ted Nelson

Chapter 10

The Operating System (Part 1)

In order to control the hardware and software resources, a computer usually comes with some type of system software, the major part of this being the operating system. The **operating system** is a collection of routines which control both the hardware resources of the system and the flow of information into and out of the computer. It also allows the application programs to handle the various input/output (I/O) devices to perform their specific tasks.

The primary objective of the operating system handling the system functions in computers is to remove direct control of the system resources, both peripherals and memory, away from the user's program. In this way the chore of writing software drivers for the hardware in the system is avoided and the standard ones incorporated in the operating system can be shared by multiple users. With these primary goals in mind, additional features such as multitasking, multiuser handling, fault tolerance, and data control may be added to the operating system, thereby increasing the power of the overall computer.

Apart from the application programs, a computer system will then typically contain the operating system kernel, system processes and utility programs. It can be seen from Figure 10.1 that a hierarchy exists within the system with the central core of the operating system sometimes called the kernel, nucleus, supervisor, executive or monitor, as the root of the tree. It is to be noted that in the past operating systems were written entirely in assembly language whereas modern practise now dictates that they should be written in a high level language since this tends to make them easier to write, understand, debug, and also move to other machines.

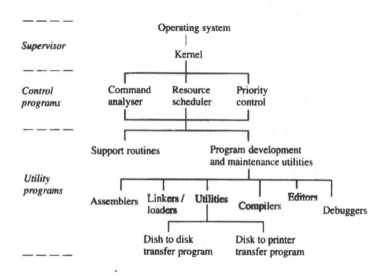

Figure 10.1

Software structure within a typical computer system.

Developments in software engineering have also resulted in operating systems that are more maintainable, reliable and understandable. Most of the modern operating systems are now written in C, a powerful high level language that is well suited to operating system implementation; however, for certain time critical parts of the operating system some of the code may still have to be written in assembly language, generally less than 10% of the code.

Operating system

In many operating systems functions like interrupt handling, task switching, I/O services and memory management may be provided in microcode.

10.1 OPERATING SYSTEM TYPES

The evolution of the computer has brought about an increased sophistication in operating systems. Table 10.1 notes the different types of operating systems that have been developed.

Table 10.1

Types of operating systems.

Operating system	Description
Batch	Single task system.
Multiprogrammed	Allows multiple tasks to use the CPU and the I/O devices.
Time shared	Allows multiple tasks and multiple users to share the computer resources. Each task is given a short amount of time on the CPU.
Multiprocessor	Controls the scheduling of task onto multiple CPUs.

10.1.1   BATCH PROCESSING

The operating systems used in the early first generation computers of the 1950s were in fact the human operators who were involved in the maintenance and particularly the supervision of the computer system. Typically these operators were also the programmers and it was then their job not only to write the programs for these early machines (in machine code) but also to manually enter this code into the computer and then monitor the results as the program ran.

With the wide variety of computers available in the late 1950s the input/output requirements became more varied and I/O devices more sophisticated. This led to a growing desire among computer users that processing systems should provide more and more functions and that these functions were in some way automatic. Hence, simple operating systems called job–to–job monitors were devised to control the I/O devices and manage several user programs stored in memory, thereby alleviating some of the work of the operator. These early operating systems worked on a batch basis.

Batch processing is a method of processing within a computer system in which the programs or data items to be fed into the computer are collected and coded prior to processing, and then forwarded to the computer in a group called a batch.

Figure 10.2 shows the basic batch operation used in the computers of the early 1960s. During batch processing, the computer system would read in one job, pass control to it and then, when it was completed, read in the next job, and so on. With this type of system one person, the computer operator, controlled the collection and loading of all jobs.

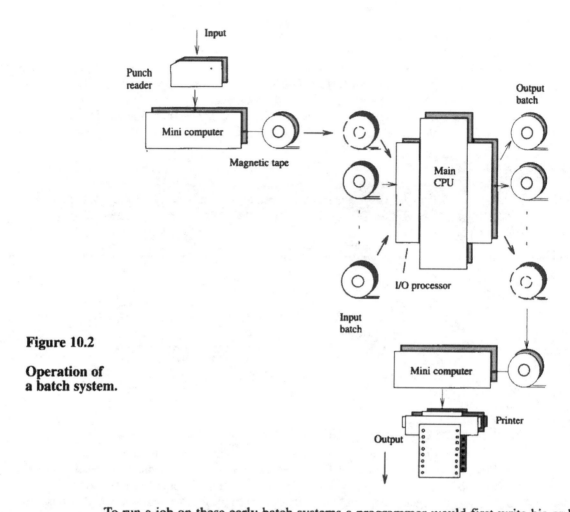

Figure 10.2

Operation of a batch system.

To run a job on these early batch systems a programmer would first write his or her program on paper in either a high level language or assembly language then punch it onto a series of cards. He would then bring the card deck down to the input room and hand it to an operator. The operator would transfer the job along with a batch of other programs onto a magnetic tape or disk, usually making use of a second small computer. The tape or disk would then be transferred over to the mainframe which would run the jobs in sequence, transferring the results for each job onto another magnetic tape or disk which would be passed to a second small computer for printout.

Batch processing

Batch operating systems were directed by control statements submitted at the beginning of each job, in the form of one or more punched cards or program instructions. These statements were part of a control language called a **job control language** (JCL). Typically the user had to specify:

1. the language translator needed, e.g. FORTRAN
2. the amount of memory space needed
3. time requirements
4. the data to be used
5. the I/O devices required
6. job accounting information.

One of the major features of the early batch operating systems was their ability to control the different I/O devices in the computer system. However with the initial batch processing systems the central processing unit (CPU) was often idle for large periods of time, while the I/O transfer was taking place. An improvement was later added to these systems by overlapping the system input, job processing and system output. This type of overlap was called spooling (simultaneous peripheral operation on line).

In order for spooling to function within a processing system it required the addition to the system of I/O **channels**. These channels were specialized hardware devices functionally similar to small computers which were designed to move input from I/O devices to main storage and output from main storage to I/O devices, independent of CPU intervention. A CPU instruction was usually required to initiate a channel operation, but once started the slow transmission and buffer operations themselves were done by the channel. This freed the CPU from the tedious business of I/O transfer, allowing it to carry out more useful operations.

To make the most use of spooling techniques the operating system buffered the system input and output through an intermediate level of direct access storage, such as magnetic tape or disk. In this way the operating system was able to interleave the output of several jobs on (for example) a printer, such that when a job tried to write to the line printer, a spooling program intercepted the request and converted it to a write to secondary memory. This had the effect of converting the physical line printer into many logical line printers through the use of the secondary memory. The spooling technique was also applied to any slow input device. By disassociating a running program from the slow operation of devices like printers and instead directing the I/O to fast devices like disks, the operating system attempted to keep the card reader(s) and printer(s) operating concurrently, by applying them to different jobs at the same time.

It was possible that during certain times, the job output could accumulate faster on the direct storage than it could be printed. If this condition prevailed for too long a period, the processing that was causing the accumulation had to be suspended. Spooling, however, had a major effect on improving the turnaround time of a job, i.e. the time from when the job was submitted to be processed to the final result being output. Spooling was in fact a primitive form of a technique called multiprogramming.

The executive control routines within the operating system would then read the statements and check that the required resources were available; if they were not, an error was signalled and the program was not executed. Early forms of batch systems used to call the operating system routines from auxiliary memory; however, as programs came to depend increasingly on system routines, it became more practical to retain them in the central main memory. Such operating systems were described as resident.

In a batch system the operator's tasks were the following.

Batch processing

1. Starting the system through the initial preloading procedure.
2. Feeding input jobs in the form of punched card decks or paper tapes into the computer system and routing the final output, such as printed listings, to particular user stations.
3. Starting and stopping certain system programs such as the reader or interpreter, which read jobs into the system from a card reader or tape drive, and the initiator terminator, which selected jobs for processing by moving their programs to main storage.
4. Mounting magnetic tape reels or disk packs as they were required during job processing, as directed by console messages.
5. Attempting recovery procedures in case of system failure.

Batch operating systems nowadays are only found on small computers with few input devices and no time sharing.

10.1.2   MULTIPROGRAMMING

The first generation computers, up until about 1963, used batch operating system. However, during the early 1960s the requirements for a more efficient operating system wer identified.

1. To relieve the programmers from certain detailed and tedious tasks, such as the direc naming and programming of the physical I/O devices and the mechanics of storage an device allocation.
2. To reduce the efforts required by the human system operators in job setup and termination. Batch operating system required a large amount of operator interventior
3. To schedule jobs in a manner that would provide good user service according to som criterion, for example highest priority jobs would be scheduled first.
4. To utilize hardware resources as fully as possible, such as the central processing un (CPU), the main storage, the I/O channels and the I/O devices.
5. To permit multiple users, or jobs, to currently share the hardware and software systen resources. This was the major innovation compared to batch systems which could onl' process one computer job at a time.

In an attempt to satisfy these requirements and increase the efficiency of compute operation, multiprogrammed operating systems were introduced. Multiprogramming refer to a method of running two or more applications programs on the same computer suc that the operating system allows these application programs to share the CPU, the mai memory and the system I/O devices.

The use of the main, or primary, memory for storing the jobs to run on the system als means that these jobs can be rapidly switched into and out of the CPU. This is much quicke than a batch system can do, due to the fact that the batch operating system typically use secondary memory for storing and retrieving the tasks. The main reason for the greate use of the central memory in multiprogrammed systems compared to batch systems wa due to the decreasing cost and increasing size of random access memory in the newe computers of the mid–1960s.

A diagram of the operation of a multiprogrammed operating system is shown in Figur 10.3 where the input jobs are initially held in a job queue and returned to the queue whei they require for example I/O operations.

Figure 10.3

Operation of a multiprogrammed operating system.

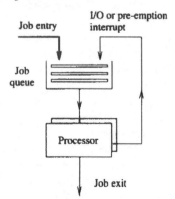

Once the jobs in a multiprogrammed system have been assigned a priority each job car be allocated memory and I/O devices accordingly. By sharing memory and I/O device among different programs, the computer can then run several applications program

Glossary 10.1

Multiprogramming

BACKGROUND PROCESS – In a computer system, a background process is a low priority process that is executed when high priority processes do not need the central processing unit.

BOOTSTRAP LOADER – Bootstrapping refers to a technique to bring a system into a desired state by means of its own action. Within computer systems a bootstrap loader program is a short program which is initialized when the computer is started and is used to get the processor to load a larger program, usually the operating system, from disk into the main memory ready for use. Typically the bootstrap program mechanism consists of firmware holding a small program, which in turn reads the core of the operating system from either a floppy disk or from a central hard disk into the main memory. Most of the operating system code must be kept on disk because the routines are too big to be stored in a read only memory. In addition the processor, memory and peripherals are initialized to the extent necessary to accept the first external inputs. Finally, the operating system creates a special log–in process connected to each terminal of the system to allow an operator to direct the processor. The instructions within the initialization routines may be regarded as forming part of the operating system.

JOB – A job is a set of programs, files, data and instructions to a computer, that collectively constitutes a unit of work to be done by the machine.

JOB CONTROL BLOCK (JCB) – During the operation of a multiprogrammed computer, as each processing job is read in and filed onto disk, a job control block is created by the high level scheduler within the operating system of the computer. The JCB contains all the important information about the job such as its resource requirements, its priority, the amount of memory it requires, the time the job needs and any data files requested.

JOB CONTROL LANGUAGE (JCL) – In the early 1960s the IBM/360 mainframe computers were the first commercial computers to have acceptable multiprogrammed operating systems. With these machines the user could inform the operating system of his required operating parameters by utilizing a special set of commands called job control statements which prefaced each program. The statements indicated to the computer such things as the I/O resources needed and the compiler required. Together, these job control statements constituted the job control language of the machine. A JCL was then a type of language which was used to indicate to the computer various features of the job to be processed. The statements in the job control language, simply called JCL statements, could be seen to identify a job and to specify its requirements for system resources. This helped the multiprogrammed operating system decide whether it could accept a program into the job queue, dependent for example on whether the I/O peripherals the program required were available, and how to allocate time on the central processing unit (CPU) to each program.

MUTUAL EXCLUSION – In a large complex processing system where multiple tasks are in operation, tasks desiring exclusive access to a physical resource such as a disk drive, or a virtual resource such as a system table, must generally compete for it. Since it must be guaranteed that only one task has access to any resource at any one time, exclusive access to the resource must be assured. Hence one of the jobs of the operating system routines is to ensure that nonsharable resources are accessed by only one task at a time. This exclusive access to the resources is called mutual exclusion between the tasks.

VIRTUAL RESOURCE – In multiprogrammed or time shared computer systems there may be a large number of programs and users trying to access the various hardware and software resources of the computer simultaneously. The operating system hides this fact by providing a virtual picture of the computer to each user of the system in which each user or program has the impression that they have access to all the physical resources of the computer. Hence what appears to the user to be a simple object may in fact have to be simulated by a complicated underlying mechanism, controlled by the operating system. By implementing useful virtual objects, an operating system can save the programmer from a considerable amount of programming.

As an example, a file may be considered a virtual object since it appears to the programmer simply as a set of records, which he or she may update or read on command from their program. However, the operating system must format the records and store them in various media, it must set up buffers, and provide subroutines to control the transmission of records between the main memory and the secondary memory, and it must protect files from unauthorized access or accidental damage. Also, by hiding the internal details of the object, the operating system can protect the computer from misuse, and so present a more reliable system for all users.

Other examples of virtual objects are the following.
1. **Processes** – processes are short programs in operation.
2. **Virtual Memory** – virtual memory refers to the set of addresses which can be generated by a process that can exceed the available main memory.
3. **Semaphores** – semaphores are channels through which processes can send signals.
4. **Message Queues** – message queues are channels through which processes can exchange messages.

With this perspective, an operating system can be viewed as a piece of software that turns the real computer, including all resources into one or more virtual machines composed of virtual resources.

concurrently, thereby maximizing the utilization and operation of all the hardware elemen
in the system. This minimizes the cost of computing, by distributing the cost of hardwar
use across a large set of concurrent users. This type of operating system therefore provide:
the mechanisms to control the patterns by which a population of active programs can gai
use of the central processor in the system.

The efficiency of a multiprogrammed operating system depends on maintaining a larg
population of available jobs, of heterogeneous resource needs, within the job queue. Th
heterogeneity is important since, in order to run the hardware well, it is necessary to hav
a mix of programs that are I/O bound and programs that are CPU bound, to keep all resourc
Multi-   in use. It can be noted that when such a multiprogramming mix is formed, a particula
programming program A, for example, can be arranged to be using the CPU, when program B is using I/
channel 1 and program C is using I/O channel 2 etc. If A, B and C never wish to use the san
resources at the same time then no contention for the resources will develop and eac
program can run as if it were alone on the machine. The net effect of achieving a goo
multiprogramming mix is to enable the operating system to fill in the gaps in the utilizatio
of the equipment during times when running a single program would not be using th
equipment. In reality such a mix is very difficult to sustain and contention for resource
inevitably occurs. Glossary 10.1 describes several terms relating to multiprogramming.

10.1.3   TIME SHARING

As multiprogrammed operating systems were being developed in the late 1960s, it wa
also realized that rather than allow application programs to have access to the CPU until the
required I/O, as in the initial multiprogrammed systems, a timed interrupt could be used t
force the computer to switch to another job. A job could then be allowed to execute for
certain period of time, called a **time slice**, after which it would be interrupted by a timer. Th
operating system would subsequently select another job which could run until stopped b
I/O or the timed interrupt, and so forth. Thus, a long job with little I/O would be forced t
execute concurrently with other jobs, and since job execution was no longer strictl
sequential, a short job could have good turnaround time – even if it started after a long job
The technique of relying on a timed interrupt to swap jobs in a computer is called tim
sharing.

Users of time shared operating systems are provided with distributed terminals linke
by transmission lines to the central processor. These terminals are buffered so that any inp
entered by the user can be held temporarily until the CPU is ready to receive it. Within th
CPU the time shared operating system utilizes a cyclical search pattern, effectively scannin
around all the terminals connected to the CPU allocating processing time in short bursts t
the different users in turn. Each of the users in the system can then construct, debug and us
programs interactively with delay times that are comfortable for direct man to machin
interaction. Provided that the number of users is not excessively high, the speed of th
rotational search and the resultant processing are so fast that a virtually immediate respons
is experienced by any user having data to input through his or her terminal.

Time shared systems very often profit by maximum homogeneity of use. This impli
that a set of terminal users who are all performing the same task will considerably improv
the responsiveness of the system, because sharing of the code can be maximized and th
resource management burden minimized. Movement of system functions into and out c
memory will also be minimized and very stable operating environments can be achieved

The use of time sharing gives the illusion that a number of users are using the CPU simultaneously. As a result, each user sees what is called a virtual processor, giving the impression that he or she has sole access to the complete machine. In this way, by multiplexing the use of every real resource within the computer, the operating system turns individual real resources into a number of virtual resources. In order to provide this view of the computer, the operating system must hide the details of the management of the CPU and the other system resources from the user. To the user these virtual resources exhibit a behaviour much like that of the real resources but they do not need to worry about the complex control mechanisms that may be required to operate the resource, since these mechanisms are handled by the operating system.

Time sharing is probably the commonest form of operating system for large modern computers. In such systems there may be a large number of concurrent tasks, the majority of which are executing interactively in communication with user terminals, while other background batch programs may also be executed in the short idle intervals.

Time sharing systems typically have a very high volume direct access backing store to hold the large amount of user programs and files. They also require a very high speed data access and transfer mechanism since, when a terminal comes on line to request processing from the CPU, the program requested by the new user has to be read from the secondary memory into the main memory and after processing it has to be transferred back to the direct access store and the next program demanded called in. A sophisticated set of hardware and software mechanisms is therefore required to implement time sharing techniques. These mechanisms control the search for and the transfer of data to/from the terminals and also organize access to the wide range of programs and files held in secondary memory.

10.1.4   REAL TIME OPERATING SYSTEM

A real time operating system (RTOS) is normally used to handle real time applications where jobs with higher priorities are executed before jobs with lower priorities. It can also be designed to handle batch and real time tasks concurrently, e.g. a typical mainframe computer might process a stream of batch programs fed into the computer from a disk or terminal while at the same time being able to respond promptly to a request from a number of user terminals.

The RTOS for this type of mainframe is basically a two–mode system where the foreground mode operates as an on–line program to control the data entry stations, while the background mode is batch orientated, compilation, file update, sorting and so on. Finally, an RTOS can be programmed to schedule jobs at some preset time, for example the preparation of an inventory report at the end of each working day.

10.2 DISK OPERATING SYSTEM (DOS)

The operating systems used in small micro and minicomputer systems are usually classified according to the nature of the storage devices that hold them. A tape operating system stores the main part of the program on tape, while a disk operating system (DOS) stores the bulk of its commands on a magnetic disk. All modern computers use disk operating systems. There are a number of popular disk operating systems including CP/M, MS–DOS, OS/2 and UNIX.

10.2.1   CP/M OPERATING SYSTEM

CP/M stands for Control Program for Microcomputers, or control program/monitor, and is a popular single user microcomputer operating system developed by J. Torode and G. Kildall in 1974. It has become a de–facto standard primarily for 8–bit microprocessor based computers. The structure of CP/M, shown in Figure 10.4, is typical of a disk operating system in that it is made up from three parts the BDOS, BIOS and CCP.

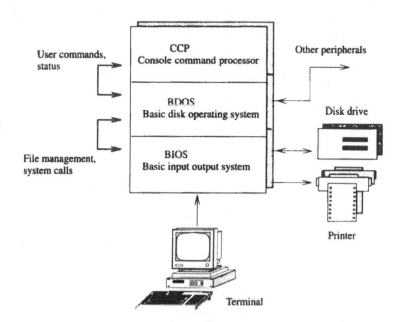

Figure 10.4

The structure of the CP/M operating system.

1. Basic Disk Operating System (BDOS) – The basic disk operating system allows the user to see the floppy or hard disk as a number of files of varying length identified by symbolic names. It also provides a hardware independent I/O interface to a console device such as a visual display unit (VDU).

2. Basic Input/Output System (BIOS) – The basic input/output system is a hardware dependent mechanism whose main task is mapping the hardware independent interface of the BDOS to the actual hardware of the system. The BIOS will therefore be different for each computer. The BIOS handles the systems calls from the BDOS.

3. Console Command Processor (CCP) – The console command processor is responsible for the user interface. It operates in an endless loop reading a command from the keyboard obeying it then returning to the start to read another command. In operation it is typical for the CCP to call the BDOS which subsequently calls the BIOS to access a file on disk. Most command programs are stored on files and CCP obeys a command by reading the appropriate file into a memory transient area and executing the program loaded.

CP/M is a sequential operating system ie. there is no overlapping of input and output and once a program requests an I/O transfer it loses control of the processor until the transfer is complete. CP/M is primarily for 8–bit microprocessors; however, in 1980 Digital Research announced the CP/M–86 operating system, intended for 16–bit micros.

10.2.2   MS–DOS OPERATING SYSTEM

The software company Microsoft was formed in 1975 by Bill Gates and Paul Allen and produced the first microcomputer BASIC interpreter. In 1980 Microsoft developed the operating system PC–DOS for the IBM personal computer (PC). The wide acceptance of the IBM–PC has made the Microsoft Disk Operating System (MS–DOS) which is an upgraded version of PC–DOS, the most popular disk operating system written for the IBM PC and other computers using the Intel 8088 microprocessor and its successors.

Table 10.2 illustrates the historical development of MS–DOS, noting the additional features added to the new versions compared to previous versions. Windows support for MS–DOS is covered in Article 10.2.

Table 10.2

Features of the different versions of MS–DOS.

Date	Version	Features
August 1981	MS–DOS 1	The original operating system was based on approximately 4000 program lines. It contained a few utility programs, several elementary functions, a text editor and a disk formatting program.
May 1982	MS–DOS 1.1	This version supported double sided floppy disks.
March 1983	MS–DOS 2.0	This version supported hard disk drives and a hierarchical disk structure with subdirectories.
August 1984	MS–DOS 3.0	This version supported IBM AT diskette format for 1.2 MB floppy disk drives and 20 MB hard disk drives.
March 1985	MS–DOS 3.1	This version introduced network compatibility.
December 1983	MS–DOS 3.2	This version introduced support for 3.5 in/720 KB floppy disk drives.
April 1987	MS–DOS 3.3	This version introduced support for 3.5 in/1.44 MB floppy disk drives.
November 1988	MS–DOS 4.0	This version provided support for larger hard disk drive partitions than previous versions. It also supported expanded memory and a DOS shell.
June 1991	MS–DOS 5	This version moved the DOS drivers into expanded memory, provided a DOS shell with limited multitasking, provided support for 3.5 in/2.88 MB floppy disk and came with a full screen editor.

10.2.3   OS/2 OPERATING SYSTEM

In 1987 IBM announced the OS/2 operating system, targeted at its range of personal computers containing the Intel 80286 and 80386 microprocessors. It has a number of different features, these are shown in Table 10.3.

Table 10.3

Features of the OS/2 operating system.

	Feature	Description
1.	16 MB address space	This is the maximum allowable space on the 80286.
2.	Virtual memory	Use of virtual memory requires a large amount of hard disk space for the swap files.
3.	Multitasking	It supports full multitasking including background and foreground processing and sophisticated interprocess communication features that allow several tasks to be integrated.
4.	Ease of use	When the OS/2 system is initially booted a full screen user interface is normally loaded. This uses a menu based command selection.
5.	Presentation manager	OS/2 supports a single set of functions for windows, graphics and other features that govern what the user sees at both the system and application level. The windows can be moved, scrolled and clipped.
6.	Graphics	OS/2 controls a powerful set of graphic utilities which allow rapid drawing and text manipulation.

WINDOWS (Microsoft)

The user interface to the operating system of a personal computer has changed over the years. It is possible to identify four types of interfaces dependent on how the user enters his or her commands.

1. **Command–orientated interface** – This prompts the user to type in a single letter word or line that is translated into an instruction by the operating system.
2. **Menu–driven interface** – This offers a list of choices or options called a menu as a way of letting a user choose from several different commands or functions.
3. **Natural language interface** – This allows a user to input simple English statements that are translated into commands by the operating system.
4. **Graphical user interface (GUI)** – This uses pictures and graphic symbols to represent commands, choices or actions.

Of the four, the graphical user interface has become the most popular for PCs. Since its appearance in 1985, Microsoft's Windows has rapidly become the operating environment for the IBM compatible computer. Windows is neither an operating system nor an operating environment, it is both – termed an operating extension. It is partly an operating system in that it facilitates features like multitasking but, nevertheless, requires a disk operating system either MS–DOS or DR–DOS to be loaded on the computer first. In 1990 Microsoft released Windows 3.0.

Windows has an easy to use graphical user interface, multitasking facilities and access to large amounts of memory, integrated applications, networking, extensive hardware support, device independence and DOS compatibility. The most obvious feature of Windows is the GUI. It provides a standard approach to common operations such as starting an application, copying and moving files, in this way the user does not have to remember all the DOS commands for these tasks. It makes use of icons, menus and windows.

An icon represent an application, directory, file or other object. A list of selectable functions that are available is called a menu. A menu drops down from a menu bar when a function is selected is called a drop down or pull down menu. One that appears when an item is selected is called a pop up menu. A window is an area on a screen used to display icons, menus or to run an application. A window consists of a title bar that represents the name of the window, a menu bar that lists several option and scrolls bars that allow you to move matter up and down or left or right inside the window. Most windows have buttons in one or more corner for re–sizing the window, moving it from place to place on the screen and for closing it when done.

With Windows 3 multitasking can be cooperative, for Windows applications, and pre–emptive for DOS applications. When a Windows application is running it is periodically given control of the processor and it is the responsibility of the application developer to relinquish control of the processor to another application. Hence the term cooperative multitasking (although it is possible for badly written programs to monopolize the CPU). For DOS applications the Windows kernel is responsible for allocating processor time slices to each DOS session; this method is known as pre–emptive multitasking, more commonly associated with real–time programming.

Windows also has two basic operating modes, standard and enhanced, each having advantages and disadvantages.

Standard mode: This mode uses the protective mode of the Intel 80286 and 80386 processors, allowing access to as much as 15 MB of extended memory above the 1MB limit. Standard mode does not multitask DOS applications but uses a task swapper to swap between applications. It is normally the fastest operating mode in Windows 3.

Enhanced mode: This mode is selected automatically for 386 or 486 processors with more than 2 MB of memory. It also uses protected mode and allows the hard disk to be used as virtual memory, up to 64 MB. In other words, if the user runs out of physical memory, Windows 3 can use the hard disk as additional RAM by creating temporary swap files. The 386 enhanced mode is also capable of running multiple DOS sessions, each in its own window. It does this by using virtual 8086 mode of the 386 processor. In virtual mode each DOS session believes it has complete access to the machine, and has access to extended to the machine, and has access to extended and expanded memory, on top of the normal 640 KB of base memory.

In 1987 Microsoft announced Excel for the PC, the first real application program for Windows.

Although the OS/2 operating system is also compatible with the PC–DOS operating system, DOS was created as a single tasking operating system limited to 640 KB of memory, it does make use of special features within the 80286 and 80386 and so will not run on 8088 and 8086 based machines.

10.2.4   UNIX OPERATING SYSTEM

Perhaps the most influential current operating system used in workstations, minicomputers and mainframes is UNIX (uniplexed information and computing service). It is a popular multiuser operating system for 32–bit computers designed for multiuser environments. UNIX is a reengineering of the MULTICS (multiplexed information and computing service) operating system designed during the period 1965 to 1969 at Bell Laboratories in the USA. MULTICS was originally intended to run on the large powerful GE 645 mainframe computer, however in 1969 Bell Labs withdrew from the MULTICS effort.

A group of software engineers led by Ken Thompson subsequently went on to develop the UNIX system for minicomputers. Although an order of magnitude smaller than MULTICS, UNIX retains most of its predecessors useful characteristics, such as process creation and termination, hierarchical file system, device independence, I/O redirection, and a high level language shell. UNIX is mostly used on computers based on the Motorola 68000. However, as the major part of UNIX is written in the high level language C, it can be easily transported to a wide variety of processors.

UNIX was designed from the outset to provide an environment which is particularly suitable for software development, since it comes with a large library of utility programs such as a text editors, programmable command language interpreters, compilers for high level languages, assemblers, debuggers, document formatters and text processing programs. UNIX also dispensed with virtual memory and the detailed protection system that such a memory required and instead introduced the pipe mechanism. The pipe is a two–way communication system where messages between processes may be inserted into one end of the pipe and extracted out the other end.

Outside the kernel of UNIX is the shell which provides the user interface. User interaction is in the form of a dialogue with the shell. The shell uses about 200 standard system commands and additional shell commands to form special interfaces. Furthermore, commonly used sequences of shell commands may be stored in a file and invoked by using the file name.

Modifications to UNIX are also possible including networking, database control, distributed processing, security provisions, user friendliness and on–line transactional processing, thereby turning the operating system into more than simply a host to application program.

Although not designed for multiple processing it has been possible to modify the kernel of the operating system to include the following.

1. Mutual exclusion in order to prevent different processes from accessing the same data structure.
2. The distribution of interrupts to all processors in the system.
3. Multiple process scheduling.
4. A virtual memory environment for all the processors.

10.3 OPERATING SYSTEM FUNCTIONS

The basic goals for an operating system are to provide a programmer interface to the machine; provide mechanisms to implement programs; and to facilitate the matching of applications to the machine.

Figure 10.5 shows the relationship between the operating system, the system hardware, the application programs and the user. The various functions of a typical operating system are also noted.

The **kernel** of an operating system is the central part of the operating system. It is also known as the nucleus, monitor or executive, and is the part of the operating system which provides a series of basic utility routines, to observe, supervise, control and verify the operations of the computer system. Since the kernel is the most commonly used piece of software in any processing system it is typically held in primary storage, i.e. it is the memory resident portion of the operating system.

The functions incorporated into the kernel are shown in Figure 10.5. There are three system processes directly under the control of the kernel, these are a command analyser, a high level resource scheduler and a priority controller.

Figure 10.5

The interaction between the operating system and the hardware and software components of a computer.

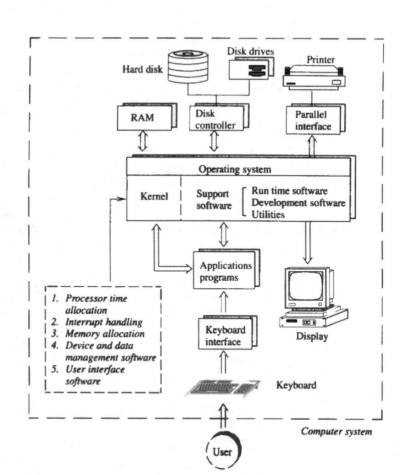

10.3.1   COMMAND ANALYSER

The command analyser is a program that monitors the keyboard and interprets any instructions entered by the operator for access to particular system programs. It then operates on these instructions.

10.3.2   RESOURCE SCHEDULER

In general the resource scheduler must look after the system's resources, where a resource in a computer system may be looked upon as something needed to complete a task. These resources are of two types: hardware such as physical objects, processors, peripherals, memory locations and software such as the utility programs, compilers/interpreters, assemblers, loader programs and data file manipulation programs.

The resource scheduling within an operating system can be regarded as being handled by a series of major system functions. These functions in turn are handled by a number of routines which may be grouped together, sometimes referred to as managers, that essentially control the different tasks that the system must deal with. Five separate managers are usually found within a typically multiprogrammed operating system. These are shown in Figure 10.6:

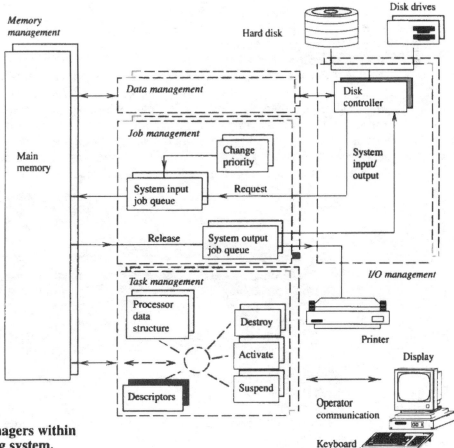

Figure 10.6

The different managers within a typical operating system.

1. **The Job Manager**: This is also known as the high level scheduler. The routines in the job manager are responsible for controlling which jobs are read in and written out to disk.

2. **The Task Manager**: This is also known as the low level scheduler. It is responsible for deciding which tasks or jobs are actually run on the CPU.

3. **The Data Manager**: The data manager controls the movement of data between the disk units and the main memory.

4. **The Memory Manager**: As well as managing the control of the CPU by the different tasks in the computer, the operating system must also control the use of the computer's memory. Within an operating system it is the memory management routines, often simply called the memory manager, that are responsible for assigning memory to specific tasks.

5. **The I/O Manager**: The I/O manager coordinates the communication with the external devices, eg. the disk drives and the printers, and must handle error detection and handling and interrupt handling.

Operating system functions

The resource managers interact with one another in order to handle the complete operation of any computer system. This interaction is handled by the use a number of lists or tables associated with each resource manager and it is via these lists that the resource managers request servicing from one another and exchange data. Resource management then involves the monitoring and control of the various system resources required by application programs as they are being executed.

10.3.3   PRIORITY CONTROL

The third of the major system processes is involved with priority control. In a multiuser computer system the operating system must have a policy for choosing the order in which the users competing for the system resources are serviced, and for resolving the conflicts resulting from the simultaneous requests for the same resources by competing users. Different priorities must be assigned to the tasks that wish to run on the CPU to coordinate the execution of different programs that may be sharing the CPU so that each separate program will execute properly under all conditions. In such a computer the operating system must also prevent one program from damaging another by intrusion into one another's memory space. Operating system software is not normally written as a general package for all systems but instead it must be tailored to fit a specific machine configuration.

10.4 KEY TERMS

Background process	Mutual exclusion
Batch processing	Operating system
Bootstrap loader	Operating system functions
Channels (I/O)	Operating system types
CP/M operating system	OS/2 operating system
Disk operating system (DOS)	Real time operating system
Job	Spooling
Job control block (JCB)	Time sharing
Job control language (JCL)	Time slice
Kernel	UNIX
MS–DOS operating system	Virtual resource
Multiprogramming	Windows (Microsoft)

10.5 REVIEW QUESTIONS

1. Describe the main function of an operating system and list several different types.
2. The operating system is referred to as the system controlling software. What does this mean?
3. Why is the definition of an operating system as 'the software that controls the hardware' inadequate? Can you give a better definition?
4. Why was a JCL necessary?
5. Detail the duties of an operator in a computer system using a batch operating system.
6. What is the primary objective of a spooling system?
7. What is a channel?
8. What advantages did a mutliprogrammed operating system have compared to a batch system?
9. Generally the more programs in memory, the greater the utilization of the processor. Explain why this is so.
10. When used to describe operating systems what does 'mutual exclusion' refer to?
11. What is a bootstrap loader?
12. What is meant by the term 'virtual' machine?
13. What does the term 'multitasking' refer to when used to describe an operating system?
14. What is time slicing? Why is it necessary on a time sharing system?
15. State some of the effects each user would notice at his terminal as more and more users are entered on a time shared computer system.
16. Are modern operating systems command or menu driven?
17. On what type of computers is the UNIX operating system mostly used?
18. What are the features of a graphical user interface?
19. What is the kernel of an operating system?
20. List the five main 'managers' within a typical operating system.

10.6 PROJECT QUESTIONS

1. Compare the functions of control programs, utility programs and application programs in a computer system.
2. Distinguish between multiuser and multitasking operating systems?
3. Can you give some practical examples of situations where a real time operating system is required on a computer.
4. What are the drawbacks with command driven operating systems?
5. What are the three software components of the DOS operating system and briefly describe each of their functions?
6. Briefly note the main features of batch processing, multiprogramming, time sharing and multiprocessing in a computer system.
7. Investigate the features offered by the Windows user interface.
8. By suitable reference find out about the functions offered by the UNIX operating system.
9. State several key differences between personal computer operating systems and mainframe computer operating systems.
10. Why does it make sense to refer to the operating system as a resource manager?

10.7 FURTHER READING

This chapter has given a brief introduction to the topic of operating systems. The next chapter expands this with more details of the working of the software routines involved. There are a large number of books on operating systems several of the more readable ones are listed below.

Operating Systems (Books)

1. *Operating Systems Design and Implementation*, A. S Tanenbaum, Prentice Hall, 1987.
2. *Operating Systems Concepts and Examples*, W. Stallings, Macmillan, 1992.
3. *Operating Systems*, P. A. Janson, Academic Press, 1985.
4. *Operating Systems*, H. Lorin and H. M. Deitel, Addison Wesley, 1980.

Operating Systems (Articles)

1. Operating systems, D. A. Anderson, *IEEE Computer*, June 1981, pp. 69–82.
2. Advanced operating systems, R. L. Brown et al., *IEEE Computer*, October 1984, pp. 173–190.
3. Operating systems for micros – too much for too little?, G. MacNicol, *Digital Design*, January 1985, pp. 78–87.

Chapter 11

The Operating System (Part 2)

Chapter Structure

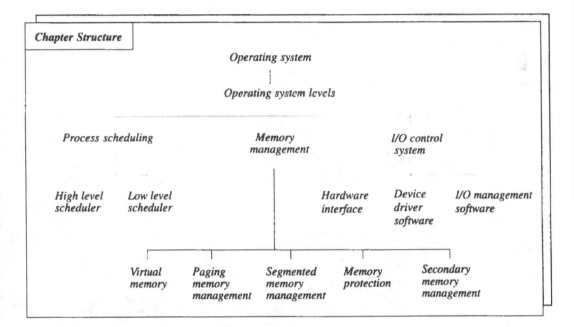

Operating system

Operating system levels

Process scheduling   Memory management   I/O control system

High level scheduler   Low level scheduler   Hardware interface   Device driver software   I/O management software

Virtual memory   Paging memory management   Segmented memory management   Memory protection   Secondary memory management

Learning Objectives

In this chapter you will learn about:

1. The functions of the operating system kernel.
2. The use made of the high and low scheduling functions within an operating system.
3. The role virtual memory plays in increasing the effective size of the main memory.
4. The use of segmentation and paging in virtual memory systems.
5. The functions performed by the file system within a computer.

'Unix is a demon – it is too dangerous and complicated for civilians. It does not solve the problem of protecting the innocent. But programmers love it.'

Ted Nelson

Chapter 11

The Operating System (Part 2)

The previous chapter introduced a number of terms used in describing operating systems. This chapter will expand on the initial introduction by considering the main functions in more detail.

11.1 OPERATING SYSTEM LEVELS

Figure 11.1 presents a block diagram of the different levels within an operating system, noting the functions and features of the levels. It can be seen that in moving down through the **operating system levels** we are moving from user oriented functions, such as providing information through a graphical interface, to hardware oriented functions, such as controlling the disks to manipulate data and program files.

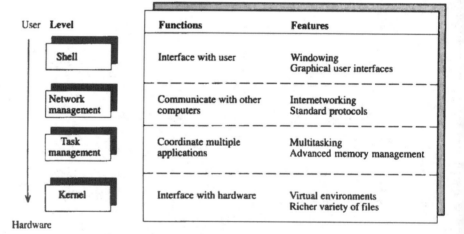

Figure 11.1

Operating system levels.

As the operating system has evolved with the computer, the software for it has tended to grow in layers of procedures. The different software routines within an operating system are often shown diagrammatically as a series of shells, like the different layers within an onion. These layers relate to the different types of functions and services provided by the system. This is shown in Figure 11.2 with an example of the XINU operating system. This operating system was originally developed for a distributed computer network based on DEC LSI 11/02 computers.

Each of the layers in an operating system handles specific features of the system, with the lower layers tending to handle the more basic tasks. The number of layers in an operating system depends on the level of control of the system and the hardware complexity of · the processing structures. For example, the operating system for a large mainframe configuration will be more complex and therefore require more layers of software than for a small minicomputer.

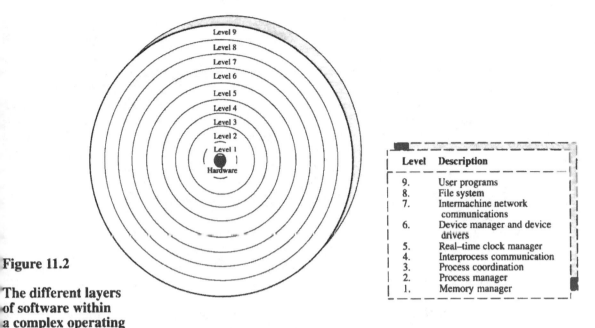

Level	Description
9.	User programs
8.	File system
7.	Intermachine network communications
6.	Device manager and device drivers
5.	Real–time clock manager
4.	Interprocess communication
3.	Process coordination
2.	Process manager
1.	Memory manager

Figure 11.2

The different layers of software within a complex operating system.

The ability to decompose the functions of an operating system into a strict hierarchical structure as shown in Figure 11.2 has a number of advantages for the development of an operating system because of the possibility of developing the levels independently. With structured programming techniques the operating system designer can take reasonable steps to verify that each level of the operating system meets its specifications and that the need for a procedure at a higher level to invoke a lower level service does not introduce unacceptable delays in the progress of the higher level procedure. It is also normal for layered systems to be constructed such that the time required to perform functions at each level is smaller as one conceptually moves closer to the hardware level.

Operating system levels

The layering of procedures may be seen as a form of functional hierarchy where a procedure at one level may directly call any visible operation of a lower level. However it must be noted that the level structure shown in Figure 11.2 is only a diagrammatic representation of the procedures within the overall operating system and it does not mean that information must be passed down through all the software layers to get to a lower level. In fact with this form of functional hierarchy no information need flow through any intermediate level but instead each level is able to locate local objects by their internal names without having to rely on some kind of central controlling program.

In Figure 11.2 the outer layers typically have to call the procedures in the inner layers to help them fulfil their functions. At the centre of the layers is the **kernel** of the operating system which will typically provide the following routines.

1. Process Management: The kernel controls the overall scheduling of the computer's operations. It pulls required programs from the resident storage devices, e.g. floppy or hard disks, and loads them into primary memory, keeping a list of the jobs run on the processing unit. It handles process scheduling, process manipulation, process control block manipulation and process synchronization.

2. Communication: The kernel is responsible for communicating with the system operator. The first type of call to the operating system is usually through the computer operator, typing a request through the system console, which the command interpreter within the operating system interprets, and subsequently causes the operating system routines to perform some series of actions. Within any computing system there is also a certain amount of communication from the operating system to the operator, e.g. messages indicating various events such as error conditions and input/output (I/O) devices that need attention. By processing faults through this central control in an orderly way it is possible to reduce the time lost due to an error in a particular part of the system. This communication is usually controlled by some form of scheduler which routes all system messages to the operator, usually printing them on a printer or on the visual display unit of the system console. In addition the kernel handle interprocess communication, ie. passing messages between processes or tasks.

Kernel

3. Interrupt handling: The kernel is responsible for handling interrupts from either a single or multiuser situation, depending on the system.

4. Resource Management: The kernel handles I/O and memory management, support of I/O activities, storage allocation and deallocation and support of the file system.

Glossary 10.1 describes some of the terms used when discussing operating systems. The three major functions of the operating system are process scheduling, memory management and I/O management.

11.2 PROCESS SCHEDULING

Scheduling within a computer system is the organization of the sequence in which tasks are run on the system. It is also used to refer to the allocation of resources to these tasks. The scheduler is the mechanism within the operating system that controls the access of the different programs to the central processing unit (CPU). It can essentially be split into the high level scheduler and the low level scheduler.

11.2.1   HIGH LEVEL SCHEDULER

The high level scheduler is often called the job or task manager. It is the collection of software procedures within the operating system that controls the movement of processing jobs from the time that they enter the system, usually from a disk device, until they complete, usually with a printout. The high level scheduler governs the policies of resource allocation and influences the order in which tasks are selected for the processor. Some of the major functions of the high level scheduler are as follows.

1. Task input: In a time shared computer system the high level scheduler must read the tasks and jobs from the many users and place them on the job queue for access to the CPU.

2. Scheduling tasks: Before a job can gain access to the CPU, the high level scheduler must request information from the memory and I/O management routines to ensure that all the resources needed by the job are available. If all the resources are available, the job is then placed in a wait state in the central memory by the memory manager and attached to the service request list of the low level scheduler; the job is now ready to be processed by the CPU. If all the resources needed by the job are not ready, the high level scheduler will select the next highest priority job for scheduling and repeat its analysis of the job's requirements. It then schedules jobs in the input queue by allocating the data sets, main storage space, and

Glossary 11.1

Operating System

CONTEXT SWITCHING – Context switching is the operation within a processing unit in which the processor switches from dealing with one task to handling another task. In operation the processor must save the state of the currently operating task and restore the state of another operating task. This usually required the saving of all the central processing unit's registers and the return address.

DISPATCHER – The dispatcher is a software mechanism within the operating system of a computer which carries out actions necessary to swap a processor from execution of one process to the execution of another. It is sometimes known as the low level scheduler.

PRE–EMPTIVE SCHEDULING – Pre–emptive scheduling is a technique whereby a process running on a computer is suspended in order that a higher priority process may have the use of the central processing unit.

PRIVILEGED INSTRUCTION – In a time shared computer system many user–programs effectively share the system's resources, hence it is important to keep the different users from interfering with one another or with the operating system. In order to ensure this, most time–shared systems include monitor, or privileged, states for the operating system and user, or restricted, states for the user. These states are such that when the computer is in the monitor state, information can be read from any memory address, while in the user state only particular memory addresses can be accessed. This prevents one application program from interfering with another. In the restricted state, an interrupt to the operating system occurs if a user tries to use the privileged commands or access certain areas of memory. To be able to use the monitor state it is necessary to include an additional set of commands called privileged instructions into the operating system.

In certain computer systems the access to the privileged instructions can be accomplished by a register within the central processing unit (CPU) containing one bit, called the monitor bit, that designates whether the current CPU program may execute privileged instructions. This bit may be ON when the CPU is executing part of the operating system and turned OFF when the CPU is executing an application program. The bit is turned ON early in the response to most interrupts and it is turned OFF by the operating system before it turns over control of the CPU to an application program. If the monitor bit is OFF, indicating the operating system is not executing, any attempt to execute a privileged instruction is aborted and an interrupt is generated to the operating system.

RESOURCE – A resource is any separate unit within a computer system which can be allocated by the operating system for the use of a specific task.

RUN TIME OVERHEAD – Most tasks when invoked within a processor have associated with them a run time overhead. This overhead includes task activation and termination, task scheduling and dispatching, context switching and allocation of task control blocks.

SYSTEM PROCESS – A system process is one of the software programs or routines within a computer system which provides a service to the operating system of a computer that is not time critical, eg. start an I/O transfer or send an interprocessor message. It is also usually arranged so that it may be run concurrently with the other processes in the system.

Since a system process is less vital to the performance of the complete processing system, compared to the kernel of the operating system, it is not so vital that it is resident in main storage and is usually held on secondary storage, eg. the hard disk, and is called into main memory as it is required. By arranging the system processes to be non resident means that the operating system can perform many complicated tasks without necessarily tying up the primary memory with infrequently used processes. Deciding on which system programs and tables are to be made permanently resident in main storage is a user–installation responsibility that can have a major impact on system performance.

TRANSIENT PROGRAM AREA – Due to limitations in program storage usually only the most frequently used portions of an operating system are retained in the main memory. Other routines called transient routines are transferred into and out of the main memory when required. These routines occupy an area of main memory called a transient program area.

I/O devices needed by each job during execution.

3. Service request: To access services from the other major routines in the operating system the high level scheduler places requests in a list used by the required routine and then invokes the routine. For example as each job is read in from the various users in a time shared computer system, the high–level scheduler requests any necessary space within the main memory from the memory manager. Should the high level scheduler require information concerning the availability of an I/O peripheral, then a request is placed in the I/O manager's service request list and the I/O management routines called up. The I/O manager subsequently examines its request list and performs the appropriate processing to service the request.

High level scheduler

4. Priority assignment: The high level scheduler is the process within the operating system which is capable of altering the priority of other applications or system processes

5. Output control: The high level scheduler also has the responsibility of handling the output control. When a particular job or application program has been executed, the resources that have been used are released and any output from the program usually spooled to a disk. This means the scheduler must control buffering of the output, by storing the output generated by each job in the system onto hard disk, and later transferring it to an output device, such as a printer or a plotter.

High level scheduler

Within the overall operation of the operating system, the high level scheduler must have a high priority. In this way when a user wishes to enter a new job into the system the currently executing program will be suspended and the scheduler routines started to bring in the new program. This leads to a fast response for most users wishing to gain access to the system.

11.2.2   LOW LEVEL SCHEDULER

The low level scheduler, otherwise known as the task manager or the dispatcher, is the operating system function that determines which task among those in the main memory storage should next receive the services of the CPU. It also controls the rate of progress of a task running on the CPU, determining the length of time it is allowed to maintain control of the processor. From this description we can appreciate that the low level scheduler is then the mechanism that takes the CPU away from the control of one task and gives it to another. Low level scheduling is regarded as one of the main functions of the kernel of the operating system.

Whenever the currently executing task enters a wait state, or is unable to continue execution, the low level scheduler is activated. The scheduler can then select, according to some priority procedure, another task to continue execution on the CPU. When such an interruption occurs the low level scheduler will also record in one of the tables it uses the fact that one task has been suspended and the reason for that suspension; the activation of a different task; and any other changes of the execution status of the tasks waiting in the main memory.

The information on each task is stored in a task control block (TCB; Article 10.1). It can be noted that in most cases the transfer of processor control is undertaken by the low level scheduler without any logical direction on the part of an active task. Most systems also include a means for driving the low level scheduler at the request of a running task.

The low level scheduler has to consider the various priorities of the different tasks in the system. These priorities may have been set to control the processing of a complete job in the correct sequence or they may have been set due to time restrictions on certain tasks. In order to perform its function rapidly the low level scheduler must have a top priority within any operating system.

In large computer systems which make use of time sharing, the low level scheduler must also control the time each task is allowed to run on the CPU. In this case before the low level scheduler passes control to a particular task it first loads a value, some integer number of time intervals equivalent to the **time slice**, into a register in the timer which defines the maximum permitted period of execution for the task.

During the execution of the task this value is constantly reduced and when it reaches zero the timer interrupts the processor which then automatically suspends the executing task to resume the low level scheduling routine. The low level scheduler can then schedule some other task. In this way a certain amount of the time is assigned to each task and the hardware

TASK CONTROL BLOCK (TCB)

Within a large multitasking or time shared computer system, when a task is scheduled for execution and brought into the main memory, the operating system creates a small record called a task control block (TCB) that will be used by the low level scheduling routines to record the status of the task throughout its execution. The structure of a typical TCB is shown in Figure 11.3.

The low level scheduling routines in the operating system maintains a queue of these TCB's to help it decide on its scheduling activities. Each of the system tables stores one part of the task descriptor for each task

As a job within a multi task system may contain many tasks so the information within a TCB will look similar in format to that of a job control block (JCB), though they will not be identical. A task control block contains such data as the task's name, its priority, its location in memory, the address of all tables used by the task, and the state vector. The state vector is the location in memory where all the contents of the central processing unit (CPU) registers used by the task are stored when an interrupt to the processing of the task occurs. In this way when the CPU returns to the task which was interrupted, the contents of the CPU's register can be restored. It is to be noted that some tasks may also create their own subtasks which subsequently have their own TCBs for dispatching. The TCB is therefore used as a repository for status and capability data about a task.

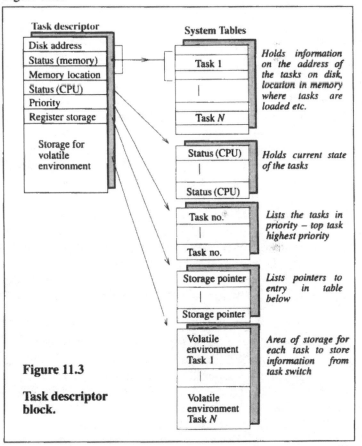

Figure 11.3

Task descriptor block.

is set to generate an interrupt after each task time has elapsed, the task entry is then deleted and the CPU is assigned to the next task on the low level scheduler's list. When the last task on the list has been serviced, the low level scheduler returns to the top of the list, this loop is repeated until there are no further requests. This method of sharing time between tasks is known as a round robin arrangement. The low level scheduler therefore allows tasks that are not aware of each other's existence to share the use of a single processor in an interleaved fashion.

Low level scheduler

11.3 MEMORY MANAGEMENT

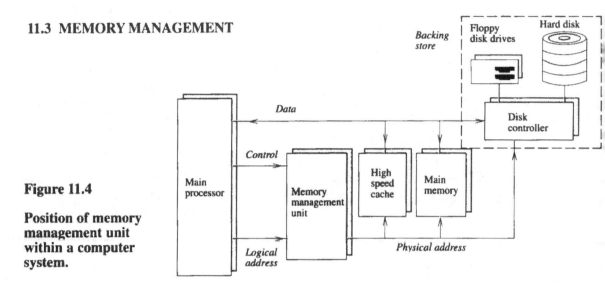

Figure 11.4

Position of memory management unit within a computer system.

Memory management within a computer system is the process of mapping the address field generated by the processor's instruction onto a physical memory location. The way a memory management system performs these functions varies widely from architecture to architecture, in order to meet the requirements of the individual system. Memory management may be handled by operating system software or some of the responsibility for directing memory transfers within a system can be given to special integrated circuits collectively called a **memory management unit** (**MMU**), as shown in Figure 11.4. The MMU acts as an interface between the CPU and the main memory. Some 32–bit microprocessors have this memory management facility built into them.

Depending on the type of computer system, the memory manager can have a number of different tasks.

1. Memory Allocation: When a programming task is initiated the memory management routine will allocate to it only sufficient space to contain those parts of the task which are needed immediately. As execution proceeds and parts not currently in the main memory begin to be required, the memory manager will read the parts of the task in from disk after allocating space for them. At the same time, it will transfer out parts which are not currently being used when the space they occupied is required. This is the technique of memory overlaying (Article 10.2).

2. Virtual Memory Control: Where the available address space of the CPU is greater than the available main memory, the memory manager may use virtual memory techniques to control the memory distribution. To perform this task it uses tabular information to associate virtual addresses generated by the processor from the program with the physical addresses allocated in the hardware, using relocation or translation.

3. Memory Protection: The memory management within a large computer system must also handle memory partitioning to protect the address space of one task from the others. This type of memory management is required to prevent problems that result when multiple tasks within a given application contend for a limited amount of physical memory, or when users share common data or employ common programs. It does this by enforcing a separation of address space among the different tasks. It uses tabular information, which when combined with state information can then allow it to restrict certain kinds of access to selected memory blocks.

Article
11.2

OVERLAYING

During the active life of a program, there are usually parts which are executed for only brief periods of time. This is also true for a large data array, since in most cases only a small amount of the data is actually needed at any particular moment. Thus, instead of having all the sections of a program and all the data it references in the main memory throughout the duration of the program, it would be more efficient in terms of the main memory usage to only put each part of a program, or its data, into memory when that part is needed. This would have the advantage of allowing large programs to only require a small amount of main memory. This is referred to as overlaying.

In this context an overlay is then a section of a computer program, or a part of a large data block, that is actually resident in the main memory of the computer at a particular time. A large program can be divided into many separate overlays and run on a computer having a limited main memory by continually swapping each overlay into and out of the main memory. The overlays not present in the main memory are held on secondary storage, such as a hard disk.

In the early computers, up until the mid–1960s, memory overlaying was handled by the programmer dividing his program and data into blocks which were as far as possible independent of one another, ie. they were never needed in the main memory at the same time. When the program was run, several blocks could share the same space in memory but at different times. In this way a block which was no longer needed would be written out onto magnetic disk, and a new block which was required could be brought into the main memory to take its place. By so doing it was possible to use a small amount of memory to execute programs, whose total space requirements were fairly large. With overlaying, programs or sections of programs could then be transferred into the primary random access memory as required, overwriting existing programs. The overlay method however placed a heavy burden on the programmer, since interaction between sections in different overlays caused continual and excessive disk transfers (disk thrashing) and errors could be difficult to locate. In general then effective memory overlays were not easy to manage.

The operating systems for the second generation of computers had memory management facilities incorporated into them to handle the task of memory overlaying. Automatic overlaying is also used in microprocessor operating systems such as CP/M and MS–DOS.

Memory management

4. Secondary Memory Control: The memory manager makes use of various file directories to enable it to keep track of the use of the main and secondary memory. The file directories are usually held on disk and contain a list of all files, their location and their size. The memory manager will also allocate a file name to a new job and place the file name in a file directory.

In small computer systems memory overlaying may be used to handle the primary memory, while in larger systems virtual memory techniques are used. For multitask systems memory protection is required. We will consider virtual memory techniques, memory protection and secondary memory management in more detail.

11.3.1   VIRTUAL MEMORY

Virtual memory is the concept of providing a memory space that may be viewed as a large addressable main memory to a computer user, but is actually made up from a small amount of main memory and a larger secondary memory, usually a hard disk. This means the system memory is logically extended beyond the confines of the available physical main memory by using mass storage to store data and code sections not needed by currently running tasks. The actual amount of virtual storage is limited by the addressing scheme of the computer.

A combination of hardware and software techniques must be provided by the computer to give the user the impression that the primary storage is very large. To handle a virtual memory environment the **memory management system (MMS)**, which usually consists of a hardware memory management unit and software routines in the operating system, must map the program, or virtual, addresses generated by the CPU onto the physical, or real, memory. The MMS then provides a mechanism to perform the translation of the virtual addresses into the physical addresses which can be used by the memory units.

In most computer systems a large address space exists across a hierarchy of memory devices with different storage capacity cost and speed ratios. In a system employing virtual memory it is the function of the MMS to organize these devices such that it appears to the user that the whole of the memory behaves like a large main random access memory (RAM) any word of which can be accessed at any time. In this way the number of random access addresses available to the programmer is substantially greater than the number of locations in the physical RAM.

*Virtual
memory*

As an example, a very small computer may have a main memory of about 64 KB of random access memory but the allowable logical address range within the system may be 1 MB of memory, the 1 MB of addresses refers to the total address space of the primary and secondary memory. Application software can issue references to the 1 MB logical memory and rely on the MMS both to automatically ensure that the sections required are moved from the hard or floppy disk into the 64 KB of physical RAM memory and then to translate the logical reference into one that addresses data or code within the section once it is in the RAM memory. In this way the MMS handles all relocation making user software addresses independent of the physical memory addresses, thus freeing the user from specifying where information is actually stored in physical memory.

In computer systems which use virtual memory management, the automatic memory transfer methods used rely on having two addresses associated with each memory word stored, a program, or logical, address which is produced by the processor when executing the user program and an actual physical address used by the memory units to reference the memory words. **Memory address translation** involves taking the address emitted from the processor, the virtual or logical address, and converting this address to a real, or physical, address. There are two main methods for achieving this translation, paged memory systems and segmented memory systems.

11.3.2   PAGING MEMORY MANAGEMENT

In a paged memory system the physical memory space is divided into equal page sized blocks and an address is then composed of a block number and a line number.

· A **page** is generally an addressable memory area of fixed size, frequently used page sizes are 2 KB and 4 KB. A page is then a set of consecutive bytes within a computer's memory map, the first byte of each page being located at a storage address that is a multiple of the page size. By using this grouping of memory locations, complete pages may be selected by their higher order address bits. In a virtual memory environment a page is the logical portion of a program associated with a real physical memory blocks.

With virtual storage management, portions of a program may then be kept in secondary storage and only loaded into real storage as a paged block when they are needed during processing. In this context a block of memory is a set of physical locations in memory to which a page is assigned. In a paging system, the actual address in physical memory is called a physical (or real) address and the program address is called a logical (or virtual) address. The virtual memory manager makes use of an address translation mechanism to convert the program virtual addresses to real physical locations.

To translate a virtual page number into a real block number it is necessary to make use of the hardware MMU to generate the correct physical address. The hardware mechanism maps the program virtual addresses into physical locations through the use of a mapping table, usually referred to as a page table, which contains a set of memory registers with each register holding the starting physical address of a block in memory. The page table may then be seen as a small area of memory that holds a list of the virtual pages and associated memory blocks. A machine address can subsequently be formed when the hardware uses the incoming virtual address in connection with this page table.

In this way the translation mechanism is fundamentally a table that relates program virtual addresses to physical locations in terms of individual page/block relationships. The conversion of the virtual addresses to real addresses is called dynamic address translation. A fault occurs whenever the page, or segment, referenced is not already in the primary memory, in which case the referenced page (or segment) must be brought into the main memory from the backing store.

Paging memory management

The concept of using a paged memory with a secondary backing store was first introduced in the late 1950s when it was used in the Atlas computer, developed at Manchester University. Although the virtual memory concept has been expanded in the intervening time, the basic theory underlying it has changed little. What has changed is the size of the processors on which this support is provided; functionality that was previously only available on mainframes and high end minicomputers is now available on microprocessors.

An example of a simple paged memory management unit in operation is shown in Figure 11.5. This example also gives a further illustration of the technique of using a paged virtual memory system. The memory management unit holds all the information about the size of the main and secondary memory and thereby manipulates the movement of the pages between these two types of memory. The total programming space shown in Figure 11.5 is 16 KB, made up from 4 KB of main memory and 12 KB of secondary memory. The page size chosen is 0.5 KB. The address from the processor is 14 bits wide with the five high order bits representing the page number and the lower nine bits representing the offset within that page. The memory management unit performs the translation between the virtual and real address by manipulating the five high order bits, leaving the lower nine bits unchanged. Figure 11.5 shows the page table with the entries at a particular moment in operation. We can see, for example, that there are only seven pages presently in the main memory with, for example, block 0 holding virtual page 9.

In practice the page number given to a particular page of a program usually points to the register within the page table, direct mapping, which contains the starting physical address of this page in memory, or contains its address in auxiliary storage if the addressed page is not in main memory.

Figure 11.6 shows the use of the page table. Any word in a program can then be identified by an address made up from a page number and a displacement within the page, called an offset. The physical address of the addressed element is produced by adding the contents of

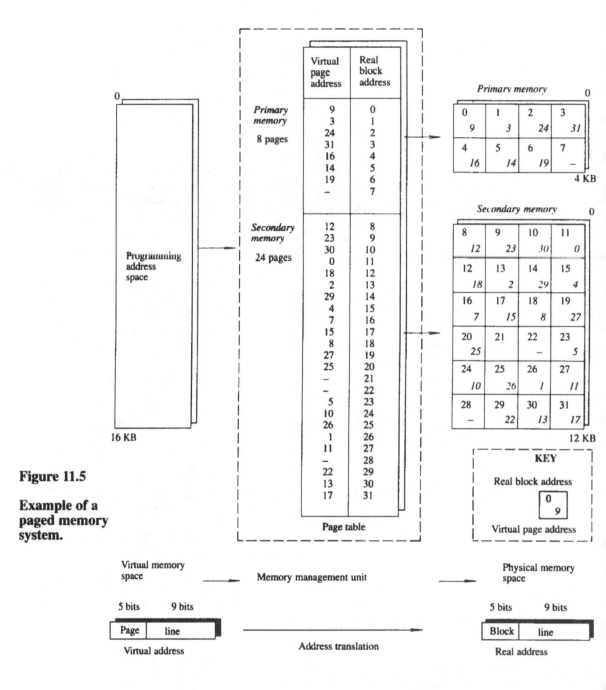

Figure 11.5

Example of a paged memory system.

the page number register and the offset. From Figure 11.6, the virtual address from the processor requests data from page 24 at line 345 within the page. This is translated with the aid of the page table to indicate that page 24 is in fact already stored in the main memory at memory block 2. The real memory address can then be put on the address bus to access the main memory.

Paging memory management

If a program instructs the CPU to access a location not present in the main memory, a hardware generated page fault occurs. For the example shown, if page 2 had not already been

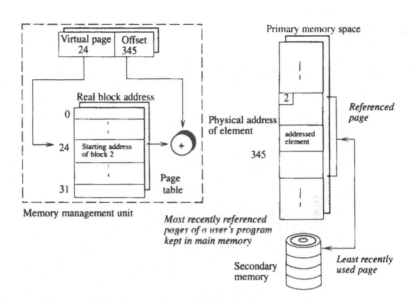

Figure 11.6

Translation example for a paged memory.

stored in the primary memory then a page fault would have been indicated. This in turn would have caused the CPU to invoke the low level scheduler within the operating system to process the next task on its service request list, and the I/O controller software to bring in the requested page from secondary memory to main memory by direct memory access, before the memory address could be referenced.

After a page fault occurs it is usually necessary to remove a page from the main memory to make room for the new page. Many systems which control memory in this way have registers associated with each page in main memory to keep track of the number of references made to each page, such that when a page fault occurs these registers are examined and the pages that have been used least are transferred back to disk, and the new pages are loaded in their place. This is the least recently used (LRU) paging strategy.

Paging memory management

When no attempt is made to predict the order of page usage in the main memory and memory is only replaced on request from a running process we have the technique of **demand paging**. In operation, when a program is loaded into memory the loader does not usually assign one register in the page table to each specific page of the total program. Only when the page is referenced at run time is it placed into a main memory block of contiguous memory locations; its starting physical address is inserted into the page table and the index of that entry is assigned as its page number.

It can be seen that paging solves the logical address space allocation problem on a computer, since loading a program for execution on a processor does not bind the program to positions in logical address space and the registers of the page table are only assigned to pages at execution time. In other words, only when a page within a program is to be executed do the page addresses become absolute within the logical address space, i.e. binding to logical address space takes place. Thus on different occasions, a page can be in different places in the program's logical address space. Furthermore the portions of a program that appear to be contiguous in memory may in fact be scattered throughout the memory in fixed size pages with the MMU handling all the necessary address translations. This ability to scatter program elements and allocate locations without regard to their contiguity in physical memory makes it easier to avoid the occurrence of unusable areas of memory as programs enter and leave a system.

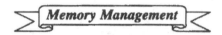

Glossary 11.2

GARBAGE COLLECTION – Garbage collection is a term given to describe the compaction of primary or secondary memory, i.e. moving all occupied areas of storage to one end of the memory to leave a single large free storage areas instead of numerous small areas. The garbage collection process has a negative impact on data integrity and availability. Since data is moved in group to group storage fragments into larger usable pieces, this movement decreases data integrity and significantly complicates data directories that point to where the data is located. Throughput is reduced because significant system resources are required to execute the garbage collection algorithms and update the impacted directories. Data availability is also reduced because users cannot retrieve or edit data during the execution of garbage collection algorithms.

SEGMENTATION – Segmentation is a method of implementing virtual memory storage by dividing a program up into variable size units called segments. A program that is written in segments is such that a number of its segments can fit into the main memory of the computer at any one time with the remaining segments being held in secondary storage. A segment is a block of memory allocated to a particular program.

STORAGE COMPACTING – The technique of storage compacting is used in random access memory (RAM) to assign programs and data to memory locations so that the largest possible area of contiguous locations remains available to other programs that are to be run. It may also be used with secondary memory to remove small areas of unused memory which can occur when files are deleted and others created which do not fit into the disk space vacated by the deleted files.

11.3.3   SEGMENTED MEMORY MANAGEMENT

Most programs intended for a processing system generally consist of a number of logical segments, such as data areas and program functions. Since all the segments in a program do not have to be in the primary memory at one time, virtual memory techniques can be used to handle the movement of segments between primary and secondary memory. Segmented memory management is a scheme similar to paged memory management, the difference is that instead of being of a fixed size, like pages, segments can be an arbitrary size. Segmented memory management reduces the fragmentation problem encountered by using a paged memory system, i.e. different successive pages of a program could be scattered round the memory by allocating the main memory in physical blocks of varying sizes, with each segment containing a complete section of a program.

The translation mechanism employed in a segmented memory system is such that the logical address supplied by the CPU consists of a logical segment number and an offset, the offset specifies the number of locations from the beginning of the segment. The MMS then uses the logical segment to select a physical segment from its segment table, where this physical segment holds the address of the first location in the physical memory. In this way, the segment table converts the logical address from the CPU to a physical address in memory. As well as providing a base value which defines where the segment begins in physical memory, the segment table usually also provides a limit, which identifies the size of the segment. The offset in the logical address must then not exceed this limit and protection is provided simply by defining the ownership and read/write access rights for the segment. If a program tries to access a segment which is not in primary memory a segment fault occurs, and the MMS then intervenes to bring the segment in from the secondary memory and the segment table is updated.

1.   **Advantages** – Segmentation possesses several advantages over paging, in that memory is not wasted since variably sized segments accommodate program divisions of any size. In programs consisting of different segments such as data blocks and program blocks, the use of segmentation can be appropriate since each block can correspond to one segment. On the other hand in programs with a number of different tasks each task may also be designated a segment. Hence when each of these blocks is being operated on by the processor the complete segment is loaded into main memory at one go.

2. **Disadvantages** – The disadvantages of segmentation compared to paged systems comes in the difficulty of managing variably sized segments in the primary and secondary memory. Physical deletion of areas of the main and secondary memory does make it possible to reuse this memory for new insertions but there is a price to be paid for this saving. Segmented memory suffers from holes of unused memory which must be compacted occasionally. The memory management routines must then manage the free space in main memory and disks by using various algorithms to fill up the memory space vacated by programs moved to lower memory levels. This is the process of garbage collection (Glossary 10.2). The use of garbage collection techniques can deteriorate the system's performance.

Segmented memory management

Due to the disadvantages noted above most time sharing systems use pagination and a block allocation of space at all memory levels. In addition, the movement of information between memory levels is performed on a block basis rather than on a segment basis. In large multiprogrammed systems where there are a large number of users and a large amount of both primary and secondary memory, a combination of pagination and segmentation can be used with segments being divided into independently relocatable pages. This is claimed to gain the benefits of both segmentation and paging by subdividing each segment into pages and using a three–part address consisting of a segment number, a page number and an offset.

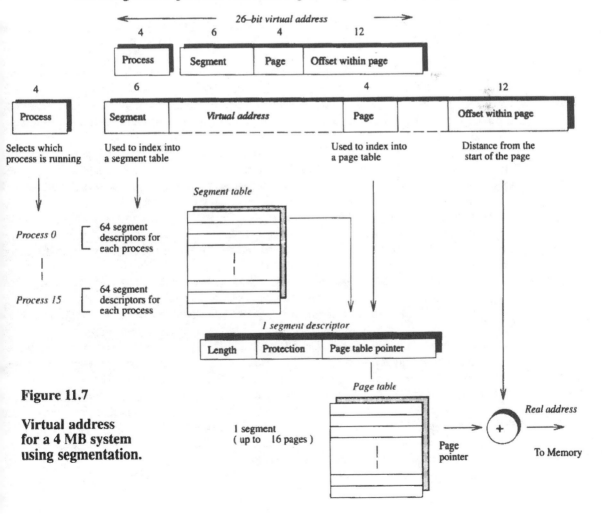

Figure 11.7

Virtual address for a 4 MB system using segmentation.

Figure 11.7 shows a segmented system intended for a large computer. The way the 26–bit virtual address is structured into different parts is shown, each of which can be seen to operate on the different memory management tables.

The mapping of the virtual, or logical, address of the user into the smaller physical memory is invisible to the user and need not be considered when programs are written, i.e. it does not require explicit actions from user programs.

A conclusion that can be drawn from this is that with a virtual memory system the addresses that are available to a program have no direct relationship to the physical locations on a machine, i.e. there is no way to directly infer the physical memory locations represented by an address from an inspection of the virtual address itself.

11.3.4   MEMORY PROTECTION

In large time shared or multiple tasking computer systems memory protection is required to prevent illegal use of the logical pages, or segments, of one task by another task. In order to do this the operating system software and MMU check a number of attributes. Some of these are now described.

1.   User/Supervisor Differentiation – It is necessary for a large processing system to have two operating modes, a normal mode which is for ordinary computer users and a system mode dedicated to the operating system. Generally when in the normal mode, there will be certain instructions which cannot be executed by a user and the only method to enter the system mode is through a system call to the operating system, either intentionally via a software or hardware interrupt, or an error condition. This allows functions such as input/output to be totally controlled by the operating system without interference from user programs.

In some systems, protection is achieved by maintaining two page tables, one for application programs and the other for system programs, and the main memory is divided in such a way that the system programs occupy an area of RAM with low order addresses while the application programs are loaded into the high order addresses.

2.   Size of the Segment – In computer systems using virtual memory techniques with variable sized segments, a limit (segment length) field may be provided to enable a maximum segment size to be specified. If an input offset within a segment is greater than the limit size an error is flagged.

3.   Class of Ownership – Since processing jobs are usually divided into several distinct pieces of storage it is important to protect these segments from each other. By maintaining some class of ownership it is possible to dictate which tasks are permitted access to a particular block, or page, in memory. With this kind of storage protection no user program can then store information in a storage space occupied by another user program or the control program so protecting the sections of a task from access and corruption by other tasks in the system.

4.   Mode of Access – This states whether a page, or segment, in memory is:

(a)   Read only: Assigning a page or segment as read only allows the data to be protected from alteration as a write operation would then be an invalid operation. In addition to this write protection some systems also provide read protection which prevents user jobs reading beyond the area of main storage assigned to them.

(b)   Read/write: This allows the most flexibility in dealing with the memory.

(c) **Execute only**: An execute only page can only be referenced during a the fetch cycle within the CPU, which prevents for example, unauthorized copying of the page since execute–only code is unable to be read as data.

(d) **System only**: This mode can be used to protect the operating system routines from improper access by any of the application tasks.

To implement memory protection, requires that if there has been a violation of an accepted memory operation any address, or address strobe, signals are prevented from being passed to the memory units. Hence storage protection of this kind requires a check for authorized access each time storage is referenced. This check must usually be done by hardware, or the speed of the computer system would suffer intolerably.

11.3.5   SECONDARY MEMORY MANAGEMENT

Before a program starts to run in a processing system it must be brought into the main memory usually from the secondary memory, and its pages assigned to memory blocks. The MMS, comprising the MMU and other software routines, determines what memory blocks may be made available to the pages of the program and then assigns and loads each page individually. When not in main memory the program pages may be returned to the disk.

To perform its tasks the MMS must know the location on the secondary storage devices of each page of the program in order to fetch it into memory. This relies on the support of the I/O elements in the operating system since the page addresses on the disks must first be translated into I/O orders and device addresses (in order to cause the movement of the pages from a disk to the main memory). The MMS also controls the access, storage and movement of files among the computer storage media. This requires it to keep track of the location and size of the available free memory space, allocating it on demand and recovering it when processes are completed.

The operating system makes use of secondary memory management routines to allocate mass storage space and keeps files stored on disks or magnetic tape in an organized manner known as a directory. A **file** is an organized collection of related data or instructions which may be treated as a unit and stored in secondary memory under a generic name. Files are usually held on a hard or floppy disk or on magnetic tape, with the data being organized into a series of records.

The **file handling** routines in the MMS must provide the capability needed to keep track of the information stored on the magnetic disks and to make it available to the various software tools in the system. The file handling facilities must allow a number of tasks to be performed.

1. Formatting of a hard or floppy disk.

2. Creation, deletion and copying of files. The initial job of the file handling routines in a processing system is usually the creation of files and the provision of a mechanism whereby programs may be loaded into primary memory for subsequent execution. Once created, the files may be further manipulated by other application programs.

3. Allocation of meaningful names consisting of a number of characters to files. As they are created each file is assigned a name that consists of two parts, a file name and a name extension. In small IBM AT computer systems the extension may be up to three characters long and is separated from the primary filename by a period. The name extensions used with each file name are usually chosen to easily differentiate files associated with each module or procedure. In DOS, for example, there are several common extensions such as shown in Table 11.1.

Glossary 11.3

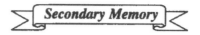

ARCHIVED FILE – An archived file is one which has been stored on a backing medium that is not held permanently on–line to the main memory.

DIRECTORY – A directory is a list of file names together with information enabling the files to be retrieved from backing store by the operating system; a directory is usually held in the form of a file on the backing store to which it refers.

FILE STORE – A file store is a means of holding bulk information within a computer system, usually through the use of a number of magnetic disks. The file store is split into a number of named files each of which can be individually processed by just referring to the name of the file. A file is a collection of related records. In a file store maintained by a computer system the contents of the files can be organized in many ways, for example, sequential, indexed sequential and random.

INDEXED SEQUENTIAL FILE – An indexed sequential file is similar to a sequential file except that some space is left for the addition of records. This type of file is then split up into smaller sections, each of which is maintained in sequential order and the organization is augmented by an index which maintains the range of records kept in each section. The use of an index allows the search for a particular record to be started at some point within the file thus reducing the time to search for a particular record in the file. To access a record therefore involves searching the index to find in which section the record resides and then searching the indicated section for the desired record.

RANDOM FILE – A random file is a collection of data records stored on a random access device, such as a hard disk. Algorithms are then used to define the relationship between the record key and the physical location of the record on the device. A random file utilizes the fact that every record location within the file has a unique address, where this address is usually an integer ranging from one to the number of records in the file. When a record is to be processed, the key value of the record is taken and processed by means of a hashing algorithm. The effect of the hashing algorithm is to map the range of key values on to the number of records which can be stored in the file.

RECORD – A logical record is a collection of related items of data, treated as a unit for processing. It is constructed from data values from which information can be found for the use processing of the file. The structure of the records can vary between files since different files hold different information to which different record structures are appropriate. A physical record on the other hand is the record held in computer storage. It may include extra control information not present in the logical record.

SEQUENTIAL ACCESS FILE – In a sequential access file the records are stored and maintained in a predefined order specified by the user, usually they are located one after another in an order indicated by a special tag called a key number. It is then not necessary to use any index or algorithm to locate a particular record as it can be quickly referenced with the aid of the key. However, within such a file there is no means of adding records into the middle of the file except by copying into another file, the records of the file up to the position of the new record, adding the new record and then copying the remainder of the original file. A similar process is needed to delete a record from a file.

THRASHING – The memory management system (MMS) within a computer system usually consists of both a hardware memory management unit (MMU) and software memory management routines in the operating system. The MMS allows the swapping and overlaying of a piece of main memory to be done automatically, ensuring that only actively used blocks, or pages, of programs and data are kept in the primary memory. Automatic management of memory generally works well if the program running on the central processing unit (CPU) normally accesses consecutive words within a page such that once a collection of words (page) is brought into the main memory one can expect to read or write the next few words in the fast main memory. A problem however arises if the data sought constantly requires the memory management system to access the slower secondary memory such as the hard or floppy disks. Not only does the performance of the computer suffer but the system might find itself swapping between memory levels, continuously exchanging pages between the different levels in the memory hierarchy without doing any effective computations. This phenomenon is called thrashing. Thrashing occurs when every page fault forces the replacement of a useful page, perhaps belonging to another job, and as a result operating system overhead increases severely since most of the processor time is spent servicing page faults.

To minimize thrashing and effectively utilize the CPU in a multiprogrammed or time shared computer system, it is normal to keep a number of pages for each job in the main memory at any one time. These pages are sometimes referred to as the working set. The number of pages in a working set depends on the pattern of execution over a period of time. By making use of its tables the MMS can then determine which pages are free such that when the job is scheduled for execution more than one of its pages may then be loaded into the main memory.

File handling

4. Opening and closing files.

5. Protection of files from accidental erasure.

6. Listing of the complete directory or files whose names contain a specified character string.

7. Handling of any errors associated with the file system.

COM	DOS external command or executable program
EXE	DOS external command or executable program
BAT	Batch file
SYS	System set up file
ASM	Assembly language program
BAS	BASIC language program
PAS	Pascal language program
TXT	Text file
BAK	Backup copy of some other file
DOC	Document file of some word processing programs
DBF	Data base file of some data base managers

Table 11.1

Common DOS extensions.

The file handling routines must manage the file structures contained within a variety of hardware devices, e.g. hard disk, floppy disk and magnetic tape. They must map logical file structures onto physical devices and provide access methods whereby particular records within a file can be accessed without having to specify at the user level the details of the file structure and disk layout. This form of file management is important since there may be a variety of sequential, random, contiguous and indexed files in the system.

To aid in their function the file handling routines use a file directory of some type, where the various file names can be listed. The directory table for each disk or tape lists all the information of the files stored and their addresses – this information can be displayed by giving a directory command followed by the device designation. Other typical commands which the file handling routines will respond to include such things as find, save, load, delete, rename, open and close.

File handling

The file handling routines must also provide information about the location of referenced files within the computer system as part of its function of allocating devices and device space to data files. As an example of its operation consider when a request is made to store a file that a user has been working on. The directory must first be examined to see if a file of that name already exists. If one is found, an error is indicated, if not then a list of all possible disk sectors is examined and one or more are assigned to the file. The data for the file is subsequently stored into the assigned disk location, and the directory is updated to include this new file.

By efficiently allocating the main memory and the disk memory the memory management hardware and software allow the basic operations to be invisible to the user's program or task. It is to be noted that with automatic memory management the programmer no longer has control over the exact location of data and program blocks, since the memory management routines places these where it is most convenient. There are several reasons why it is better to let memory management mechanisms manage primary and secondary storage automatically rather than to count on user programs to tell the system when to move information between memory levels.

1. Memory management is not the objective of a user program. If programmers have to worry about it their programs get more complex, because they have to explicitly call specific operating system procedures. Also, such programs are less easy to port from one system to another, since they include calls to the procedures of the specific operating system they are designed for.

2. Every time a user program or the memory configuration it runs on is modified, the program has to be re–optimized with respect to memory management. This can be costly particularly if the person who modifies the program is not the original programmer and does not know how the program was designed to manage memory.

3. While a good programmer can probably optimize memory management for his program better than any memory management routines could, the global result of all programmers optimizing their programs in a large shared system would be worse than what an memory manager could do, because of the diversity and potential conflicts among the optimization methods used by individual users.

Secondary memory management

4. The use of memory management hardware/software does not prevent programmers from optimizing their programs if they insist, by organizing them in ways that suit the memory management system.

5. In a typical time shared system, no matter how memory management is implemented, most information ends up having to move into memory when a user interaction requires it and migrate back to secondary storage when the interaction is over. Neither an operating system nor user could improve memory management much in view of such a clear cut scheme. Thus, one is well advised to relieve the users from the burden of managing memory and let the operating system do it.

11.4 INPUT/OUTPUT CONTROL SYSTEM

The last of the major functions that the operating system handles is that of I/O control. The I/O devices within a computer system are controlled by the I/O control system (IOCS). This is a combination of the hardware interfaces necessary to control the different I/O devices, the device driver software for each I/O device, and the I/O routines in the operating system. The IOCS may then be seen as a hardware and software package designed to remove the burden of handling I/O details from a user.

We will consider each of the three parts of the IOCS.

11.4.1   HARDWARE INTERFACE
The hardware interface for each I/O device contains the necessary logic circuits to control the device itself. I/O interfaces have already been described in Chapter 5.

As the hardware interface to most I/O devices is relatively crude it requires complex software to control and use it. The software comes in two parts, device drivers and the I/O manager.

11.4.2   DEVICE DRIVER SOFTWARE
A device driver, or device handler, is a processor controlled software module that manages a specific I/O device in a computer system. It is a program segment which may be called upon by an operating system to perform specified I/O operations on an I/O device. All device dependent code goes into the device driver software with each device driver handling one device type. In general the device driver must accept abstract requests from the device independent software above it, issue the commands to the registers in the I/O device and then check to see that these commands have been carried out properly.

Because each type of device has its own particular handling procedure, it is convenient to prepare device driver programs for each I/O device and later integrate them into a more complete I/O control system. Although device handlers may be written fairly easily, their integration into an I/O control system is a nontrivial task. Difficulties arise due to the need for concurrent processing, arising from the mismatch in the speeds of the processor and the I/O device. Device handlers may be supplied by hardware manufacturers or prepared by the user in accordance with manufacturer specifications for the microprocessor.

11.4.3   INPUT/OUTPUT MANAGEMENT SOFTWARE

The management of the I/O devices in a computer system is one of the functions of the operating system. Although the interface circuits and device driver software take most of the burden of control and data acquisition from the CPU, the CPU itself and in particular the operating system software within the computer which is still in charge of the I/O devices.

The I/O management routines in the operating system are sometimes simply called the I/O manager or I/O supervisor. It is necessary for the I/O manager to control the I/O devices in a processing system for five reasons.

(a) To provide a uniform interface for application programs.

(b) To protect shared resources.

(c) To allocate the requesting tasks or processes to the devices according to policies that make access to them fail safe and fair.

(d) To ensure maximum performance of the computer system. The management of the I/O peripherals in a computer system is very important since if it is not handled well the performance of the system will be degraded. Hence careful and efficient scheduling and operation of input/output devices is essential if a computer is to run at maximum processing speed and without interruption.

(e) To provide fault tolerance such that if an I/O device is not functioning it substitutes another I/O device so that the processing will not be interrupted. The I/O manager must also take care of any error handling.

For any I/O device, whether it has only basic control mechanisms built into it or whether it is a complete I/O processor the primary functions of the I/O manager dedicated to providing a uniform interface are the following.

1.   Device Accessing – The I/O manager must permit users to perform high level operations without knowing about the actual configuration of the machine. It can be noted that the I/O functions associated with presenting a uniform interface to the programmer are directly concerned with providing the methods of access to the computational resource. To do this it must locate the I/O devices by transforming the symbolic device names encountered in a program into real device names. In order to perform this task the I/O manager maintains a table of all I/O resources showing such things as the address of the I/O device (where it can be accessed) and the status of the device (i.e. is it busy or free).

2.   I/O Preparation – The I/O manager must submit the I/O requests from a program to the I/O devices and prepare the devices for use by opening the circuitry that permits data to flow between the devices and the computer. If a request is made to a device which is busy, the I/O manager sets up a queue of requests such that when the device is free, requests are serviced on a first in, first out basis.

3.   Data Transfer – To transfer data, the I/O management routines translate commands at the interface to the I/O device into the instructions required for operating that device. When an I/O instruction is issued it involves a series of events between the I/O manager, the programming language and the hardware. Each I/O instruction must designate:

(a) the function to be performed

(b) whether the transfer is an input or an output operation

(c) the device it is to be performed on

(d) the number of words to be transferred

(e) the set of memory locations to which the data is to be sent or from which it is to b obtained.

4.   Buffering – When performing an I/O operation, the I/O manager usually reads data int a buffer, an area of storage, where the data can be held temporarily to facilitate transfe between devices operating at different speeds, or on different time cycles. There is a trade–off between the buffer size and the speed of the I/O operations, such that increasin the buffer size increases the memory requirements and therefore the cost, but it also ma reduce the total operation time. Double buffering is used when data is input from one devic and output to another, i.e. data is read into one buffer from an I/O device until that buffer i full when it can be read out by the I/O manager and at the same time as this first buffer i being read the I/O device can fill a second buffer. In this way input and output operation can be carried out almost simultaneously, i.e. buffer A is being emptied while buffer B being filled and when both operations are complete the roles of the two buffers ar exchanged.

*I/O
Software*

5.   Interrupt Handling – At the end of an I/O operation the calling program must b informed, usually by the I/O device signalling the processor with an interrupt. This require the I/O management routines to respond to interrupts from the devices to indicate that th I/O operations have been completed. Some I/O managers do not use interrupts but instead they continually check the status of the I/O operation and do not return control to the callin program until the operation is complete.

In operation the I/O manager checks the I/O program and its parameters to ensure tha both the I/O devices and memory segments containing the I/O buffers are accessible to the user. It also requests translation of virtual addresses in the I/O program to physical addresse using memory management functions. Where a processing system uses dedicated I/C channels, or I/O processors, the I/O manager must oversee their operation by using message passed to them to indicate what they are to do. These messages or channel status words the direct the operation of the data channel or I/O processor to begin.

It can be seen from this description that the I/O management routines then provide constant uniform and flexible interface to all I/O devices in the system, thereby allowin users to write programs that refer to the various devices in the system by name. This involve handling the basic I/O requests, submitting I/O instructions to devices, handling interrupt from the devices and allocating devices and channels for the use of particular programs o the residence of particular data sets.

11.5 KEY TERMS

Archived file
Context switching
Demand paging
Device driver software
Directory
Dispatcher
File
File handling
File store
Garbage collection
High level scheduler
Indexed sequential file
Input/output control system
Input/output (I/O)
Management software
Kernel
Low level scheduler
Memory management
Memory management unit
  (MMU)
Memory management system
  (MMS)
Memory protection

Memory address translation
Paging memory management
Pre–emptive scheduling
Privileged instruction
Operating system levels
Overlaying
Page
Random file
Record
Resource
Run time overhead
Scheduling
Secondary memory management
Segmentation
Segmented memory management
Sequential access file
Storage compacting
System process
Task control block (TCB)
Thrashing
Time slice
Transient program area
Virtual memory

11.6 REVIEW QUESTIONS

1. In drawing an operating system as a series of software levels which functions are at the centre of the diagram?
2. Why is the name 'kernel' used to describe the memory resident part of the operating system?
3. What are the four main functions of the kernel?
4. Why have the terms 'high' and 'low' been given to the scheduling functions within an operating system?
5. Decide which level of scheduler should make the following decisions.
   (a) Which ready process should be assigned to the CPU when it next becomes available?
   (b) Which of a series of waiting batch jobs that have been spooled to disk should next be initiated?
   (c) Which processes should be temporarily suspended to relieve a short term burden on the CPU?
6. What information is stored in a TCB?
7. What advantages does allocating memory management to hardware rather than software have on the performance of a computer system?
8. What is 'overlaying'?

9. How does virtual memory storage increase the usable computer memory?
10. What does the term 'memory address translation' mean?
11. What size is a page?
12. Detail how an address in a computer system is used to control memory segmentation.
13. In terms of memory management distinguish between the use of segmentation and paging.
14. What is a page fault, how is it detected and what is done about it?
15. Why is segmentation more difficult to implement on a computer than fixed partition memory management?
16. What role does virtual memory play in memory protection?
17. Why is memory protection important in a time shared operating system?
18. What is a file?
19. What is a file name?
20. What is a file directory used for?
21. List some of the functions associated with file handling.
22. The operating system in the most part hides the detail of memory management from the user. Why?
23. What do the following terms refer to:
    (a) device driver software,
    (b) I/O control system?
24. What are the functions of device driver software?
25. List the five tasks that the I/O management software in an operating system is responsible for.

11.7 PROJECT QUESTIONS

1. An operating system is often shown as a series of layers. Is this a valid picture? Justify your answer.
2. How does the design of an operating system differ from the design of an application package?
3. Give several reasons why it is effective to place major portions of the kernel in microcode within the CPU.
4. Why does the kernel normally run on a CPU with the interrupts to the CPU disabled?
5. Distinguish between resident and transient operating system modules. Why does it make sense to have some modules transient? Why must other operating system modules be resident?
6. Is it preferable to have a large operating system transient program area?
7. What types of instructions would be included in a bootstrap loader program?
8. What types of operation could cause a task to enter a wait state.
9. Define each of the following terms: program, job, procedure, process and task.
10. Why is the term 'virtual memory' a misnomer?
11. Discuss how fragmentation manifests itself in each of the following types of virtual storage systems: paging, segmentation.
12. Detail some of the advantage/disadvantages of using a large page size in memory management.
13. How does garbage collection operate in a memory system?

14. What is disk thrashing and why does it degrade the performance of a computer system?
15. Distinguish between sequential, random, contiguous and indexed files.

11.8 FURTHER READING

The following books and articles present reference material for operating system investigation.

Operating Systems (Books)

1. *Operating Systems Design and Implementation*, A. S Tanenbaum, Prentice Hall, 1987.
2. *Operating Systems Concepts and Examples*, W. Stallings, Macmillan, 1992.
3. *Operating Systems*, P. A. Janson, Academic Press, 1985.
4. *Operating Systems*, H. Lorin and H. M. Deitel, Addison Wesley, 1980.

Operating Systems (Articles)

1. Operating systems, D. A. Anderson, *IEEE Computer*, June 1981, pp. 69–82.
2. Advanced operating systems, R. L. Brown et al, *IEEE Computer*, October 1984, pp. 173–190.
3. Operating systems for micros – too much for too little?, G. MacNicol, *Digital Design*, January 1985, pp. 78–87.

Networks

Part Five

Chapter 12

Computer Networks

Chapter Structure

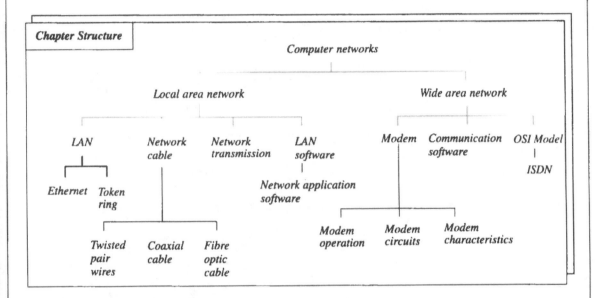

Learning Objectives

In this chapter you will learn about:

1. The importance of computer networks.
2. The components of a local area network of computer
3. The Ethernet and Token ring communication networks
4. The difference between analog and digital transmissions.
5. The use of a modem and its operating characteristics
6. The OSI model and its implementation within ISDN networks

> *'Using a computer should always be easier than not using a computer.'*
>
> Ted Nelson

Chapter 12

Computer Networks

The last decade has seen the increased use of minicomputers and small personal workstations dispersed throughout many large companies. This diversification in computing resources has come about due to the improving performance and the decreasing cost of the small computers. Hence it has become more economically viable to disperse computing resources to the point of need within a company rather than to rely on a centralized mainframe computer site. However, in order to provide a coherent company resource and provide access to dispersed data bases has meant the need to develop many communication techniques and methods to link such systems. This interlinking of computers has produced systems referred to as **computer networks**, multicomputers or distributed processors.

The decentralized resources and dedicated servers in computer networks create a structure that is a collection of loosely coupled processing elements tied together by a communications network. Such multiple processor systems use message passing as their data transfer mechanisms. Some of the various topologies that can be used to interconnect small computer networks is shown in Figure 12.1.

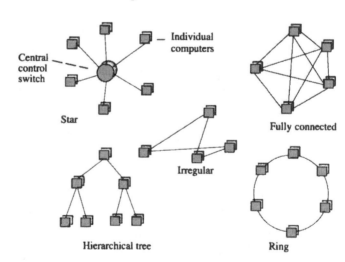

Figure 12.1

Computer network topologies.

Table 12.1

Different types of communication networks.

	Network	Description
1.	Wide area networks	Resources distributed over a large area such as a city or even a country.
2.	Local area networks	Resources situated in the same building or factory site.
3.	Cellular radio networks	Messages sent through various radio communication cells.
4.	Global networks	The most common of the global networks is probably the telephone system which carries the bulk of data communication traffic over a country or continent wide area.
5.	Satellite networks	Usually reserved for transcontinental messages.

FEATURES OF A LOCAL AREA NETWORK (LAN)

1.   Single Ownership: Most LANs rely on a switching communication network that is usually owned by one owner, meaning the computers and terminals are under the control of one organization, thereby easing management problems.

2.   Small Network Dimension: A LAN usually connects various devices in a communication network of small physical dimension, such as one floor of an office building. It may then be seen as an intraoffice or intrabuilding communication system that supports some type of communications processing and information transfer between users and/or electronic devices.

3.   Many Independent Devices: The major purpose of a LAN is as a data communications system to allow message transfer between a number of independent computer related devices such as computers, terminals, mass storage devices, printers, plotters, or copying machines.

4.   No Central Controller: The computers attached to a LAN can usually access the network without the aid of a central control mechanism. However, a network control centre may be necessary to monitor the health and performance of the system.

5.   Ease of Reconfiguration: In a LAN many different system architectures are possible, ranging from the use of intelligent terminals connected to a central mainframe, to placing powerful workstations on the desk of each worker within an organization. The configuration of the LAN can also be easily changed with devices being added or removed as desired.

6.   Fair Access: A LAN is usually organized through a set of protocols to allow fair access to the network for all the devices attached to it.

7.   Efficiency: In a LAN the individual processors are able not only to share system resources such as a printer but can also to cooperate on the handling of work, i.e. they can function relatively independently of one another, in that they have separate missions, but are able if they wish to communicate with each other on occasion.

8.   High Data Rates: LANs provide high bandwidth, typically 1 to 10 Mbps (million bits per second), communications over an inexpensive media, generally coaxial cable or twisted pair wire.

9.   Good Reliability: LAN networks usually feature several different pathways between individual stations so bringing about a better reliability profile to the network. The failure of any single device on the network does not then degrade the entire network.

10.   Low Error Rates: By using a number of error detecting and correcting codes the error rates in some LANs can be such that there will be no more than one unpredictable error per year.

11.   Communication Standards: To ensure that the computers within a LAN can process the data passed among or between them a strict communications standard must be enforced . Standard communication systems such as Ethernet have become extremely important in linking small computers together into LANs.

Network (computer)

There are also now many different types of networks that can be connected together to form complex interlinked communication systems. Some of these are shown in Table 12.1.

With all these different forms of networks, a computer system can no longer be regarded as a stand alone piece of equipment but, for maximum flexibility, must be designed to incorporate facilities to allow access to these open communication highways.

12.1 LOCAL AREA NETWORK (LAN)

With the decreasing cost of minicomputers it is now more economically viable for large organizations to base their computing system on a distributed philosophy rather than on a large central mainframe. This allows the computing power to be dispersed to the point of need, thereby giving the user instant access to powerful computing facilities without the inherent slow response of a large time shared mainframe.

LAN

The most widely used form of communication network to link locally dispersed computers is the LAN. The first LAN, called the ARCnet, appeared in 1977 and was released by Datapoint Corporation. It could transmit data at 3 Mbps over a coaxial cable (Article 10.1 gives the main features of a LAN).

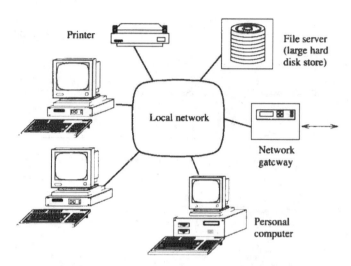

Figure 12.2

Example of the types of devices that can be attached to a local area network.

An example of a block layout of a LAN is shown in Figure 12.2 where the personal computers share a printer, file server and a network gateway, used to access other distant networks. The network mechanism could be a message switch, shared cable or a collection of serial lines making use of coaxial cables, twisted pair wires or fibre optics links.

We will consider the construction and features of two of the most widely used LANs, the Ethernet and the token ring. Glossary 12.1 explains computer network terminology.

12.1.1   ETHERNET

The Ethernet local network system was first developed in 1976 by Metcalfe and Bloggs of the Rank Xerox Company. It is an example of a **CSMA/CD** (carrier sense multiple access with collision detection) system where the shared communication channel is a simple coaxial cable which forms a passive broadcast medium with no central control. It is a listen while talk system, where not only do the individual stations listen on the network before they start transmitting, in order to ensure that no other transmissions are taking place, but they also continue to listen on the network while they are transmitting, comparing what they transmit with what they receive. In this way they can detect any other device transmitting on the network at the same time. Each station also performs its own packet address recognition to take in packets from the channel.

Figure 12.3 shows a part of an Ethernet system with three stations and one repeater with each station connected to the network through an interface and a transceiver. The repeater is to connect two individual Ethernet segments together.

A station wishing to transmit on the Ethernet system is said to contend for use of the common shared communication channel, sometimes called the ether, until it acquires the channel, therefore coordination of access to the bus in this multiple access bus system is distributed among the contending processors.

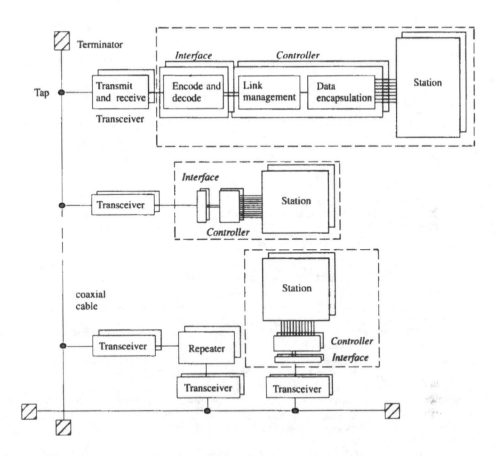

Figure 12.3

Ethernet network.

Ethernet

To acquire the channel, a station checks whether the network is busy, that is it uses carrier sense, and defers transmission of its packet until the ether is quiet, no other transmissions occurring. When quiet is detected the deferring station immediately begins to transmit a message packet onto the serial bus. This packet is heard by all the stations on the bus and, provided no interference occurs, the packet will be completely received by the destination processor. If more than one device attempts to transmit through the network at the same time it can lead to a confusion of the data on the line, or a so called collision. When a station detects a collision it sends off a burst of random data so that all the other stations on the network also detect the error. Should bus contention occur during transmission, the message will be aborted and retransmitted after a random time interval. Due to its operation Ethernet does not require to provide an acknowledgement signal.

Collision detection on the network is accomplished by a receiver detecting the signals on the cable such that if a signal exceeds the maximum voltage swing that could be produced by a single transmitter, a collision is assumed to have occurred. Attenuation, however, presents a potential problem in such a system since if two stations are far apart and are both transmitting, each station will receive a greatly attenuated signal from the other. In this situation the combined signal strength at each transmitter could be so small that the combined signal does not exceed the collision detection threshold and so neither station detects a conflict. This is one reason that Ethernet restricts the maximum length of cable to 500 m.

Glossary 12.1

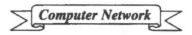
Computer Network

BACKBONE NETWORK – A backbone network is a digital transmission facility, or arrangement of such facilities, designed to interconnect lower speed distribution channels or clusters of dispersed users or devices.

BAUD – The baud is one of the units chosen to indicate the transmission speed of a serial data communications device. It is the maximum number of signalling elements or symbols that are generated per second. Most types of modem communications device are usually able to encode several bits from a digital source into one symbol, or baud, by the device making use of advanced encoding techniques. Hence the bits per second rate of a digital input is not equivalent to the baud output of the communications device.

BITS PER SECOND (bps) – This is the basic measure for serial data transmission capacity. It is used to indicate the speed of a digital communication device.

BLOCK – A block of information in a communication system is a quantity of transmitted information which may be regarded as a discrete or logical entity, either by its size, or more commonly by its own starting and ending control characters. A block also normally contains its own control routing and error checking information. Blocks within a computer system can either be of fixed or variable size.

CCITT – The Consultative Committee for International Telegraph and Telephone Standards (CCITT) is an international consultative committee that sets the usage standards for international communications.

CHECK BIT – A check bit is a single binary digit used to determine the parity status of a byte of data. Often it is called a parity bit. It is used in digital data communications to allow a receiver to determine if any error has occurred in the transmission of a data byte.

CODE CONVERSION – Code conversion is the process of changing the bit grouping for a character in one code into the corresponding bit grouping for the character in another code.

DATA LINK – A data link is any serial transmission path which can communicate data, generally between two adjacent nodes, or devices, without intermediate switching nodes.

DATA TRANSFER RATE – The data transfer rate is the average number of bits, characters or blocks transferred per unit time from a data source to a data sink. It indicates how fast data can be moved into or out of a system.

DECIBEL (dB) – In the area of communications the decibel is the relative strength of a signal, when the signal is compared to a value of one milliwatt. As an example a 0 dB signal delivers one milliwatt to a line load while a –30 dB signal delivers 0.001 milliwatts to a load.

DESTINATION FIELD – Within data communications the destination field in a data message contains the address of the station within the communications network to which the message is being directed. It is usually incorporated into the message header.

ENCODING / DECODING – In the area of communications, encoding/decoding is the process of putting information into a format suitable for transmission, encoding, and then reconverting it after transmission, decoding.

FRAME – In the area of data communications, a frame is a group of data bits similar to a block which can be sent serially over a communication channel. The frame usually contains its own control information for addressing and error checking.

HEADER – The header in a data block, or packet, is the control information and codes that are appended to the front of the block of user data, for control, synchronization, routing and sequencing of the transmitted data packet.

HSLN – A high speed local network (HSLN) is a data communications network designed for use in a computer room. It is used at centralized data processing sites to interconnect mainframe computers with possibly smaller satellite processors. Such a network is intended to deliver very high data rates but at very small distances.

ISO – The International Organization for Standardization (ISO) is based in Switzerland and is concerned with setting international standards primarily in the area of data communications.

LOCAL INTERFACE – In a local area network (LAN) of small computers each computer requires a local communications interface to be able to access the network transport system. The local communications interface is a special I/O interface which is able to handle the different protocols that the communication systems may require.

MEDIUM (for transmission) – In the area of communications, a transmission medium is any material or substance that can be used for the propagation of signals. These signals are usually in the form of modulated radio, light or acoustic waves. Typical mediums are for example optical fibers, cables, wires, dielectric slabs and air.

PBX – A private branch exchange (PBX) is a computer controlled device capable of providing a switching capacity to interconnect telephone lines to telephones, computers and terminals. In this way any telephone, terminal or computer connected onto the PBX can be switched to any other telephone, terminal or computer that is also connected onto the PBX. This means that it is possible to share an expensive local resource among a different number of diverse and dispersed user. PBXs in turn may be connected together across a distant network by special leased lines thereby interconnecting an even larger number of devices.

When correctly operating, a collision on an Ethernet system should only occur within a short time interval following the start of a transmission, since after this interval all the other stations in the system will detect the carrier and defer their own transmissions. This time interval is called the collision window or the collision interval and its length is determined by the round trip propagation delay time between the farthest points in the system.

Ethernet

To ensure that all parties to a collision have properly detected the collision, any station that detects a collision invokes a collision procedure that briefly jams the channel. It is however impossible to guarantee that all the packets transmitted will be delivered successfully and error free. For example, if a receiver is not enabled, an error free packet addressed to it will not be delivered. Higher level protocols which can insert and interpret error codes and which can monitor stray messages must therefore detect these situations and cause the message to be re-transmitted.

12.1.2   TOKEN RING

A token ring communications network consists of a complete loop of cable which leaves from, and returns to, a ring monitor or control station. In this type of system each active communication station is connected in series with the cable, such that it receives a stream of bits on one side, retransmitting them on the other. The general configuration of the ring is such that the communicating stations, connected to the ring, are identical to one another with the exceptions of the special stations on the ring for monitoring and logging the system transactions.

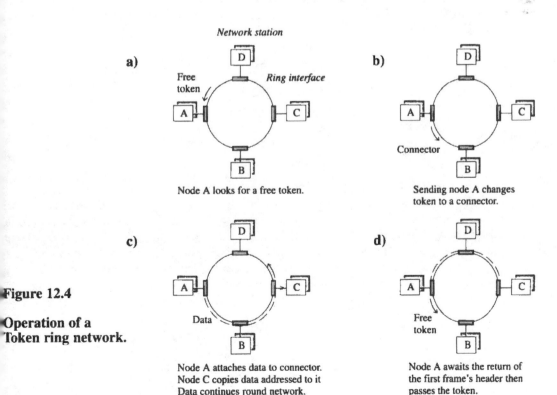

Figure 12.4

Operation of a Token ring network.

Node A looks for a free token.

Sending node A changes token to a connector.

Node A attaches data to connector. Node C copies data addressed to it Data continues round network.

Node A awaits the return of the first frame's header then passes the token.

In a token ring computers are normally connected to the ring through an access box station unit and repeater, the whole combination forming a communication station. The purpose of each of the units is as follows.

(a) The repeater manages the flow of bits on the ring.

(b) The station unit implements the protocol of the transmission packets.

(c) The access box provides the conversion between the network specific functions of the station unit and the attached computer. The interface between the computer and the access box is usually provided on the computer's own internal bus. Channels are coupled to the station logic by transformers and the electronics are so arranged that if a station fails the repeater acts as a purely passive coupler.

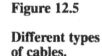
Token ring

Each of the stations on the network are usually powered from the ring on separate wires provided for the purpose. Optical fibres may also be used to carry the signal, though they cannot of course carry electrical power for the stations. When powered off each station must be by-passed.

Figure 12.4 shows the operation of a token ring system. When a user computer needs to transmit a packet, it forwards it to the station, which attaches it to the end of the free token. The message traverses the network until it is detected by the receiver station which copies the message. When the token returns to the sending station the connector is changed back to a token. (Article 12.2 describes communication devices used in a LAN).

12.2 NETWORK CABLE

In the construction of communication networks there are different forms of cabling which can be used (Figure 12.5). The three most widely used for communication networks are described below.

Figure 12.5

Different types of cables.

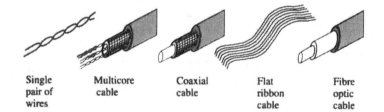

| Single pair of wires | Multicore cable | Coaxial cable | Flat ribbon cable | Fibre optic cable |

12.2.1   TWISTED PAIR WIRES

Twisted pair wires are chiefly used for low data rate digital transmission over short distances. Studies have shown that a paper–insulated twisted pair cable can transmit digital signals at rates up to 10 Mbps. It has the distinct advantage that it is comparatively cheap however, it is susceptible to crosstalk interference, particularly from adjacent wires. In order to cope with very fast transmissions within a communications network coaxial cable is required. Figure 12.6 displays two twisted pair wires in an insulating cover.

Figure 12.6

Construction of twisted pair cabling.

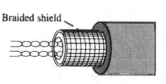

Braided shield

Two twisted wire pairs, each wire individually insulated

Insulating jacket

Article 12.2

COMMUNICATION DEVICES

SERVER – In order to keep the cost down many local networks of computers tend to share system peripherals and input/output (I/O) devices such as printers, scanners, plotters, modems and hard disk drives. In order to control the sharing of the I/O device, the small computers may be linked to a special service system, or server, whose main function is to transfer data between the I/O device and the sharing systems. In this way the server may be seen as a shared resource on a network, i.e. a device which provides a service to all the terminals, computers or workstations on it. The server, for example, may control a printer, this constitutes a printer server system, or a large capacity hard disk system, a file server.

1. A print server can provide high speed high quality centralized printing more economically than the use of multiple dot matrix and daisy wheel printers.
2. A file server provides a centralized filing service and typically consists of a Winchester disk, a processor and file management software. It is usually a network station dedicated to providing file and mass data storage services to the other stations on the network.

BRIDGE – A bridge is a device used to interconnect two similar local area computer networks (LANs). A bridge does not modify the content or format of the data packets it receives nor does it encapsulate them with an additional header. Each packet to be transferred is simply copied from one LAN and repeated with exactly the same bit pattern on the other LAN. Since the two identical LANs use the same protocols its permissible to do this. The bridge must contain addressing and routing intelligence and at a minimum it must know which addresses are on each network to know which packets to pass.

ROUTER – A router is a hardware/software combination which can be used to connect similar LANs. A router must be able to cope with a variety of devices on several types of networks.

CONCENTRATOR – A concentrator, or communications processor, is a type of multiplexing device which is able to combine the data from a number of terminals or slow speed communication devices, onto a high speed synchronous line for transmission to a host computer. In this way the concentrator can systematically allocate communication channels among several computer terminals, handling different speeds, codes, buffer storage and protocols on the low speed side. It is chiefly used within an organization to allow many users to access a mainframe computer.

Typically the concentrator is made up from a multiplexor and a front–end processor. Since the concentrator may have many different types of terminals and computers connected into it, all of which may be producing different data rates and formats, it may also have to alter the form of the data streams prior to merging them onto the high speed line. It is normal for the concentrator to be physically located close to its mainframe computer, often in fact sited next to its large host. By controlling the communication network leading to the mainframe computer, the concentrator can take some of the load from the host. While the processing of this communication traffic is not computationally demanding it can be a time consuming task, because each transmitted character needs to be checked to ensure that it conforms to the particular protocols of the system.

FRONT–END PROCESSOR – A front–end processor is a device which is placed between a communications terminal and its host computer. The processor processes the communications between the terminal and the host computer thus freeing the host computer for other work.

MULTIPLEXOR – When used in a communications system a multiplexor may be regarded as a multiport device that allows two or more users to share a common physical transmission medium. It is normally employed in pairs with one multiplexor at each end of a communications channel, such that one device performs the multiplexing of the multiple user inputs while the other demultiplexes the channel back into the separate user data streams. Typically the user data streams are interleaved within the multiplexor on a bit or byte basis, called time division , or separated by different carrier frequencies, frequency division.

12.2.2   COAXIAL CABLE

Coaxial cable provides a higher throughput, can support a larger number of devices, and can span greater distances than twisted pair wire. Coaxial cable can be used for either analog or digital transmissions and for baseband or broadband transmissions. Transmission rates of up to 1000 Mbps are possible over specially constructed coaxial cables.

Figure 12.7

A cross section through a coaxial cable.

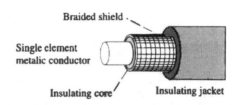

Coaxial cable consists of one or more central wire conductors surrounded by a dielectric insulator and encased in either a wire mesh or extruded metal sheathing. A cross sectional diagram of a coaxial cable is shown in Figure 12.7.

Coaxial cable comes in many varieties depending on the degree of electromagnetic shielding afforded and the voltages and frequencies accommodated. It is a popular transmission medium which protects against unwanted noise while passing a wide range of frequencies with very low signal loss.

12.2.3   OPTICAL FIBRE CABLE

An optical fibre, or light guide, is any fibre made of a dielectric material, that may be used to transmit laser or light emitting diode (LED) generated light signals. It i manufactured by drawing a large cylinder of glass, called a preform, out over a long distance until it is one piece of pure glass whose diameter is measured in millionths of a metre. The first commercially optical fibre cable was produced in 1970. At the same time at Bell labs in the USA, the first semiconductor laser diode was developed. By combining the two led to the introduction of fibre optic communications.

Figure 12.8

Cross section through an optical fibre.

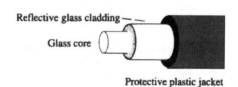

A cross section through a fibre optic cable is shown in Figure 12.8. Individual optical fibres usually consists of a narrow glass core, anywhere from 5 to 100 microns in diameter, which carries the light signals, and a cladding which surrounds the core and serves to refract the light signal back into it. The cladding is composed of glass but it is designed in the preform stage to have a different refractive index to the core. Although the resulting fibre is drawn from the same preform the two components transmit light at slightly differen velocities thus light is bent or refracted as it enters the transition between the core and the cladding. Various materials are used to encase the fibre.

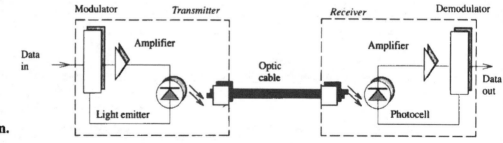

Figure 12.9

Fibre optic transmission.

The use of optical fibre as a communications medium is based on the fact that it is physically light, has a high potential capacity and is resilient to electrical interference. The fibre is used in conjunction with laser light sources and detectors to transmit and receive information. Figure 12.9 shows a fibre optic transmission. The modulator at the transmitter receives the data to be broadcast across the fibre. It subsequently modulates the current through a light emitting diode. The optical fibre carries the infrared light to a photocell at the receiver which drives a demodulator to recover the transmitted data.

Optical fibre

Table 12.2

Cable characteristics.

Feature	Twisted Pair	Coaxial Cable	Optical fibre
Date rate per kilometre	16 Mbps	500 Mbps	1000+ Mbps
Accessibility to being tapped	Easy	Easy	Difficult
Signal Radiation	Yes	Yes	No
Size and weight	Large	Large	Small
Bit error rates	1 in 10^6	1 in 10^6	1 in 10^8

Table 12.2 compares the three types of cables and notes their features. In metallic cable signals degrade more and more over unit lengths of the cable. With optical fibres this is also true but repeaters are able to regenerate the high frequency components lost in the transmission. Metallic cables carry signals that are basically radio waves hence their cable act as antennas radiating their signals freely. Fibre optic cables do not radiate at all.

12.3 NETWORK TRANSMISSION

The passing of digital signals over a network usually relies on serial transmission. The signalling can either use circuit, message or packet switching methods depending on the hardware and software at each of the stations on the network (Glossary 12.2).

Circuit Switching: In data communications, circuit switching is a method of communication where an electrical circuit between calling stations is established on demand for exclusive use of the stations until the connection is released. There is then a physical link between the source and the destination communications units which remains set up while data passes, and may then be disconnected, so that other communications units may use the same facility. After the circuit has been established every piece of information may be transferred at very high speed since there is no need to add addressing information, hence no decoding time is required.

Message Switching: In data communications, message switching is a method of handling data such that a data packet, or message, is transmitted to intermediate points, or nodes, on its communication route, stored for a short time at each node, and then retransmitted again towards its destination. In such a system the entire message is relayed from one processing node to another through the communication network. The communication network may then be seen as being constructed from processing nodes, each of which acts as a switching element to direct the messages through the network, where the destination of each message is indicated by an address integral to the message.

A **message** is any data unit which contains information arranged in an ordered format. A message may be sent by means of a communications process to a named network entity or interface.

Glossary 12.2

Network Transmission

ACQUISITION TIME – In data communications it is the time required to attain synchronism.

ARQ – Automatic request for repetition (ARQ) is a term used in communications to stand for the retransmission of a message across a communication network whereby the receiver asks the transmitter to resend a block or frame of data. This may be necessary because errors had been detected in the first message transmitted to the receiver.

BASEBAND – In a baseband transmission the cable carries one data signal at a time at data rates of 10–50 Mbps, for coaxial cables. It is chiefly a method of sending digital data over communications lines that is suitable for short distances and low transmission speeds. For network transmission the signal is modulated as an analog carrier frequency. It does allow for time division multiplexing; however, as the entire spectrum of the cable is consumed by the signal it does not allow frequency division multiplexing.

BLOCKING – Blocking occurs in a communication network when the network switch, or access node, is unable to grant service to a requesting user, due to the unavailability of a transmission channel.

BROADBAND – In a broadband transmission the coaxial cable can carry many signals at a time, with a different analog, or radio, frequency band being allocated for each signal. To transfer digital data across a broadband network the digital signals are passed through a modem and transmitted over the cable using one of the frequency bands. It can allow 20–30 channels on the same coaxial cable with data rates of 1 to 5 Mbps for each channel. Broadband transmission equipment and media can therefore support a wide range of electromagnetic frequencies and are suitable for transmissions over long distances with high data rates. In the area of data communications, broadband transmissions usually use radio frequency carrier signals in the 50–500 MHz range. Amplifiers can be used to extend the distance of broadband transmission. Broadband can carry voice, data and video information. The term broadband is sometimes used synonymously with the term wideband.

BROADCAST – Broadcast transmissions use a protocol mechanism whereby group and universal addressing is supported. Transmissions can then be sent to two or more communication stations at the same time, such as over a bus type local network configuration.

CARRIER – In the area of communications, a carrier is a continuous frequency transmission capable of being modulated or impressed with a second data–carrying signal.

CONTENTION – Contention on a communication channel is a condition when two or more user stations vie for the right to use the transmission channel. The channel in this instance may refer to a private branch exchange (PBX) circuit, a computer port, or a time slot within a multiplexed digital facility.

DIGITAL SWITCHING – In data communication networks digital switching is a process of establishing and maintaining a connection, under stored program control, where binary–encoded information is routed between two communicating devices within the network.

FDM – Frequency division multiplexing (FDM) is a technique used in communication systems for sharing a transmission channel among a number of users by dividing the available transmission frequency range of the channel into narrower frequency bands. Each of these narrow bands can then be considered as a separate channel for each user. In this way carrier signals of different frequencies can be transmitted down a channel or across a network simultaneously. It is used in broadband transmission. Modems which utilize FDM techniques use one frequency band for transmission from A to B and another frequency band for transmission from B to A. Within each band, two frequencies represent the mark (logic 1) and space (logic 0) signals. These four frequencies are selected to be far enough apart for filters at the modem to separate them, and prevent mutual interference. Consequently, each modem can use only a portion of the full bandwidth of the transmission line and therefore only a portion of the full transmission capacity.

QUEUE – A queue is any line, or list, of items, such as computer jobs or messages waiting for service.

SEQUENCING – When applied to the area of data communications, sequencing is the process of dividing a user data message into smaller frames, blocks, or packets for transmission. Each of these frames has an integral sequence number for re–assembly of the complete message at the destination end. (In large networks individual packets may be routed differently, due to changes in network operating conditions, and arrive at their destination out of sequence).

SWITCH MATRIX – In data communications a switch matrix is that portion of the switch architecture where the input and output communication channels meet. The matrix allows any pair of channels to be connected to establish a through circuit.

VIRTUAL CIRCUIT – In the area of communications, the term virtual circuit indicates the use of a packet switched network facility which gives the appearance to the user of an actual end to end circuit. Sequential user data packets may then be physically routed differently through the network during the course of a virtual connection but the appearance to the user is still of a dedicated link between the source and the destination.

Packet Switching: In a packet communication environment the nodes in the system communicate asynchronously over an interconnection network by sending fixed size packets of data. In this context packet switching can be defined as the transmission of data by means of addressed packets, whereby a communication channel is occupied for the duration of transmission of the packet only. The channel is then made available for use by packets being transferred between different data terminal equipment. The data to be transmitted across the network must be divided and formatted into a number of packets for transmission and multiplexing purposes.

Figure 12.10

Typical packet format used in a packet switched network.

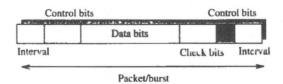

In data communications, a **packet** is a group of binary digits including data and control signals which is switched as a composite whole, with the data, call, control signals and possibly error control information being arranged in a specified format. The format of a typical packet used in packet switched networks is shown in Figure 12.10.

12.4 LOCAL AREA NETWORKS: SOFTWARE

In order to use a LAN efficiently it is important to have network operating software. The software controls the data traffic flow in the network, converts transmission and reception speeds, translates different data codes produced by different computers, detects and corrects errors, forms data messages and performs user access functions. In its simplest terms the network software may be seen to behave like a form of operating system which makes the resources dispersed across the network appear to be local to each user's workstation.

The network software is in fact a set of programs, or utilities, that resides for the most part on the network server. In this context a network server is usually a workstation or computer that handles special functions. For example a **file server** is used to store shared program and data files on a large capacity high speed hard disk, a **print server** is used to control a printer shared by other workstations on a network and a **communications** server allows network users to communicate with computers outside the network via serial ports and a high speed modem. If a workstation or computer is reserved solely for one or more of these tasks it is called a **dedicated server**. In large networks there may be several servers. The network software has four main tasks.

1. File Sharing: A major function of any LAN is to provide a means for users to share and exchange data and information files. The major flexibility in using a computer network is that it allows more than one user to have access to particular files, thereby allowing the sharing of information. In most LANs all the data and program files tend to be stored at a central location on a large hard disk store, under the control of the file server software. The file server computer may contain several hard disk drives, each of which in turn is subdivided into sections called volumes, each of which will have a distinct name. The network operating software controls the different hard disks and makes the data files available to users.

It is worth noting that files stored on the file server can only be accessed by authorized users. Each user is assigned a unique user name and an optional password which identifies that user to the server. Most commercial network software provides a sophisticated security system that allows a user to decide which of the other network users should have access, or rights, to his or her files.

2. Program Sharing: In most companies, networks are used as a central repository for shared files and as a central connecting point for shared resources, such as printers and CD–ROM drives. Since all network users can access groups of files on a file server, it makes sense to also use the network as a storehouse for application programs. Keeping the application software used by an organization on a file server within the network offers two major advantages.

LAN
software

(a) It eliminates the need to store a single copy of each application program on every PC in the network. Any user can then access the application software they need using little or no local storage space.

(b) It ensures that each user in an organization is using the same version of an application. Using application software in this way requires the purchase of a site licence for each application package as it is usually illegal to share a single copy of a proprietary piece of software among multiple users without permission from the software manufacturer.

3. Printer Sharing: The third major function provided by the network software is the ability for many users to share printers. The network software handles the queueing of the different jobs in the system to ensure that printing services are provided to the users on the network on a first come first served basis.

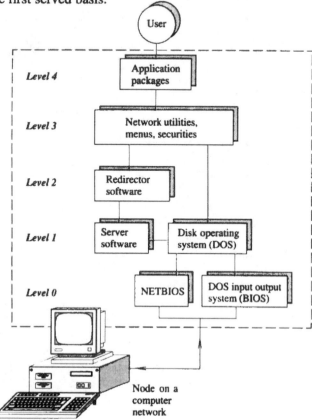

Figure 12.11

**Software layers
in Network software.**

4. Electronic Mail: Network software allows users on the network to send and receive messages to/from other users on the system.

From this description it can be seen that the network software manages the flow of data in the system and controls access to files, programs, printers and other services, collectively called resources, that are attached to the network. The software is also usually sophisticated enough to be able to control a network constructed using a variety of cable types, network configurations and network protocols.

The actual structure of the network software is constructed in layers (Figure 12.11). The function of each of the layers is described below.

LAN
software

Level 4   At the top level are application packages like word processors, spreadsheets, and data base managers.

Level 3   Network utilities, menu systems and security programs make up the level 3 layer of software. These are programs that users can all upon to set up workstations to run on the network, send messages to other workstations, send files to the print server to be printed, gain access to disks and directories on file servers and see network status reports.

Level 2   The Redirector layer acts as a traffic controller for the data and messages transmitted over the network.

Level 1   Level 1 software works beside the DOS operating system. It allows workstations to share disks and peripheral devices across the network as well as communicate with other workstations.

Level 0   NETBIOS routines are added at the lowest level of the network software to send and receive data to and from the network adapter board just as ordinary BIOS routines control the serial and parallel interfaces.

Table 12.3

Different types of utilities available with Novell NetWare.

	NetWare Utility	Description
1.	Command line utilities	These are executed from the DOS command line. They can be used to view listings of files, directories, users, copying and printing files and logging into and out of file servers These utilities perform tasks directly, without leading you through menus and screens. If you are familiar with the command line utilities they are faster and easier to use than the menu utilities.
2.	Console commands	These are entered at the file server console to monitor and control various file server activities. They can be used to change memory allocations, to monitor how the file server is being used and to control the way workstations can use the file server's resources.
3.	Loadable modules	These are used to link file server management and enhancement utilities with the operating system.
4.	Menu utilities	These can be used to perform network tasks by choosing options from menus. You can perform most tasks by using either menu utilities or command line utilities. Some menu utility tasks can be performed only by supervisors or users who have operator status.
5.	SERVER executable file	This file boots the NetWare operating system on the file server. After SERVER executes, LAN and disk drivers can be loaded to complete the operating system configuration.

UTILITY	DESCRIPTION
1. Command Line Utility	
GRANT	This allows access to be granted in a specified directory to another user or to a group of users.
HELP	This allows access to customized 'Help' menus the user has created.
LOGIN	This connects a user's workstation to the network and runs a login script.
MENU	This runs the NetWare Menu program.
NDIR	This is a network replacement for the DOS DIR program. It shows the access and other network specific information for each file.
NPRINT	This prints a file to the network queue.
SETPASS	This is used to change a user's password.
USERLIST	Lists all users currently connected to a server.
2. Console Command	
BROADCAST	This can be used to send messages to all users logged in to the file server.
CLS	This clears a user's VDU screen.
MEMORY	This is used to display the total amount of installed memory that the operating system can address.
SEND	This can be used at the file server console to send messages to all users logged in or attached to the file server.
TIME	This displays the date and time kept by the file server clock.
3. Loadable Module	
EDIT	This allows a text file to be created and modified.
INSTALL	This is the main menu command for installation.
PSERVER	This is used to locate the print server on the file server.
4. Menu Utilities	
COLORPAL	This sets the colours used by the menu driven NetWare utility programs.
FILER	This is a menu driven program that allows you to move, copy and delete files and directories.
PCONSOLE	This allows print jobs to be viewed, deleted and changed.
SESSION	This is a menu driven multifunction program which can be used to attach an additional file server, view and change drive mappings, send messages to other users, see who is logged in to the network and log out from a file server.
SYSCON	This is a multipurpose, menu driven program which can allow a user to, attach to an additional server, select a server as the current server, list the groups on a server, change a user's password, create or modify a login script, grant full rights to users, view a users account balance.

Table 12.4

Example of the different utilities available with Novell NetWare.

LAN software

Most networks have a system administrator who will have installed the network and monitors the network operation. However once the network software has been installed on the servers and computers attached to the network, the system can then be accessed by various users. To access a LAN it is usually necessary to LOGIN to the network. The user will then enter a user name and a password to identify the user to the file server.

If these items have been entered correctly the user is granted access to the different disk drives and printers installed on the network and to certain files on the network.

When the network is operating, the network media, interfaces, topologies, protocols and software layers are generally transparent to the end user. However, the user can interact with the software through commands or requests entered at the command line in their workstation or PC. Most of the commands operate in the same way as DOS operating system commands on a stand alone machine except the user has additional disk drives and peripheral devices which can be accessed on a network.

LAN software

On entering a command, or requesting a utility, the network software recognises the user, associates their pre–programmed privileges with their identities and then re–routes their commands and utility requests to the appropriate server for action. Running an application programs on a network is also similar to running them on a stand alone computer. Many database management systems, such as dBase III+ for example, are designed to run on networks and so using them within a network should pose few problems to users. Such software programs employ a variety of file locking, record locking and other security measures to prevent users from destroying each other's data.

As has been stated the network software tends to consist of a number of utility programs which allow each user on the network to perform various tasks. Table 12.3 presents an example of the five types of utilities available with NetWare, a local area network operating system produced by Novell Inc. NetWare operates LANs using standard IBM personal computers running MS–DOS or OS/2 and for Apple Macintosh computers. It was first introduced in 1983, since when there have been a number of versions each offering additional features. An example of some of the utilities is given in Table 12.4.

12.4.1 NETWORK APPLICATION SOFTWARE

All the utilities noted in Table 12.4 are invoked from the command line on the PC. While it is very powerful it does make access to the many network facilities awkward.

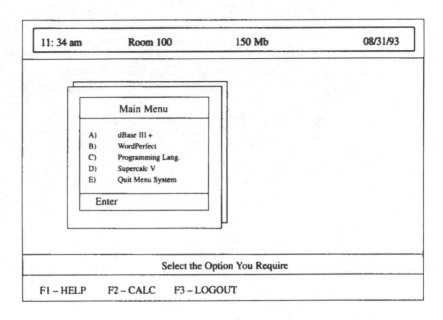

Figure 12.12

Example of a menu screen created with Saber software.

There are, however, a number of third party software vendors who have developed application programs which can interface to network software such as NetWare. These packages can make certain facilities available with the LAN software more user friendly.

1. Saber Menu: Saber Software Corporation's Saber Menu program picks up where Novell's MENU program leaves off. It allows sophisticated professional looking menus to be created for network users. An example of the type of menu that can be created is shown in Figure 12.12. The Saber Menu software includes dozens of features that are not available with NetWare's own menu system. It also supports mouse interaction and overlapping menu window displays.

Network software

2. ShareFax: This application software allows the transmission and reception of FAX messages via a hardware FAX board within the network. It allows one FAX board to be easily accessible by many users on the network.

3. Network HQ: This application software package allows a system manager to maintain a database of information about each network workstation. It allows detailed information on the machine and resident operating system software, CPU and amount of memory and disk space to be displayed to the system manager.

12.5 MODULATOR/DEMODULATOR (MODEM)

In order to communicate over long distances it is necessary to make use of long distance or wide area communication networks. In most cases these networks are usually the common telephone system.

A wide area communication network is simply one which spans large geographic distances. The processors in a wide area network, or **distributed processing system**, are physically dispersed with each processor having its own memory and I/O devices. The individual computers are independent units which although they rarely coordinate to perform any system task, may from time to time access one another's data bases. To facilitate this interaction the various units within the distributed system will have a communication network linking them, which each computer in the network is eligible to control. In this way the distributed computer system is a configuration part way between a centralized and decentralized organization. A **decentralized computer system** is one in which the computer support, both computer equipment and staff members, is placed in separate organizational units and dispersed at different locations and departments throughout an organization. There is limited or nonexistent communication among units.

Figure 12.13 shows the typical connections required to interface a computer to the telephone network. There are essentially two parts to this, the DTE and the DCE.

Figure 12.13

Modem connections.

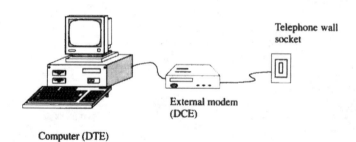

Telephone wall socket

External modem (DCE)

Computer (DTE)

Data Terminal Equipment (DTE): DTE is equipment that originates or receives digital data. DTE is used in the field of data communications and generally refers to the end–user devices, such as terminals or computers, that manage the interface. The input data from a communication network is usually processed initially by data communications equipment (DCE) before the message is sent to the DTE.

Data Circuit Terminating Equipment (DCE): DCE is the equipment which transforms the transmitted digital data from the DTE into a suitable form for transmission over a telecommunication network. It is the equipment which provides an access point to a communications network by establishing, maintaining and terminating a connection between two communicating parties. It must also perform the signal conversion and coding required for the communication between the DTE and the data circuit. For computer communications the DCE may or may not be an integral part of the computer.

Figure 12.14 is of a data communication link showing the DCE and DTE equipment. The digital data from the DTE cannot be transmitted directly onto a standard analogue telecommunications network as the digital signals contain high frequency components which would be attenuated by the 300 to 3300 Hz band limited telephone voice network.

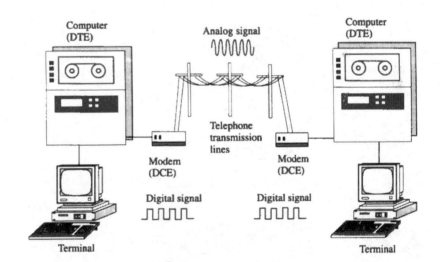

Figure 12.14

Operation of DTE and DCE equipment.

For passage over such a network it is necessary for the DCE device to convert the digital signal to a modulated time–varying carrier signal. The DCE therefore modulates a carrier signal in the proper frequency band for transmission over the network and then demodulates the signal at the receiving end of the network. The DCE which performs the signal transformation for analogue telecommunications is often called a **modem**, a contraction of modulator/demodulator. The DCE equipment also includes signal conversion devices, error protection devices, automatic calling devices and so forth.

The connection between the DTE and DCE is called the digital interface. In interfacing digital networks onto the long distance telephone network there may only be one DCE to supply the needs of the whole digital network. In this case the modem may be connected onto a multiplexor which can provide data compression in the high speed communications channels to/from the digital network stations.

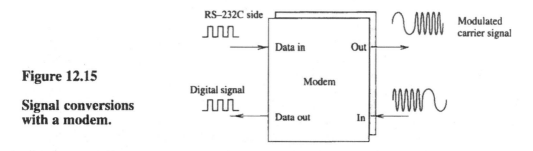

Figure 12.15

Signal conversions with a modem.

In operation the modem can act as both a transmitter and a receiver, allowing the conversion of the digital logic level signals into a modulated carrier frequency at the transmitter, and the decoding of this carrier at the receiver back into the logic levels. This is shown in Figure 12.15 where the digital data into the modem gets converted from an RS-232 signal to a modulated analog output and vice versa. Article 12.3 describes one of the first types of modem, the acoustic coupler.

12.5.1   MODEM OPERATION

The most important modem types include the 300 bits per second (bps) full duplex asynchronous modem and the 9600 bps synchronous type. Figure 12.16 shows a schematic diagram of a 300 baud full duplex modem using frequency division multiplexing.

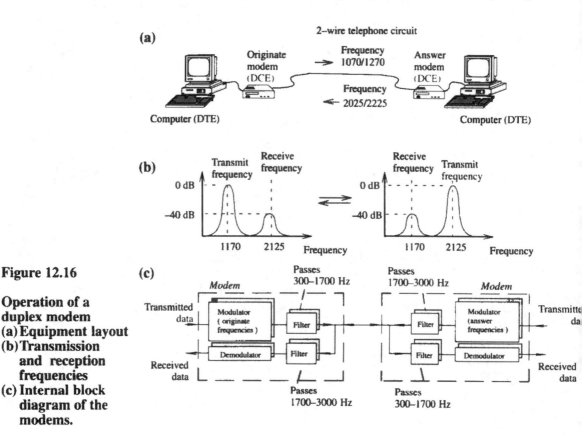

Figure 12.16

Operation of a duplex modem
(a) Equipment layout
(b) Transmission and reception frequencies
(c) Internal block diagram of the modems.

An acoustic coupler is an older type of modem fitted with a loudspeaker and a microphone in which the coupler takes digital signals and modulates them as sound waves before transmission over the telephone lines. A diagram showing the connection of the device to a telephone is given in Figure 12.17.

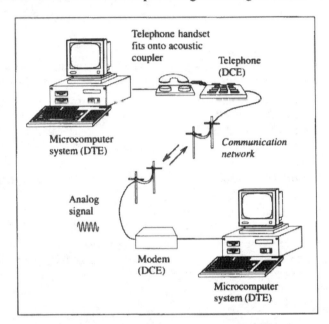

Figure 12.17

Connection of an acoustic coupler modem to a DCE and DTE.

The acoustic coupler allows a telephone handset to be used for access to a switched telephone network and is primarily used for data transmission between computers. The handset of the telephone terminal is fastened to the acoustic coupler via rubber cups which help to prevent external noise from being picked up and transmitted. In the transmission mode the acoustic coupler generates tones which are transmitted through the network to a receiving modem, while in the reception mode the tones from the transmitter modem are picked up by the acoustic coupler and decoded back into data. The standard modulation method used with acoustic couplers is frequency shift keying.

Data rates are typically limited to about 300 bps; however, some are able to transmit at rates up to 1200 bps. Acoustic couplers therefore provide a simple data transmission facility over a dial up circuit without the need for a special electrical connection to the network.

Modem operation

From the figure it can be seen that there are essentially two frequency blocks: one block (the message originator) transmits data using a frequency of 1170 Hz ± 100 Hz and receives data at a frequency of 2125 Hz ± 100 Hz, while the other (the answer modem) operates in the opposite way. The frequency spectrum of both modems is shown in Figure 12.16(b) while part(c) shows a block diagram of the system when it is functioning. Figure 12.16(c) illustrates the fact that both modems can operate at the same time, allowing for duplex transmission.

12.5.2   MODEM CIRCUITS

In a modem there are many different parts ranging from the circuits that perform the modulation and demodulation to those that perform the timing and control functions. The interchange circuits of the digital interface consist of a number of different circuits.

Data Circuits – The data circuits handle the outgoing digital data, to be transmitted by the modem and the incoming data that has been digitized by the modem.

Control Circuits – Control signals are required to cause the modem to be connected to or disconnected from the transmission line, to indicate the direction of transmission, and to note whether or not a connection to a remote modem exists. Other control signals may need to be produced to indicate the speed of the modem, the quality of the received data, and the results of self–test operations.

Timing Circuits – The timing circuits in the interface are used only when the modem is operating in a synchronous mode to provide a clock signal that is generated at the sending station and sent over the line, for recovery by the receiving modem. This clock signal controls both the transmitting and the receiving serial circuitry, thus ensuring synchrony between the two ends of the link.

Error Detection Circuits – When modems are used within a communication network, care must be taken to prevent the transmission of errors through the system to the DTE reception unit. Since the level of noise in telephone networks is high, error detecting circuitry must be incorporated into the modem.

Interface Circuits – In order to achieve compatibility between modems and computers there is a need to standardize the interface. The adoption of standard interfaces makes it possible for a computer manufacturer to provide an interface which will permit communication with many types and makes of terminals, peripherals, instruments and other computer types. The EIA RS–232C interface standard is almost universally used for the interconnection of terminal equipment that receives or transmits serial asynchronous data.

12.5.3   MODEM CHARACTERISTICS

It is possible to produce many different types of modems, depending on certain factors that can be altered within the device. Among the factors that can be altered are

1. the type of modulation, e.g. amplitude, frequency or phase;
2. the transmission protocols available, e.g. simplex, half duplex or full duplex;
3. the data transmission mode, e.g. synchronous or asynchronous;
4. the speed at which the modem can send and receive data;
5. the international standards for establishing communication.

1. Modulation – Modulation is a technique used to systematically change the properties of an analog signal in order to encode and convey information across a communication medium. The analog signal can be varied in one or more of three characteristics – amplitude, frequency or phase. It is the technique which a modem uses for transforming the digital data from the computer into an analog form suitable for the telephone network. Demodulation is the extraction or decoding of the transmitted digital signal information from a modulated analog carrier signal.

There are a number of different forms of modulation such as amplitude, frequency and phase. Figure 12.18 notes the three types of modulation.

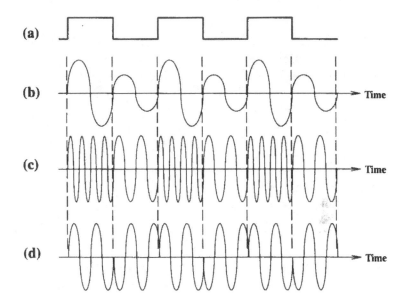

Figure 12.18

Types of modulation
a) Digital input
b) Amplitude modulation
c) Frequency modulation
d) Phase modulation.

(a) Amplitude Modulation (AM): AM is a communications transmission method in which variations in the magnitude of a voltage, or current, carrier signal, indicates the encoded information. When amplitude modulation is used to convey digital information the carrier signal is usually switched through on a logic 1 and cut off on a logic 0. Figure 12.18(b) shows an example of amplitude modulation.

In terms of quality, modems using the amplitude modulation method alone are not usable on a packet switched transport network (PSTN). The fact that the receivable information depends totally on the amplitude of the received analog signal means that any changes in the analog amplitude affects the information. PSTN channels do not however always maintain constant attenuation. This, plus the high level of impulse noise which occurs on PSTN channels would result in a high error rate if AM alone was used.

(b) Frequency Modulation (FM): FM is a communications method in which the frequency of a transmitted analogue carrier wave is altered to correspond to changes in the input information. **FSK (Frequency Shift Keying)** is a specialized form of frequency modulation used to encode digital data, where a data 1 bit is represented as one frequency and a data 0 bit as another, different, frequency. It is frequently used in modems to transmit data between computers. Figure 12.18(c) presents an example of FSK. It can be seen from the figure that a logic 1 corresponds to four cycles of one carrier frequency while a logic 0 corresponds to two cycles of a second lower carrier frequency.

Modulation

Modems using frequency modulation principles are in common use, despite two main drawbacks. Many frequencies on a packet switched transport network are allocated for other purposes, eg. signalling, and therefore cannot be used as carrier frequencies. At least half of the carrier frequency must be received in order that a modem can demodulate the analog signal, since it is not technically possible to demodulate it otherwise. This means that the digital message signal must not change faster than twice the lower output frequency – a limiting factor on the digital signaling rate.

(c) Phase Modulation: Phase modulation is a technique of encoding digital data into th different phases of a modulated analog signal. There are basically two types of phase modulation, phased shift keying and differential.

Phased shift keying is a form of phase modulation in which a logic 1 in the digita message signal causes an output carrier signal of a set frequency to be produced by the modem, while a logic 0 causes the carrier signal to be set 180° out of phase with that for logic 1. Figure 12.18(d) shows an example of phase modulation being used on a digital inpu signal. It can be seen that each change in the input signal causes a shift in phase of the outpu signal.

Differential phase modulation is a modulation technique in which each bit to b transmitted alters the phase of a carrier frequency by an amount depending on which logi state is represented. For example a bit 0 may be coded as a $+90^\circ$ phase change, whereas 1 bit may be coded as a $+270^\circ$ phase change. As only two phase changes are used to modulate the carrier, this method is referred to as differential two phase modulation. However there is nothing to stop more phases being used to modulate the carrier, each phase representing say, a combination of bits. For example, differential four–phase modulation is a method i which four different combinations of two bits modulate the carrier with four different phase changes. With such encoding it is worth pointing out that the modulation rate, baud, is no the same numerical value as the signaling rate in bps. With differential two phase encodin there are in fact two bits per baud.

Modulation

(d) Quadrature Amplitude Modulation (QAM): QAM is a sophisticated modulatio technique using variations in signal amplitude and phase to allow data encoded symbols t be represented as a number of different states. It is sometimes known as amplitude phase modulation. In its most popular form it generates an eight–level encoded analog signal, a each different group of three bits is represented by one of eight different combinations o carrier amplitude and phase.

2.   **Modem Protocols** – The term protocol when used in data communications refers to set of formal rules governing the format, timing, sequencing and error control of exchange messages on a data network. It may also designate procedures for managing communication links and contention resolution. By using a standard convention for the transmission of data devices are then able to communicate easily. A protocol may be orientated toward data transfer over an interface between two directly connected logical units, or on an end to end basis between two users within a large a complex network.

Protocols are required to control the orderly flow of data from one computer to anothe at maximum efficiency and with the minimum number of undetected errors. These protocols can have many different means of expression. For example, in the transmission of information among computer systems the protocols to be used can be either bit–orientated byte–orientated or character–orientated. The last term refers to a communication protoco or transmission procedure that carries control information encoded in fields of one or more bytes.

One of the most important protocols for modems is whether a modem uses simplex, hal duplex or full duplex methods of transmission.

(a) Simplex: Simplex communication allows only transmission in one direction e.g. from a computer to a terminal.

(b) Half Duplex (HDX): In a data communications system, HDX is an operational mode of serial communications, where transmission may occur in both directions but only in one direction at a time. The transmission directions between two communicating parties are then alternately switched to accommodate any twoway data flow.

(c) Full Duplex (FDX): A full duplex communications link is one in which data transmissions between two communicating devices can occur in both directions at the same time. Although this form of transmission is particularly useful when large amounts of data are to be transmitted, it is more expensive than half duplex and the interfacing required for computers and terminals is much more complicated. Full duplex modems use either frequency division multiplexing (FDM), or echo cancelling techniques to achieve bidirectional data transfers over a two wire link.

The distinctions between these types of communication depend on the line, the software, and the hardware used to connect the units to the transmission at either end.

3. Modem Transmission Mode – The timing of the transmission and reception of the data to and from a modem is accomplished through the use of oscillators or clocks located in either the DTE equipment or in the modem itself. In the case of high speed synchronous modems both the DTE and the modem need timing information. This is achieved by using a single clock in either the terminal or the modem and then copying the timing pulses to the other device over the interface.

With the synchronous system the clock then determines the rate of transmission and at the transmission end it tells the modem when to take a bit from the DTE and modulate it onto the line; at the reception end it tells the modem when to sample the line for an incoming bit; at both ends it facilitates the orderly passing of bits over the interfaces.

An asynchronous modem utilizes an asynchronous mode of data transmission. It characteristically operate at speeds of 1200 bps or less, but some can achieve higher asynchronous data rates, up to 9600 bps, by using asynchronous to synchronous converters. These converters allow asynchronous DTE units to be interfaced to synchronous modems by synchronizing the incoming data with the modem's clock. However asynchronous modems are still best suited to low speed low volume data.

It is also normal in asynchronous modems that when the transmission line is not carrying data it will transmit a continuous carrier. If the carrier is regarded as a logic 1 then the start of a character in this form of transmission will be indicated by the transmission of a start bit of zero, which causes the receiver to activate its clock, the opposite condition of the carrier. Immediately following the start bit, or start bits, will be the data bits, usually seven of them.

4. Modem Speed – The speed at which a modem can transmit and receive data can either be expressed in terms of bits per second (bps) or baud. These two need not be the same since in phase modulation, for example, a single change of phase in the analog signal can represent a group of two or more data bits. Hence the actual bit rate can be higher than the baud rate, depending upon the modulation technique used and the number of bits transmitted per baud.

The narrow bandwidth, 300–3300 Hz, of public switched telephone lines does not allow baud rates greater than 2400 to be used. To speed up data transmission it is therefore necessary to devise techniques to allow multiple bits to be encoded as single–baud items. The desire to transmit digital data over the telecommunications network at higher transmission rates has led to the design of a number of relatively complex frequency, phase and amplitude modulation schemes.

Until 1980, modems had a maximum transmission speed of 9600 bps, however in th
early 1980's modems with transmission rates higher than 9600 bps appeared using trelli
coded modulation (TCM). These modems can operate at rates up to 19 200 bps by addin
error correction to the transmitted sequence of signal points. This is close to the widel
regarded theoretical maximum of 20 000 bps. By adding data compression technique
modems can transmit at up to 38 400 bps using the V42bis standard. However, the mo
widely used standard bit rates for analog modems are probably still 300 bps and 9600 bp

5. Modem Standards – The three main organizations which issue standards related t
electrical and electronic equipment are the Electronics Industries Associations (EIA), th
International Consultative Committee for Telephony and Telegraphy (CCITT), and th
International Standards Organization (ISO).

The data rates of typical two wire full duplex modems are summarized by the CCITT
committee in the V series standards as shown below:
1. CCITT V.21 relates to a 300 bps modem
2. CCITT V.22 relates to a 1200 bps modem
3. CCITT V.22bis relates to a 2400 bps modem
4. CCITT V.23 relates to a 1200 bps/75 bps modem
5. CCITT V.32 relates to a 9600 bps modem.

These modems are generally used over the public switched telephone networks, th
actual speed of operations depending on the quality of the particular telephone network.
The CCITT recommendations for data transmission over digital networks are known as th
X series. Some of the more popular are X.21 and X.25.

X.21 is a general purpose synchronous DTE/DCE interface to public digital dat
networks. Recommendation X.21(bis) contains the complete specification whe
interconnecting a synchronous terminal to a modem. It is almost identical to the mor
common RS–232 interface used in many systems though it uses a smaller set of wires, eigl
wires are used, within a 15–pin connector.

The recommendations which specify the DCE/DTE interface onto packet switche
networks is recommendation X.25. The X.25 standard is defined in terms of the lower thre
layers of the seven–layer reference model of the Open System Interconnection from the ISO
It is often referred to as the three–level protocol of the X.25 network.

12.6 COMMUNICATION SOFTWARE

Communication application software is different from a LAN operating system in tha
it is only responsible for establishing and maintaining communication between tw
computers and it is not responsible for managing a local area communications network. I
most cases it is used to establish the communication protocols and transmission between tw
computers usually across a long distance telephone network. From this description it can b
seen that the communication software has several tasks, these are noted in Table 12.5

The use of communication software allows computer users to keep in touch with on
another across large distances. They are then able to exchange ideas with other users of th
same type of computer equipment through bulletin boards and electronic mail and acces
information services and commercial data bases. More advanced communication softwar
can also control teleconferencing, gateway services and videotext facilities.

Initialization	The communication software allows the serial interface and modem attached to the local host computer to be initialized. Before communicating with another computer both must agree on the file transfer protocols. These are the set of rules for the transfer of data. The parameters which govern these rules are: (a) baud rate of the transmission (b) number of data bits used to create each character (c) number of stop bits (d) type of parity used for error checking during transmission.
Establish communication	It lets you dial the requested computer's number to be dialled and in certain cases informs the distant computer of the communication parameters.
Control data transfer	Once a connection is made it acts as the vehicle with which a user can work with the host computer. It also provides ways for the user to transmit and receive data to and from the distant computer by directing outgoing data from the keyboard or disks through the communication port and into a modem and directing incoming data from the communication port to the screen or disk.
Terminate communication	When finished working with the distant computer, the communications program terminates the session, hangs up the phone and return to the operating system of the local host computer.

Table 12.5

Communication software tasks.

Examples of this type of software are Crosstalk, PC–TALK, Procomm and SmartCom. Figure 12.19 notes the various features available through the Procomm communication software. As well as being able to initialize various parameters, the software allows the uploading and downloading of files from and to the local host computer. Uploading occurs when data from the user's disk are sent to another computer while downloading occurs when data are received from another computer and saved on the user's disk.

Figure 12.19

Features to be found in Procomm software.

ProComm Help

MAJOR FUNCTIONS		UTILITY FUNCTIONS		FILE FUNCTIONS	
Dialling Directory	Alt–D	Program Info	Alt–I	Send Files	PgUp
Automatic Redial	Alt–R	Setup Screen	Alt–S	Receive Files	PgDn
Keyboard Macros	Alt–M	Kermit Server Cmd	Alt–K	Directory	Alt–F
Modem Parameters	Alt–P	Change Directory	Alt–B	View a File	Alt–V
Translate Table	Alt–W	Clear Screen	Alt–C	Screen Dump	Alt–G
Editor	Alt–A	Toggle Duplex	Alt–E	Log Toggle	Alt–F1
Exit	Alt–X	Hang Up Phone	Alt–H	Log Hold	Alt–F2
Host Mode	Alt–Q	Elapsed Time	Alt–T		
Chat Mode	Alt–O	Print On/Off	Alt–L		
DOS Gateway	Alt–F4	Set Colors	Alt–Z		
Command Files	Alt–F5	Toggle CR–CR/LF	Alt–F3		
Redisplay	Alt–F6	Break Key	Alt–F7		

PIL Software Systems

12.7 OPEN SYSTEM INTERCONNECTION (OSI) MODEL

There are many different types of computer and communications networks, such as wide area networks, packet switched networks, local area networks, satellite networks, circuit switched networks, cellular radio networks and global networks. Unfortunately many different protocols have existed across the broad spectrum of communication pathways, hence internetwork communications has proved difficult and expensive, requiring the services of complex gateways and other protocol conversion devices.

In 1977 the International Organization for Standardization (ISO) in Geneva, Switzerland, began to develop standards that would allow open system interconnection. It set out to produce a standard which when followed by manufacturers would mean that all networks could interconnect easily. In 1983 the basic reference model for the Open System Interconnection (OSI) standard ISO 7498 became an international standard. The model finally agreed upon by the ISO defines a seven layer model to be used as a framework for defining the communication process between systems. It was intended to be helpful when designing open system networks and to act as a framework for coordinating the development of OSI standards.

The OSI model refers to a seven layer architectural reference model and to the set of standards that describe how to provide communications among computers and terminals. It uses a layered structure to break up the communications problem into manageable pieces and identifies the functions at each of the seven level required for computers to transfer information and to internetwork for achieving common distributed goals. Later ISO standards defined the implementations for each layer to ensure full compatability is achieved at each layer. Using the OSI model it has become possible to design comprehensive communications networks that perform both packet and circuit switching.

Within the OSI model messages originating in an upper layer are transmitted down through the lower layers to the physical transmission medium. The reverse procedure occurs at the receiving device. The OSI model does not give the specific details for implementation of each layer but provides a framework for standards to be developed for each layer. The details of the functions of each of the layers is given in Table 12.6.

Now the combination of global, wide area and local area networks makes it possible to create multi–network systems that could offer economically feasible processing of huge data sets. Also as a result of the development of a large assortment of integrated circuitry and the enhancement of data transmission techniques, various forms of information other than computer, data and voice communication have begun to be transmitted in digital form.

It has then become possible to create comprehensive communications networks that perform both packet and circuit switching. Such networks have come to be known as Integrated Services Digital Network (ISDN).

12.7.1   INTEGRATED SERVICES DIGITAL NETWORK (ISDN)

An ISDN may be defined as a network that provides end to end digital connectivity to support a wide range of services, including voice and nonvoice services, to which users have access by a limited set of standard multipurpose user/network interfaces. It is the communications industry's vision of the public telephone network of the future and may be seen as an evolving body of international telecommunication standards being established by the CCITT.

OSI LEVELS	GENERAL CHARACTERISTICS
7. APPLICATION	User written programs
	File transfers
	Resource sharing
	Network management
	Access to remote files
	Database management
	Planning network operations
	Control of network
	Operation of network
6. PRESENTATION	Defines I/O procedures
	Controls network functions of application layer
	System dependent process to process communication
	User application connections
5. SESSION	Definition of connected session services
	Dialogue coordination
	Management of network resources
4. TRANSPORT	Provides location–independent transport of packets
	Provides this end to end communication control, once data path has been established
3. NETWORK	Distributed control policy that can allocate network management to different messages on the network
	Addresses messages
	Sets up paths between nodes
	Controls message flow between nodes
	Provides control and observation function for network planning and operation
2. DATA LINK	Frames message packets
	Locates channel capacity
	Determines station addresses to receive data
	Error detection
	Establishes access to physical link
1. PHYSICAL	Provides electrical transmission of information, 1s and 0s, over the physical medium
	Encoding
	Decoding
	Physical connection
	Signaling

Table 12.6

Layers within the OSI model.

ISDN is based on the development of digital transmission and switching technologies and their use to construct open networks for telecommunications. The main feature of the ISDN is the support of voice and non voice in the same network. For users, ISDN will provide end to end digital access for all communication applications over existing phone lines, without modems. Many software services should become available such as electronic mail, video conferencing etc. A key element of service integration for the ISDN is to provide a limited set of multipurpose user interface arrangements.

The bandwidths of the ISDN specifications and other communication networks are shown in Table 12.7.

ISDN

CHANNEL	CHARACTERISTICS
Phone line	19 200 bps
Ethernet LAN	10 Mbps
Token ring LAN	4 to 16 Mbps
Basic ISDN	64 000 bps
Primary ISDN	1.5 Mbps
Broadband ISDN	150 Mbps

Table 12.7

ISDN bandwidths.

To enable the new facilities it is intended that new technology will turn an ordinary low bandwidth telephone line into a high bandwidth multiple channel communication line. Basic ISDN splits a phone line into two 64 000 bps B channels and a 16 000 bps D channel. By creating multiple channels ISDN frees the B channels for carrying more data and makes way for two–way transmission. These digital transmission networks are expanding rapidly in many countries and as the standards set down by ISDN continue to grow in importance, digital transmission will become even more widespread. To provide conversion between analog and digital domains a **Codec (coder/encoder)** device is used. A codec is an integrated circuit (IC), or series of ICs, that performs a specific analog to digital conversion, usually the conversion of an analog voice signal to a 64 000 bps digital stream or an analog television signal to a digital format.

ISDN

It is widely regarded by many commentators that ISDN is set to revolutionize existing computer networks and may hold the most important development in the computer communications industry during the 1990s. Although ISDN is still evolving in technology and standards, a distinct visualization of the architecture, design approaches and services of ISDN is emerging.

With the development of ISDN in the future, computer networks will no longer be the stand alone items often found in many establishments but will simply become one node on a global computer network. This should allow the information in all the data banks throughout the world to be accessible from any part of the globe. Such freely available and easily accessible knowledge should benefit not only the research establishment but also business companies.

12.8 KEY TERMS

Acoustic coupler
Acquisition time
Amplitude modulation (AM)
Automatic request for repetition
 (ARQ)
Backbone network
Baud
Baseband
Bits per second (bps)
Block
Blocking
Broadband
Broadcast
Bridge
Cable
Carrier
Carrier sense multiple access with
 collision detection (CSMA/CD)
CCITT
Check bit
Circuit switching
Coaxial cable

Codec
Code conversion
Communications server
Communication software
Computer networks
Computer network devices
Concentrator
Contention
Data circuit terminating
 equipment (DCE)
Data link
Data terminal equipment (DTE)
Data transfer rate
Decentralized computer system
Decibel (dB)
Dedicated server
Destination field
Digital switching
Distributed processing system
Differential phase modulation
Electronic mail

(Continued)

Key
terms

Encoding/decoding	Modem standards
Ethernet	Modem transmission mode
File server	Modem protocols
Frame	Modulation
Frequency division multiplexing (FDM)	Multiplexor
	Network (computer)
Frequency modulation (FM)	Network application software
Frequency shift keying (FSK)	Network transmission
Front–end processor	Optical fibre
Full duplex (FDX)	OSI model
Half duplex (HDX)	Packet
Header	Packet switching
High speed local network (HSLN)	Private branch exchange (PBX)
Integrated services digital network (ISDN)	Phase modulation
	Phased shift keying
International Standards Organization (ISO)	Print server
	Quadrature amplitude modulation (QAM)
LAN software	
Local area network (LAN)	Queue
Local interface	Router
Medium (transmission)	Sequencing
Message	Server
Message switching	Simplex
Modem	Switch matrix
Modem characteristics	Token ring
Modem circuits	Twisted pair
Modem operation	Virtual circuit
Modem speed	Wide area network

12.9 REVIEW QUESTIONS

1. What do you understand by the term data communication?
2. Name and describe four types of communication topologies?
3. List five types of communication networks.
4. Give six features of a local area network.
5. What does the term distributed data processing mean when applied to LANs?
6. Why can a local area network of computers do without a central network controller?
7. What does bits per second indicate?
8. What do the following terms mean when used to refer to computer networks: PBX, data transfer rate and baud?
9. Describe the process that a computer would use in an Ethernet system to send a message across the network.
10. What does the term collision detection refer to in an Ethernet network?
11. From what does the token ring network derive its name?
12. Contrast the function of a router and a bridge when used in a LAN.
13. Which of the communication devices would be used to interface a computer to a number of terminals to facilitate the use of a large scale central computer?

14. Note the three main types of cabling used in LANs.
15. Why can optical fibres carry data at much higher data rates compared to twisted pair wiring ?
16. Name and describe the three modes of data transfer across a network.
17. What advantages does packet switching have over message switching?
18. What do the following terms refer to: broadcast, contention and carrier?
19. Which of the two types of transmission techniques can support FDM, broadband or baseband?
20. Where are broadband communication methods most often used?
21. What is the function of the control bits added to each data packet in a packet switched network?

Review questions

22. In a short paragraph describe the main functions of the network software used to control a LAN.
23. What is electronic mail?
24. What uses would electronic mail be used for in a university?
25. Novell NetWare provides five different utilities; what are they?
26. What is the main function of the Saber Menu application software when used on a LAN?
27. How does a wide area network differ from a local area network?
28. Describe the difference between analog and digital communication.
29. Describe the various types of modems.
30. What advantages does an acoustic coupler have over other types of modem?
31. Draw a diagram illustrating the three types of modulation.
32. What do the terms FSK and QAM refer to when used to describe modems?
33. What is a communication protocol?
34. Define the terms
   (a) simplex transmission
   (b) duplex transmission
   (c) half duplex transmission.
35. Why do the transmitting and receiving clocks of modems have to be synchronized during message transmission?
36. Why is a bit per second not necessarily the same as a baud?
37. Note the four tasks that communication software must support.
38. Describe what is meant by down–loading information.
39. What are electronic bulletin boards?
40. What is the significance of the OSI model in linking communication networks?

12.10 PROJECT QUESTIONS

1. Describe several applications of data communications.
2. What are the advantages/disadvantages of distributed computing?
3. Identify factors that determine the speed at which data are transmitted through a LAN.
4. What are the issues to be discussed before a company installs a local area network
5. What alterations would there be in office practice when a company changes from decentralized processing at each computer to distributed processing in a networked environment?

6. What are the managerial difficulties in operating a local area network?
7. Prepare a list of all the data communication facilities at your college or university.
8. Investigate the process of optical fibre manufacture.
9. A large business is considering buying a local area communication network. Compare the advantages of the different topologies they could choose from, hierarchical, ring, bus or star if they require the computers attached to the network to operate independently and to be able to exchange information without the need for a central controller.

Project questions

10. What are computer viruses and how do they effect large computer networks? How can a company be protected from viruses?
11. Note a number of application areas where intelligent terminals would be used.
12. What is a terminal emulator?
13. Which of the following may be regarded as the most common piece of data communication hardware: dumb terminal, multiplexor, gateway or modem?
14. What is the relationship between distributed data processing and tele–communications?
15. Do telephone systems process information? Explain your answer.
16. In computer jargon what is voice mail?
17. Explain the process of error correcting in data communications.
18. If computers and communications are converging what will the computer systems of the future look like?
19. Investigate the use of bulletin boards.
20. Will fibre optic networks make satellite transmission obsolete?
21. Describe where a FAX machine would be used.
22. Identify several advantages of teleconferencing.
23. Explain the increasing demand for communications services throughout the world.
24. What are microwave systems used for in the computer communication process?
25. With suitable reference find out how geostationary satellites are used in communication networks.

12.11 FURTHER READING

There are a number of journals, periodicals and magazines targeted at users of computer networks. There are also frequent articles in magazines such as *Byte*, *EDN* and *Computer Design* indicating the current developments in network hardware and architectures.

Computer Communications (Articles)

1. *Computer Communications*, 2nd Edn, K. G. Beauchamp, Chapman & Hall, 1990.
2. *Introduction to Computer Hardware and Data Communications*, P. A. Goupille, Prentice Hall, 1993.
3. *Local Area Network Architectures*, D. Hutchison, Addison–Wesley, 1988.
4. *Data Communications and Distributed Networks*, 2nd Edn, U. D. Black, Prentice Hall, 1985.
5. *Information Technology Electronics Computers*, A. Kirk (Editor), GEJ Publishing Ltd, 1984.

Local Area Networks (Books)

1.. *PC Magazine Guide to using NetWare*, L. Freed and F. J. Derfler, ZD Press, 1991,

Local Area Networks (Articles)

1. When one LAN is not enough, W. Stallings, *Byte*, January 1989, pp. 293–298
2. Ethernet ten years after, R. Seifer, *Byte*, January 1991, pp. 315–321.

Modem (Articles)

1. Correct modem choice requires knowledge of communication needs, J. Douglass, *EDN*, September 1986, pp. 299 – 310.

Wide Area Networks (Articles)

1. Remote connections, R. Green, *Byte*, July 1991, pp. 161–168.

OSI Model (Articles)

1. Helping computers communicate, J. Voelcker, *IEEE Spectrum*, March 1986, pp. 61–70.
2. Fiber vs metal, J. Y. Bryce, *Byte*, January 1989, pp. 253–258.

Advanced Hardware

Part Six

Chapter 13   Advanced Computer Architectures

13.1   Complex Instruction Set Microprocessor (CISM)
13.2   Reduced Instruction Set Microprocessor (RISM)
13.3   Digital Signal Processor
13.4   Computer Architectures
13.5   Multiple Processor Systems: Examples
13.6   Integrated Circuit: Future Development

Advanced Computer Architectures

Chapter Structure

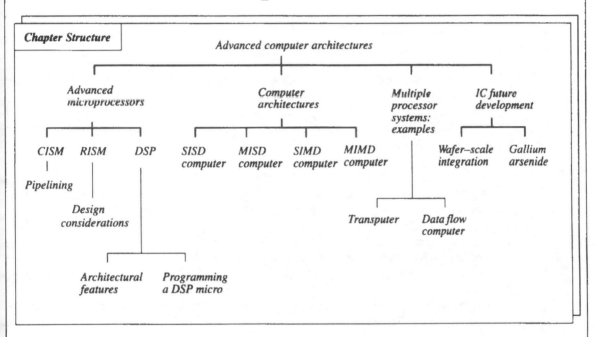

Advanced computer architectures

Advanced microprocessors · Computer architectures · Multiple processor systems: examples · IC future development

CISM · RISM · DSP · SISD computer · MISD computer · SIMD computer · MIMD computer · Wafer–scale integration · Gallium arsenide

Pipelining

Design considerations

Transputer · Data flow computer

Architectural features · Programming a DSP micro

Learning Objectives

In this chapter you will learn about:

1. The architectural features of modern microprocessors.
2. The features of RISC microprocessors.
3. The architecture of digital signal microprocessors.
4. The implementation of concurrency in computer systems.
5. The variety of MIMD computers.
6. The transputer and its application to multiple processing.
7. The projected growth in IC technology.

> *'The technological realities of today are already obsolete and the future of technology is bound only by the limit of our dreams.'*
>
> *How Wachspress*

Chapter 13

Advanced Computer Architectures

This final chapter considers the features and architectures of modern advanced computers. There are three parts to this study: the features of modern microprocessors, the architecture of high performance multiple computers, and a brief investigation into the future development of integrated circuit (IC) technology and how this will effect the computers of the future.

In deciding the major advances that have occurred in computer architecture over the past 45 years, Stallings notes six major advances that have directed the ways computers have evolved.

1. **Computer Families**: The family concept introduced by the IBM System/360 in 1964 followed shortly after by DEC. The family concept decouples the architecture of a machine from its implementation. A set of computers are offered with different price/ performance characteristics that present the same architecture to the user. The difference in price and performance are due to different implementations of the same architecture.

2. **Microprogrammed Control Unit**: This was suggested by Wilkes in 1951 and introduced by IBM on the S/360 line in 1964. Microprogramming eases the task of designing and implementing the control unit and provides support for the family concept.

3. **Cache Memory**: The use of a cache type memory was first introduced conceptually on the IBM S/360 Model 85 in 1968. The insertion of this element into the memory hierarchy dramatically improves the performance

4. **Pipelining**: Pipelining the instruction processing within a microprocessor allows a means of introducing parallelism into the essentially sequential nature of machine instruction program. Examples are instruction pipelining and vector processing.

5. **Multiple Processing**: By using more than one processing element to handle a program allows an enhanced performance from a computer.

6. **RISC Architecture**: The RISC approach rethought how the instruction sets of microprocessors were decided.

We have considered the first three of these major advances in the earlier parts of the book. In this chapter we will now consider the other three.

13.1 COMPLEX INSTRUCTION SET MICROPROCESSOR (CISM)

Since its introduction in the 1970s the development of the microprocessor has seen its hardware complexity increase dramatically, to the point where, for example, the present Intel 80486 32–bit microprocessor has over one million transistors. The number of machine code instructions required to manipulate the modern 32–bit microprocessors has also increased due to the numerous addressing modes and control signals necessary to deal with the variety of different hardware units that are contained within the devices.

A large number of machine–level routines are also required to handle the complex situations that can occur with systems utilizing them. Microprocessors with large **complex instruction sets** are often referred to as complex instruction set microprocessors (CISMs) while computer systems designed with processing units that use CISMS are referred to as complex instruction set computers (**CISCs**).

Table 13.1 presents some of the more recent CISMs. CISMs make use of a number of different architectural techniques to improve their performance.One of the most important is the use of pipelining.

Table 13.1

Examples of CISC microprocessors.

CISC microprocessor	Year released	Physical address size (bits)	Data bus size (bits)
Intel 386DX	1985	32	32
Intel 386SX	1987	24	16
Intel 860	1989	32	64
Intel 486DX	1989	32	32
Intel 486SX	1991	32	32
Motorola 68000	1980	24	16
Motorola 68020	1985	32	32
Motorola 68030	1987	32	32
Motorola 68040	1990	32	32

13.1.1 PIPELINING

Pipelining is a method of introducing some parallelism into the operation of a microprocessor by allowing several tasks to be performed in parallel within the microprocessor. There are two basic forms of pipelining: instruction pipelining and arithmetic pipelining.

1. **Instruction Pipelining** – The easiest conceptual form of pipelining is that employed in instruction pipelining, where the execution of a single instruction within the microprocessor is broken down into a number of different parts:

Instruction fetch: getting the instruction from the main memory;

Instruction decode: deciding what the instruction is to do;

Data fetch: fetching any data items necessary to execute the instruction;

Instruction execution: performing the operation associated with the instruction;

Write result: sending the result to a register or back to memory.

To use instruction pipelining, the microprocessor is designed with a number of separate hardware units which are arranged to carry out a number of the activities listed above. In particular this means that while one instruction is being fetched from memory another instruction can be decoded while another can have its data fetched from memory.

Figure 13.1 shows a five–stage pipeline which has been designed to allow three instructions to be in the process of being executed. For example, while the data for instruction 1 are being fetched, instruction 2 can be being decoded while instruction 3 is being fetched from memory. By overlapping a number of operations an increased rate of completion of instructions can be achieved.

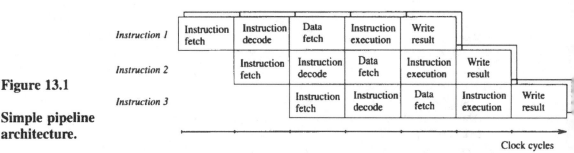

Figure 13.1

Simple pipeline architecture.

The Stretch computer designed by IBM and released in 1961 as the IBM 7030 was the first computer to employ a lookahead facility (a basic form of instruction pipelining) to pick up, decode, calculate addresses and fetch operands several instructions in advance.

Instruction pipelining

In terms of the implementation of instruction pipelining in modern CISMs there are two forms of architecture, superscaler and superpipelined.

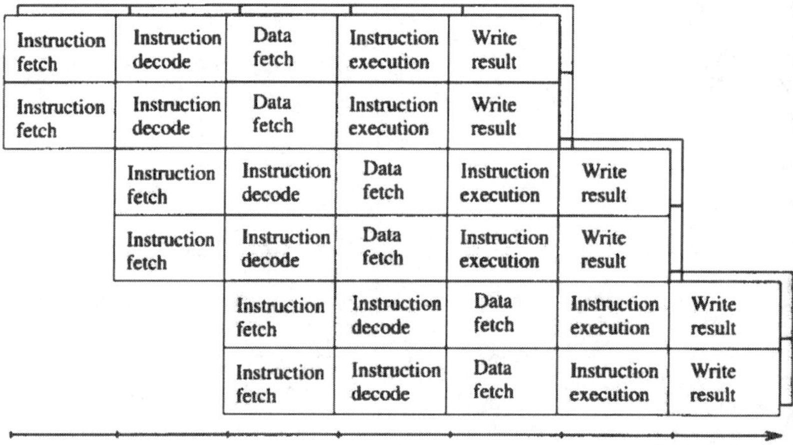

Figure 13.2

Example of the operation of a superscaler architecture.

(a) *Superscaler Techniques*: A superscaler architecture replicates each of the pipeline stages so that two or more instructions at the same stage of the pipeline can be processed simultaneously. The heart of a superscaler processor is then at least conceptually multiple pipelines where each pipeline is not significantly different from the basic pipeline of Figure 13.1. An equally important part of a superscaler processor is the instruction dispatcher that attempts to issue multiple instructions per cycle, one into each pipeline. Figure 13.2 shows the operation of a superscaler architecture in which there are two hardware units in the processor responsible for each of the five stages in the pipeline.

A superscaler architecture can be created by having different arithmetic logic units within a computer to deal with the different arithmetic tasks, such as addition, and multiplication, that may be encountered in a program. These multiple functional units can operate in parallel and so provide spatial parallelism. In this form of architecture common instructions – integer and floating point arithmetic, loads, stores and conditional branches can also be initiated simultaneously and executed independently and since almost every instruction stream involves some mix of additions and multiplications it is possible to keep many functional units busy at any time.

ARITHMETIC PIPELINING

In terms of handling arithmetic, microprocessors can either use fixed or floating point arithmetic.
Fixed Point Arithmetic – Fixed point arithmetic is the performing of arithmetical calculations without regard to the position of the radix point treating the numbers as integers for the purpose of calculation. The relative position of the point has to be controlled during calculations.

Floating Point Arithmetic – The term floating point refers to a numeric notation in which the integer and the exponent parts of a number are separately represented, frequently by two computer words; for example:

in base 10 $1024 = 1.024 \times 10^3$ the mantissa = 1.024, the exponent = 3
in base 2 $1024 = 1 \times 2^{10}$ the mantissa = 1, the exponent = 10

Using floating point arithmetic permits the computer to keep track of the individual exponent and mantissa parts of a number rather than the entire number itself. This allows the implied position of the decimal point to be freely varied with respect to the integer digits and means that numbers can be stored more economically and in wider ranges of magnitude. The four different operations required for a floating point addition using base 10 arithmetic are shown in Figure 13.3(a). Figure 13.3(b) presents the two different parts required for floating point hardware. Each half is dedicated to manipulating either the mantissa or exponent part of the calculation. In microprocessors the floating point calculations are carried out using base 2 arithmetic. These four operations can be handled by different hardware elements in a floating point unit so forming a four stage arithmetic pipeline.

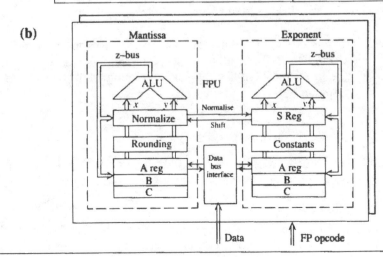

Figure 13.3

Floating point operation
(a) **Operations required for a floating point addition.**
(b) **Block layout of a floating point hardware unit.**

The Control Data Corporation CDC 6600 mainframe computer, which first appeared in 1964, was the first computer to employ functional pipelining as a major feature of its design.

(b) *Superpipelined Techniques*: Superpipelining as a term was first coined in 1988. Superpipelining exploits the fact that many pipeline stages perform tasks that require less than half a clock cycle. Thus a doubled internal clock speed allows the performance of two tasks in one external clock cycle. A superpipelined architecture is one which can make use of more, and more fine grained pipeline stages. With more stages more instructions can be in the pipeline at the same time, increasing parallelism. Figure 13.4 shows an example of the operation of such an architecture.

Figure 13.4

Example of the operation of a superpipelined architecture.

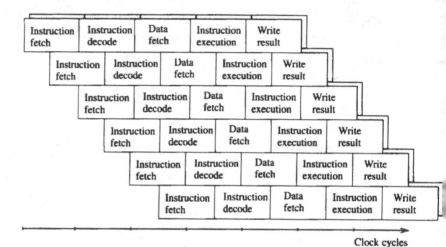

2. **Arithmetic Pipelining** – Arithmetic pipelining breaks a task into smaller pieces and assigns a separate hardware unit within the microprocessor to handle each of the pieces. The set of units are then strung one after the other in assembly line fashion and data pumped through the system such that on each tick of the clock cycle each of the units operates on its own data, i.e. each unit repeatedly executes the same operation or instruction on each successive piece of data passing through the pipeline. The most obvious example of arithmetic pipelining is in the handling of floating point numbers (Article 13.1).

The use of pipelining tended to dominate the way high performance computer architectures were developed in the 1970s. Both forms of pipelining were used in devices called vector computers. Two of the most famous early supercomputers, the Cray 1 and the Cyber 205, were vector computers.

13.2 REDUCED INSTRUCTION SET MICROPROCESSOR (RISM)

During the early 1980s a number of microprocessor designers decided to take a fresh look at the way microprocessors were designed in the light of newer implementation technology. They noted, for example, that the instructions within a complex instruction set microprocessor (CISM) required complex and potentially time consuming hardware steps to decode and execute them. This in turn meant a large microprogrammed control store had to be incorporated into the CISMs, taking up a considerable amount of the area on the

HISTORY OF RISC MICROPROCESSORS

The first reduced instruction set computer (RISC) was developed at IBM in the early 1980s. The IBM 801 project set in motion a series of research projects at both Berkley and Stanford Universities in the USA to generate reduced instruction set microprocessors (RISMs).

The RISC 1 and RISC 2 RISMs were designed and built at Berkley University in the USA. They were designed to provide efficient support for high level languages, especially C and Pascal, in a simple hardwired microprocessor. The RISC 1 microprocessor was designed and fabricated over the course of 1981 to 1982. Most of the instructions were register to register, and data memory access was restricted to the load and store instructions. RISC 2 was designed and fabricated between 1981 and 1983 by Katevenis and Sherburne using 4 μm single level metal custom NMOS technology. This 41 000–transistor microprocessor chip had a cycle time of 500 ns per 32–bit instruction. It was later fabricated in 3 μm NMOS and ran with a cycle time of 330 ns.

In comparing the two devices we can note that RISC 1 had 31 instructions while RISC 2 had 39. RISC 2 offered a higher performance than RISC 1 through a larger register file and longer pipeline. The term RISC was first coined by David Patterson who led the Berkeley RISC team. Patterson has since worked with Sun Microsystems to define the SPARC (scalable performance architecture), which can be viewed as a commercial outgrowth of the Berkeley RISC design.

The second of the RISC projects was the MIPS (microprocessor without interlocked pipe stages) device which was designed and built at Stanford university in the USA. John Hennessy coordinated the team at Stanford University who developed MIPS. This device was a pipelined 32–bit RISC processor with 2 KB of on–chip cache memory. It was started in 1981, the first chip being fabricated in NMOS in 1983 with 24 000 transistors, a basic cycle time of 250 ns and about 70 instructions. Compared to the RISC 1 chip which used a large register set of 138 by 32–bit registers, arranged in eight overlapping windows of 24 registers, the MIPS device used only 32 registers and instead relied on sophisticated software to manipulate the registers.

The basic philosophy adopted at Stanford was to design a compiler–driven instruction set that required little or no decoding, due to close its correspondence with the microcode of the MIPS device. The MIP's sophisticated compiler optimizes the performance of the microprocessor by minimizing any resource conflicts between the individual stages within the microprocessor through a reordering of the instruction sequence in the programs for the microprocessor.

Although the RISC approach to designing microprocessors has been one of the big developments in the 1980s and 1990s, it was only initially advocated by the smaller microprocessor manufacturers. Companies such as Fairchild (with its Clipper microprocessor), Acorn (with the ARM device) and Signetics (with the STC 2000) designed and developed their own ICs in a comparatively short time due to the simplicity of the RISM hardware. In the 1980s larger companies, such as Intel and Motorola, tended to continue the development of their CISMs, basically to maintain continuity in their product development. There are now many RISC processor designs and such has been their initial success that both Intel and Motorola have developed their own devices.

RISM

integrated circuit. The end result of their deliberations was a decision to reduce the number and complexity of the instructions within a microprocessor's instruction set. This would then lead to a shortening in the time to process instructions within the control unit of a microprocessor and a decrease in the size of the control unit. This in turn would produce an increase in the free space on the chip itself which could be used to hold a large number of registers or a block of cache memory. Article 13.2 provides a history of the development of reduced instruction set microprocessors (RISMs). Computers designed with RISM devices are often called reduced instruction set computers (RISCs).

13.2.1   RISM DESIGN CONSIDERATIONS

Computer designers are actually hard put to identify a true RISM because it is unclear how far an instruction set can be reduced or simplified and still allow a microprocessor to efficiently perform all of the functions necessary for a practical machine. Many commentators and researchers have also suggested that the number of instructions is not the only criterion that should be used to define a RISM. There are six elements that would seem essential to the RISM philosophy.

1. **Relatively Few Instructions and Addressing Modes** – A traditional CISM relies on hundreds of specialized instructions, dozens of addressing modes and several high–level language constructs implemented in hardware. This requires a large and complex control unit within the CISM. In such a microprocessor the compiler must consider the many possibilities inherent in each complex instruction and then perform a number of memory transfers to execute it. This requires identifying the ideal addressing mode and the shortest instruction format to manipulate the operands in memory. Yet only a small number of instruction types take up most of the microprocessor's execution time. For example load, call and branch instructions are found in compiled code more often than any other instruction type. The complex instructions are rarely used and hence are wasteful in terms of control unit space.

RISM

The instruction sets for RISC architectures are chosen after extensive study of compiler generated code such that only the most frequently used instructions are selected for hardware implementations. The others must be synthesized from groups of the basic instructions. In this way one of the main principles behind RISC is to have almost all the instructions perform only basic arithmetic logic unit (ALU) operations and shift functions, effectively implementing the complex instructions found in CISMs as subroutines. Also, only simple addressing modes are usually provided with the more complicated addressing modes again being synthesized from the simple ones. RISMs typically use 50 to 70 instructions compared to over 150 for a typical CISM and 303 for the VAX 11/780 super minicomputer.

2. **Single Cycle Operation Facilities** – How well a RISC machine achieves its enhanced performance through reduced instructions can be measured by the number of clock cycles required to execute an average instruction. The ultimate goal of a RISC machine is to come as close to 1 as is possible.

3. **Fixed Instruction Format** – RISMs try to have a common format in the way instructions are constructed. This eases the hardware decoding of the instructions.

4. **Load/Store Design** – In a RISM which does not have the benefit of complex instructions, it is preferable if frequently used instructions and data can be stored in a large number of on–chip registers, since the more instructions and data that can be kept in these registers, the faster a program should run. Hence part of the RISC implementation includes a large number of registers. To make use of the registers in the most efficient way the registers are organized in blocks, or banks or windows, with overlapping of these register windows.

Table 13.2 shows a number of commercially available RISC microprocessors together with a note of the numbers of registers they contain. The table also presents some details of each RISC device's support for cache memory and internal architecture. In Table 13.2 a **von Neumann architecture** is a conventional microprocessor architecture with a single bus for instructions and data, while a **Harvard architecture** is an architecture has separate buses for data and instructions.

5. **Hardwired Control** – General purpose CISMs have complex instruction sets which handle both basic operations and complex functions. Typically the instructions are microcoded and take many clock cycles to complete and their control unit occupies considerable silicon area. A RISM is based on the assumption that the commonly executed instructions should be processed in the most efficient way possible. Hence using RISC architectures with the CPU instruction set reduced to a few simple operations means there

Table 13.2

**Architectural
features of
several RISC
microprocessors.**

RISC microprocessor	Type of architecture	Number of registers	Cache memory support
Motorola 88000	Harvard	32	Yes
MIPS R3000	von Neumann	32	Yes
AMD 29000	Harvard	192	No
Cypress SPARC	von Neumann	128	Yes
Sun SPARC	von Neumann	128	Yes
Intergraph Clipper	von Neumann	40	Yes
Intel 80960	von Neumann	32	No

RISM

are few instructions for the control unit within the CPU to decode. The control unit can then be reduced in size and complexity, and since there are fewer instructions to decode, most of them can be decoded through hardware rather than through microcode. This leads to an overall decrease in the cycle time for the RISM over its CISM counterpart.

6. More Compile Time Effort – Due to their small instruction sets, RISMs must rely heavily on efficient compilers, since it is the compiler in a RISM which is left to optimize the program. Much of the performance of RISMs comes down to the technology of the compilers that convert the high level program code into the set of machine instructions the RISM will execute. The role of a compiler in any processing system is to break down the high level language programs into the machine code for the particular target processor. For a RISM machine the program simply has to be broken down a little further into more basic actions, before the translation to machine language can take place. With a reduced instruction set, since there are fewer choices of instructions for the compiler, the compiler can potentially be made simpler, since it does not have to keep the complex accounting required to choose between several alternative ways of doing the same thing in CISM machines. One of the drawbacks however is that the code produced by a RISM compiler may be longer than that encountered for a similar CISM operation.

RISM compilers try to remove as much work from the microprocessor as possible at compile time, such that for example they try to manipulate the data used within a program to ensure that they are always kept in the different register sets within the processor. In this way simple register to register instructions can be used without having to rely on transferring data to and from the main memory. Traditional compilers, on the other hand, try to discover the ideal addressing mode and the shortest format to handle different instructions.

Table 13.3 compares the general characteristics of CISMs and RISMs, while Table 13.4 notes three specific examples of CISC and CISMs, and RISC and RISM systems. Article 13.3 describes three RISC microprocessors.

Table 13.3

**Comparison between
CISC and RISC
microprocessors.**

	CISM	RISM
Number of instructions	Over 200	Under 100
Number of address modes	5–20	1–2
Instruction formats	3+	1–2
Average cycles per instruction	3–10	Near 1
Memory access	Most CPU operations	Load/store only
Registers	2–16	32+
Control unit	Microcoded	Hardwired
Instruction decode area of the CPU	Over 50%	10%

RISC MICROPROCESSOR INFORMATION

ARM MICROPROCESSOR – The ARM microprocessor was designed by Acorn and launched in 1985. It is a 32–bit RISM with a 26–bit address bus and a separate 32–bit data bus. It has twenty five 32–bit registers but only 16 of these are available to a programmer. The device uses 44 basic instructions almost all of which are completed in a single instructions cycle (150 ns). The upgraded device is now produced by VLSI as the VL86C020, released in 1989.

MOTOROLA MC 88000 / 88110 – The Motorola MC 88000 microprocessor was released in 1988. It comes in a 182–pin grid array package and measures 1.8 in^2. It has 32 general purpose 32–bit registers and uses register to register addressing for all data manipulation instructions. It also has both data and instruction cache memories.

The MC 88110, released in 1992, with 1.3 million transistors uses a 50 MHz clock. It is a concurrent RISC processor which achieves its performance through the use of overlapping parallel execution units. Everything about the processor is pipelined – integer and floating point calculations and data and instruction fetches. It has a superscaler architecture which issues as many as two instructions per clock cycle. It uses an 8 KB instruction cache and an 8 KB data cache. It has a performance rating of 5.4 million floating point operations per second (MFLOPS).

MIPS R2000 – The MIPS R2000 microprocessor, released in 1986, consists of two tightly coupled processors implemented on a single chip. The first processor is a 32–bit reduced instruction set central processor unit while the second processor is a system control processor containing a translation look aside buffer and control register to support a virtual memory subsystem and separate cache memories for instructions and data. The MIPS R4000, released in 1990, is a 64–bit device.

Table 13.4

Comparison of complex and reduced instruction computers and microprocessors.

Characteristic	CISC IBM 370/168	CISC VAX 11/780	CISM Intel 80486	RISM Motorola 88000	RISM MIPS R4000	RISC IBM RS/System 6000
Year developed	1973	1978	1989	1988	1991	1990
Number of instructions	208	303	235	51	94	184
Instruction size (bytes)	2–6	2–57	1–11	4	32	4
Number of general purpose registers	16	16	8	32	32	32
Cache memory (KB)	64	64	8	16	128	32–64

RISM

As we have noted the main reason for the development of RISMs was to improve the performance of microprocessors. Table 13.5 shows some of the modern RISMs and an indication of their performance in MIPS (millions of instructions per second).

COMPANY	PRODUCT	PERFORMANCE (MIPS)
Motorola	88000	20
Motorola	88100	28
Intel	80960	10
Intel	i860	33
Intergraph	Clipper	17
Cypress	SPARC	20
Sun	SPARC	9.5
Cypress	CY7C601	29
AMD	29000	17
AMD	AM29050	32
Fujitsu	MB86903	29
MIPS	R3000	20

Table 13.5

RISM performance.

Another advantage claimed for RISMs is that they are easier to design. This is shown in Table 13.6.

Table 13.6

Comparison in the development time for a number of microprocessors.

Microprocessors		Number of transistors (thousand)	Design time (person– months)	Layout time (person– months)
CISM	M68000	68	100	70
	Z8000	18	60	70
	Intel iAPx–432	110	170	90
RISM	RISC–1	44	15	12
	RISC–II	41	18	12

As a final point it is worth noting that computer system designers in the past have avoided using RISMs for real time applications for two reasons. The first is the impact of large register sets on task context–switch time – the key to dealing with large register sets is to find a way to avoid saving all the registers to memory every time an interrupt occurs. Secondly, performance penalties associated with cache misses – cache misses affect the performance of RISMs, in particular, because these processors are designed to operate with an unbroken stream of instructions; this leads chip architects to build them with the largest possible on–chip caches. These two problems have been overcome and RISM devices are presently finding their way into real time computer systems.

13.3 DIGITAL SIGNAL PROCESSOR

As well as the development of general purpose RISM and CISM devices, microprocessor manufacturers have also produced more application specific microprocessors. The most popular of these are the digital signal processors which are dedicated to handling digital signal processing (DSP) applications.

Signal processing is a form of numerical analysis that usually must be done in real time and which involves a large number of repetitive simple calculations, such as addition and multiplication, being performed on a large number of data items.

In the past most signal processing was done using complex analog circuits. However certain advantages can be identified in using a digital microprocessor design over an analog discrete approach. The advantages are noted in Table 13.7.

Table 13.7

Advantages of a digital approach to signal processing compared to an analogue approach.

1. The ability to implement advanced signal processing and control algorithms with software rather than special purpose hardware. This is advantageous since software can be updated or altered as the system performance demands change. Digital techniques offer the possibility of reprogrammable hardware with very stable characteristics capable of handling a wide range of applications.

2. Microprocessor circuits have a reduced size, weight, power and cost compared to an equivalent analog design.

3. Microprocessor circuits exhibit greater reliability, maintainability and testability than analog circuits.

4. Digital circuits generally have an increased noise immunity compared to analog ones. Conveying information is accurate and reliable with minimum noise effects, and storing and processing may be done simply by the microprocessor.

5. Microprocessor circuits can be programmed to compensate for drift and counteract electromagnetic interference problems.

6. Digital measurements may often be carried out much more accurately than analog ones.

7. Microprocessor technology allows a compatible communications link to be developed within a larger processing hierarchy.

A block layout for a DSP system is shown in Figure 13.5. The input signal, $x(t)$, is passed through low pass filter and an amplifier to clean up the input signal and obtain a signal of the correct magnitude. This signal is then sampled by the analog to digital (A/D) converter before the digital data $x(n)$ is passed to the digital signal processor. The output signal $y(n)$ from the digital signal processor is converted to an analog form by a digital to analog converter (D/A) before being smoothed by another filter and amplifier.

Digital signal processor

Digital signal processing was originally performed on CISMs; however, these conventional microprocessors were not very efficient in processing DSP algorithm. This is due to the fact that DSP algorithms tend to consist of repetitive arithmetic operations with little use for complex branching instructions. Over the past ten years, DSP systems have moved from the use of CISMs to the situation where they are now mostly built from single chip DSP microprocessors.

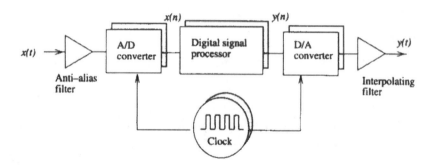

Figure 13.5

A digital signal processing system.

A DSP micro is a high speed signal processing integrated circuit which provides a very high number–crunching speed with the flexibility of a general purpose microprocessor. Such a special type of microprocessor is something of a cross between a microcomputer and a microprocessor in that it usually provides a modest amount of on chip memory, but also supports two large off chip external memory spaces.

Some of the architectural features found in DSP microprocessors are shown in Table 13.8. DSP microprocessors operate much like standard microprocessors except that they are optimized for the high speed processing of information signals such as streams of audio or video data. Their main use is to convert analog signals into digital forms according to predetermined algorithms. Their design is best suited for real time DSP and other computation intensive applications in the areas of telecommunications, graphics/image processing, high speed control, speech processing, numeric processing and instrumentation.

Table 13.8

Features found in a DSP microprocessor.

1.	Fast hardware multiplier.
2.	On–chip memories to store programs and data
3.	Extensive temporary registers to reduce unnecessary fetches of continually used data.
4.	Special registers for accelerating repetitive functions such as multiplication and addition.
5.	Floating point hardware for easier algorithm design.
6.	Wide data buses for memory, internal processor transfers, registers and on board processing.
7.	Several data buses available to reduce memory bus conflict/ transfer overhead.
8.	Harvard architecture and/or instruction cache to avoid instruction and data fetch clashes.
9.	Duplicate resources for parallel computation of real and imaginary components of complex numbers.
10.	Dedicated hardware for address calculations.
11.	Fast instruction cycle.
12.	Easily programmed.
13.	Lower power consumption with a standby mode.

Digital signal processor

The first purpose designed digital signal processing chip was the Intel 2920 which included A/D and D/A converters on the chip. The 2920 had four analog input channels, eight A/D output channels and a 24–bit processor with 192 by 24–bit read only memory (ROM) words and 40 by 25–bit random access memory (RAM) words. A selection of DSP microprocessors is given in Table 13.9. By far the most popular range of devices are those from Texas Instruments (TI). It is reckoned that TI controls 60% of the world share for DSP processors (Article 13.4).

It would appear that the success of these devices will invariably mean that their future is secure and that the continued advancement of IC technologies will fashion a platform for their increased development.

TEXAS INSTRUMENTS DSP MICROPROCESSORS

TMS 32010 – The Texas Instruments TMS 32010 digital signal microprocessor, released in 1982, was one of the first of the DSP devices. The TMS32010 contains a hardware multiplier that performs a 16–bit by 16–bit multiplication providing a 32–bit product in a single 200 nanosecond clock cycle. A hardware barrel shifter is used to shift data on its way into the arithmetic logic unit. It has an on–board 1532 by 16–bit mask programmed read only memory (ROM) and a 144 by 16–bit random access memory (RAM). The use of a Harvard style architecture within the device allows transfers between the program space and the data space, thereby increasing the flexibility of the device. It can perform a 1024 point complex fast Fourier transform (FFT) in 42 ms.

TMS 32020 – The successor to the TMS 32010 is the second generation TMS 32020, launched in 1985. It is manufactured in NMOS and has a similar architecture to the TMS 32010 but offers an increase in performance and two to three times the processing power of the TMS 32010, through the use of a single multiply/accumulate instruction with a data move option, five auxiliary registers with a dedicated arithmetic unit and faster I/O. The TMS 32020 source code has been designed to be upwards compatible with the TMS 32010 source code, meaning the device can run TMS32010 code.

The TMS 32020 contains most of the features of its predecessor with the major exception being that no on–chip ROM is provided. This is so that a greater amount of on–chip RAM (544 by 16–bits) can be provided. The program and data memory address ranges of the TMS 32020 has been increased to 64 K (16–bit words) each. The architectural design of the TMS 32020 emphasises overall system speed, communication and flexibility in processor configuration. Control signals and instructions provide block memory transfers while at the same time its general purpose applications have been greatly enhanced by its large address space, on–chip timer, serial port, multiple interrupt structure, provision for external memory wait states and multiple processor interface capability. The time for a 1024 point complex FFT is 14.18 ms.

TMS 320C25 – Texas Instrument's 3rd generation DSP device, the TMS 320C25, released in 1986, is similar to the TMS 32020 except that it has been constructed in CMOS, thus producing a faster device with a lower power consumption. It is essentially a pin compatible version of the TMS 32020 with a faster instruction cycle time and the inclusion of additional hardware and software features. The TMS 320C25 instruction set is upwards compatible with the 32020, being completely object code compatible. The processor has an instruction time of 100 ns allowing it to process 10 million instructions per second, 97 out of the 133 instructions are single cycle. The reduced instruction cycle provides double the throughput for the 320C25 over the 32020.

Control operations for the TMS 320C25 are provided by an on–chip timer, a repeat counter, three external maskable user interrupts and internal interrupts which can be generated by serial port operations, or by the timer. The device also provides 544 words of on–chip memory thereby, allowing it to be configured as a stand alone microcomputer to handle a 256 point fast Fourier transform (FFT). The TMS 320C25 supports floating point operations with a 16–bit mantissa and a 4–bit exponent for applications requiring a large dynamic range.

TMS 320C30 – See Highlight 12.1 for information on this processor.

TMS 320C40 – The latest device in the TMS range is the TMS320C40. It is innovative in that not only does it possess a high performance architecture, performing 275 MOPS with a 40 ns clock, but it also contains six serial communication links to allow multiple processing systems based on the device to be easily connected. Internally the device has several processing sections which can operate in parallel using a number of internal buses to interconnect the functional units. It has a 128 by 32–bit word cache memory, 2 K by 32–bit words of RAM and 4 K by 32–bit of ROM. The serial communication links are 8–bit bidirectional ports to provide efficient processor to processor data transfer. It also incorporates a dedicated co–processor for controlling the direct memory access operations through the communications ports. It is code compatible with the C30 and comes with a C compiler, cross assembler, linker, software simulator and an in circuit emulator.

Company	Model Number	Date introduced	Description
AMI	S2811	1978	First DSP described 12/16–bit fixed point device.
	S28211/2	1983	Update of S2811.
Analog devices	ADSP–2100	1986	16/40–bit fixed point with off chip memory.
	ADSP–2100A	1988	Update of 2100.
AT&T	DSP1	1979	Early DSP device 16/20–bit operands to multiplier.
	DSP20	1981	Update of DSP1.
Fujitsu	MB8764	1983	16/26–bit fixed point.
	MB86232	1987	IEEE 32–bit floating point device.
Hitachi	HD61810	1982	12/16–bit floating point device.
	DSPi	1988	Image processor.
NEC	μPD7720	1981	Early DSP device.
	μPD7281	1984	Data flow machine for image processing.
	μPD77230	1985	32–bit floating point.
IBM	Hermes	1981	Not marketed outside IBM.
Motorola	DSP56000	1986	56–bit arithmetic.
	DSP96000	1990	IEEE floating point device.
Texas Instruments	TMS32010	1982	NMOS, 16/32 data path.
	TMS320C10	1985	CMOS version of 32010.
	TMS32020	1985	32–bit device.
	TMS320C25	1986	Upgrade of 32020.
	TMS32030	1989	IEEE floating point device.
	TMS320C40	1991	32–bit device with serial communication links.
Zoran	34161	1986	Vector signal processor. 16–bit block floating point device.
	35325	1989	32–bit IEEE floating point device.

Table 13.9

Description of various DSP microprocessors.

13.3.1 DSP: ARCHITECTURAL FEATURES

DSP microprocessors are constructed with specialized architectures. These architectures are basically the same irrespective of the manufacturer of the device. There are several major strategies adopted within these devices to enable high performance.

1. Harvard Architecture – DSP devices differ from general purpose microprocessors in a number of ways, but probably the most important is that in a standard microprocessor based system, there is a single data bus on which three distinct types of data transfers takes place. There are instruction fetches, data bytes are moved in either direction between the central processing unit (CPU) and locations in memory, and I/O transfers occur between the CPU and the input/output (I/O) ports.

DSP devices use a Harvard architecture within the processor, such that the data and the instructions are stored and manipulated separately with each having its own dedicated bus. With two data buses and a program bus, two data elements can be moved to the multiplier and the adder in parallel for execution, while the CPU is fetching the next instruction. This keeps the multiplier and ALU continuously supplied with data, allowing the device to perform a multiplication or addition on a new set of data on each clock cycle.

Since the program memory space and the data memory space are separated, it is no longer necessary for the word lengths in the two areas to be the same. Hence, instruction word lengths may be longer than data word lengths, or vice versa with each memory space being connected to a bus of the appropriate width.

| Highlight 13.1 | EXAMPLE OF A DSP MICROPROCESSOR: TMS 320C30 |

The fourth generation DSP processor from Texas Instruments is a CMOS device utilizing 1μm processing technology. A block diagram of the device is given in Figure 13.6. The device runs at 16.7 MHz giving a 32-bit floating point performance of 33 MFLOPS. It has an addressing capacity of 16 M x 32-bit in total and has a large on-board RAM and ROM memory, a 64-bit instruction cache memory, two serial ports, an analog timer and a direct memory access (DMA) controller. The DMA channel can allow data to be read in and out of the device without interrupting the central processing unit.

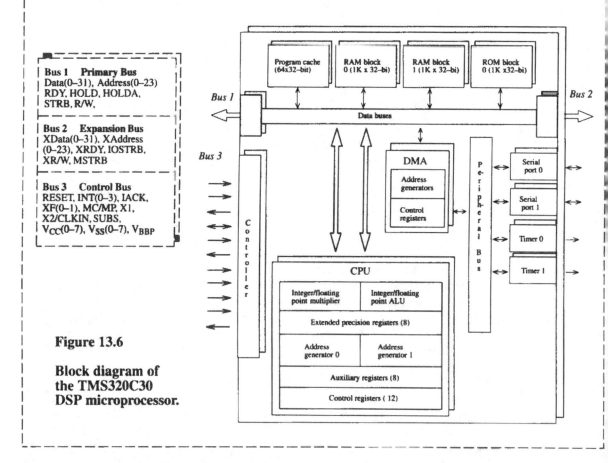

Bus 1   Primary Bus
Data(0–31), Address(0–23)
RDY, HOLD, HOLDA,
STRB, R/W,

Bus 2   Expansion Bus
XData(0–31), XAddress
(0–23), XRDY, IOSTRB,
XR/W, MSTRB

Bus 3   Control Bus
RESET, INT(0–3), IACK,
XF(0–1), MC/MP, X1,
X2/CLKIN, SUBS,
V_{CC}(0–7), V_{SS}(0–7), V_{BBP}

Figure 13.6

Block diagram of the TMS320C30 DSP microprocessor.

DSP architectural features

2. Hardware Multipliers and Adders – The hardware multiplier within DSP devices reduces the burden of multiplication from the software and allows the multiplication to be performed in hardware in a single clock cycle. Also, by using a separate multiplier and ALU in the processor, the device can have a high throughput, i.e. it is able to process many data items from input to output, for a wider range of algorithms, because it allows the processor to execute multiplication and addition operations in parallel.

3. **Pipelining** – In many DSP applications, throughput rate often represents the overriding factor dictating a system's performance. In order to optimize throughput, a different design choice is often made to that of minimizing the total processing time, or latency, of a single processing unit. By making use of pipeline techniques with many different parts within the processing unit it is possible to enhance the device's performance.

As we noted earlier in the chapter, in a pipelined architecture, computational problems are broken down into several smaller parts each of which can usually be handled in a single clock cycle. The processor itself then consists of many individual hardware sections each of which can work on a particular part of the problem. In this way for example, the multiplier portion of the pipe can be multiplying data that was specified in the current instruction, while another section of the pipe may be accessing data for use in the next multiplication operation. Arithmetic pipelining within the multiplier is acceptable because most DSP routines consist of purely sequential operations with no jumps or branches.

DSP architectural features

4. **Special DSP Instructions** – With the advent of VLSI (very large scale integrated) technology it has become possible to design a DSP processor architecture that can handle large functions supported directly in hardware, e.g. processing a fast Fourier transform (FFT). In order to use the specialized hardware, most DSP processors also include dedicated instructions within their instruction set to enable the manipulation of data between the hardware units very quickly. These may include single instruction multiply and shift operations, and other single instructions to aid in signal processing tasks.

5. **I/O Interfaces** – A requirement for DSP devices is that they implement flexible I/O structures, primarily for specialist interfaces. Hence DSP chips usually have serial I/O ports, a large data and program memory built into the DSP microprocessor and, in some cases, A/D and D/A facilities. This allows efficient acquisition and depositing of data.

13.3.2   PROGRAMMING A DSP MICROPROCESSOR

The dominating requirements in signal and image processing applications are the need for high throughput rates and huge amounts of memory. Most signal processing applications also have a well defined behaviour with no data–dependent branching. Hence within a DSP program the programmer knows which operations are independent and which require ordering or synchronization. This tends to make program production relatively straightforward.

The high level language C has become the main language for programming the modern digital signal microprocessor; however, the efficiency of C compilers has tended to be poor. In many cases the assembly language code generated by the compiler fails to capitalize on the superb architecture of the chip and just does not produce executable code from a tried and tested C language source. Hence, in practice to maintain maximum efficiency for a DSP system, a mix of assembly language and C language modules will usually be used in any program. The difficulty with modern DSP devices is that their assemblers have a relatively steep learning curve, the reason for this stems from the parallel and/or pipelined architecture of the new devices. For example, in a number of DSP microprocessors a single line of assembly code may contain four or more sub–instructions and learning how to capitalize on the architecture through efficient use of the instruction set does take time. With a pipelined architecture, learning is even more difficult since the processor can be executing parts of four lines of code simultaneously.

The software development tools available for most digital signal microprocessors includes cross assemblers, linkers, software simulators and high level language compilers.

13.4 COMPUTER ARCHITECTURES

Over the past 50 years since the introduction of the computer, the way that the architecture of the device has changed and evolved can be easily seen when we consider the different forms of processing systems that can be found throughout the computer world today. Table 13.10 notes some of them. Each of these devices has an architecture that has been designed and developed for particular applications. In the future no doubt there will be many more types of system that can be added to the list, all claiming to offer solutions to the ever increasing processing needs of the computer establishment.

Table 13.10

Different type of computer systems.

1.	Uni–processor system.	7.	Computer network.
2.	Multifunction processor.	8.	Array processor.
3.	Pipeline or vector processor.	9.	Associative processor.
4.	Multiprocessor.	10.	Systolic array.
5.	Hypercube.	11.	Data flow processor.
6.	Multicomputer.	12.	Demand driven processor.

In 1966 J. Flynn proposed a **computer taxonomy** to describe the architectures of different computers in which the instruction stream and the data stream of the computer system are divided in many ways. **Flynn's taxonomy** consists of four classes of processor systems, usually referred to by their acronyms:
1. SISD (single instruction stream, single data stream),
2. MISD (multiple instruction stream, single data stream)
3. SIMD (single instruction stream, multiple data stream),
4. MIMD (multiple instruction stream, multiple data stream).

The term stream is used to denote a sequence of items, or instructions or data, as executed, or operated upon, by a single processor. An instruction stream is a sequence of instructions executed by the machine, while a data stream is a sequence of data including input and partial or temporary results called for by the instruction stream. The four Flynn classes may be further subdivided into different categories. The categories are open to interpretation and may not be regarded as the definitive groupings within a class; they do however, afford a way of differentiating between different computers which may be in the same class. We will consider each classification in turn.

13.4.1 SINGLE INSTRUCTION, SINGLE DATA STREAM (SISD) COMPUTER SYSTEM

This is a conventional single processor system characterized by the standard von Neumann architecture. We have already considered the characteristics of such an architecture in the preceding chapters of the book.

13.4.2 MULTIPLE INSTRUCTION, SINGLE DATA STREAM (MISD) COMPUTER SYSTEM

There are no real examples of this type of system under the Flynn classification; however, vector computers exhibit some characteristics of an MISD type of system. A **vector computer** is a special form of computer in which repetitive calculations on a large number of data items, called vectors, are handle by special hardware.

The instruction set of the device contains a set of vector instructions that can cause the computer to automatically perform both the data processing and the control sequencing to handle the data vectors. Therefore, a single instruction specifies both the function to be performed and the location of the vector to which the function is to be applied. In this way executing a vector instruction requires only one instruction fetch for many data operations.

Vector instructions range in complexity all the way from the vector ADD where corresponding elements of two input vectors are added together to produce an element of an output vector, to a FFT, where the vector of data undergoes a long complex sequence of multiplies and adds to produce an output vector, all by the execution of one machine instruction.

To enable the synchronous manipulation of the data in the processing units within a vector computer, the data must be arranged in the form of a vector. This task is either carried out by a unit called a vectorizer or by a special vector compiler which is able to arrange the program code to handle vector data items. The vectorizing compiler also takes the code written in a sequential language and where possible generates parallel machine instructions from a sequence of operations. The automatic vectorization of a program, performed by this type of compiler can be difficult in languages such as FORTRAN, due to the side effects caused by the global nature of the variables. To try to overcome this and provide the programmer with some means of assisting the compiler discover the parallelism in a program, vector extensions are usually included in the high level languages that the particular vector computer supports. Exploiting vector operations certainly leads to a speed up in the operation of a computer. However, as the amount of code that can be vectorized within any general program tends to range from 10% to 90%, this may still leave a large amount of code that cannot be vectorized.

MISD
computer

Vector processing was introduced in the early 1970s with the STAR 100 computer from Control Data Corporation (delivered to the Lawrence Livermore National Laboratory in 1974) and the Texas Instruments Advanced Scientific Computer (which became operational in Europe in 1972). Two of the most famous supercomputers, the Cray 1 and the Cyber 205 were both vector processors. The Cyber 205 computer was a vector supercomputer which could trace its origins back to the Star vector computer. The Cyber was introduced in 1979 and contained both a vector processor and a scaler processor. The vector processor consisted of either one, two or four vector pipelines, each of which could perform a variety of arithmetic operations. The peak performance of the Cyber 205 with four pipelines was 200 MFLOPS. The newer ETA multiprocessor supercomputer is based on the Cyber 205. It features gate arrays of 20,000 gates and is an order of magnitude faster than the Cyber 205. The use of vector operations is now well developed in all classes of computers from the supercomputers down to the small personal computers.

13.4.3   SINGLE INSTRUCTION, MULTIPLE DATA STREAM (SIMD) COMPUTER SYSTEM

SIMD computers are characterized by systems with a single controller unit which has a large number of small slave processing elements. The controller interprets the instructions in the program while the execution of the instruction is performed by the large number of simple processing elements usually arranged in an array topology. Each processing unit executes the instructions on its own data, i.e. the execution of the instruction governs many identical operations applied in parallel to the data items. SIMD computers work on large arrays, or matrices, of data applying a single instruction to all the data in the array at one go. They therefore implement data parallelism in that they can effectively handle a large number of independent data streams. There are several types of processing systems which may be classed as SIMD: the array.processor, the systolic array and the associative processor. It is possible to distinguish between these three types depending on the complexity of the control unit, the processing power, the addressing method of the processing elements, and their interconnection facilities.

SIMD computer

1. <u>**Array Processor**</u> – An array processor is a type of SIMD computer architecture typically made up from a control processor, many arithmetic processing elements, and sometimes a separate scaler processor.

Figure 13.7 shows a block diagram of an array computer with the single master control unit, a large number of processing elements and a number of memory modules. It can be seen that the processing elements receive there instructions from the control unit and their data from the memory modules under the control of the control unit which controls the connection network between processors and memory modules.

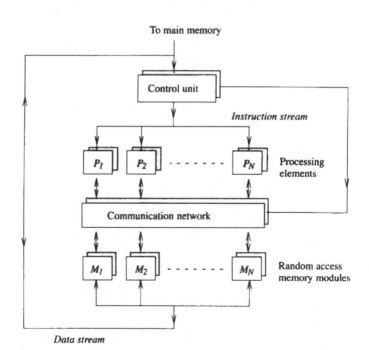

Figure 13.7

Array computer.

In operation the array control unit decodes the instructions within the single program and then broadcasts the instructions and appropriate control signals to all the processing elements in the system. All active processing elements execute the same instruction at the same time on the data in their own local memory. Thus there is a single instruction stream yet multiple data streams. In many such systems the control unit is usually a computer in its own right, with its own local memory, sometimes its own arithmetic unit, registers and control section. The crucial difference however between the control unit and the arithmetic processing elements is that the arithmetic processors lack the ability to decode instructions and, in effect, operate simply as slave units to the master control unit. Individual processing elements within such a computer usually contain an ALU and a number of registers.

ARRAY PROCESSOR INFORMATION

Four of the more well known array processors which have been developed are described below.

1. The ICL Distributed Array Processor (DAP) was proposed in 1972, commissioned in 1976 and the first machine was installed in London in 1980. It has a square array of 32 by 32, or 64 by 64, arithmetic processors linked by row and column highways for data interchange. Each of the processors has a 4 Kbit memory store. The processors simultaneously execute a single instruction broadcast from the master control unit on data values which have been previously loaded into the memory from a host machine.

2. The Massively Parallel Processor (MPP) is a 128 by 128 parallel processing array computer. It was delivered in 1983 to the Goddard Space Flight Centre in the USA by Goodyear Aerospace Corporation. It is a descendant of the Illiac IV supercomputer which was developed in the late 1960s. Each of the MPPs 16 384 processors is only one bit wide with each processor being connected to its four nearest neighbours. All 16 384 arithmetic processors execute the same instruction at one time. Eight arithmetic processing elements each containing approximately 2500 logic gates and 1000 transistors are contained within a single integrated chip.

3. The Connection Machine, devised in 1984, is an example of a large multiple processor computer. It is collection of 65 536 one–bit processors which are connected in a 256 by 256 grid. In addition, clumps of 16 processors are also interconnected by a packet–switched 12–dimensional hypercube network for routing messages. Peak performance is 100 MFLOPS. It can be configured as either an array processor or an MIMD hypercube computer.

4. The BBN Butterfly Array Processor produced by BBN Advanced Computers Inc. was operational by the early 1980s. It contains up to 256 processing nodes with each node containing a Motorola 68020 processor, 68881 arithmetic coprocessor, a microcode processor, node controller and 1 MB of memory.

With the increasing ability of integrated circuit technology to incorporate large numbers of digital logic gates onto an IC it is now possible to produce array type ICs. The NCR geometric arithmetic parallel processor (GAPP) is a two–dimensional array IC chip containing 72 single–bit processors each with 128 bits of RAM. Larger arrays may be constructed from multiple GAPP chips for image processing, signal processing and database applications. The GAPP processes data words on parallel working on words of varying lengths by sequentially processing each bit, word parallel, bit serial processing. The GAPP is classified as an array processor rather than a systolic array because it has the ability to broadcast data to all cells.

Array computer

This type of system then works on large arrays, or matrices, of data. To use the array computer the system's compiler must first detect any parallelism within the program to be run on the computer and then generate the required object code suitable for execution by the multiple processing elements and the control unit. If an instruction within the program does not involve a matrix operation, then the control unit or a separate scaler processor will execute it itself, whereas if the instruction does involve a matrix operation, the instruction can be passed on to the arithmetic processors, where it can manipulate many data items in parallel (Article 13.5).

2. Systolic Array – A systolic array is a regular arrangement of a large number of very simple processing elements, typically add/multiply cells, which are constructed on VLSI circuits. In such a system, data flows from the computer memory in a rhythmic fashion, passing through the many processing cells within the systolic device before it returns to the memory. Each of the cells in the array performs only a short computation on the data but they can all operate concurrently making use of pipelining both within and between the individual cells, e.g. overlapping I/O and computation operations. The complete VLSI systolic device is intended to be used as the processing engine within a more general purpose computer system.

Figure 13.8 shows an **example** of a systolic array with an array of 16 processing cell
and a central control unit responsible for entering data into the array. The input and outpu
buffers are temporary storage areas for the input and output data items.

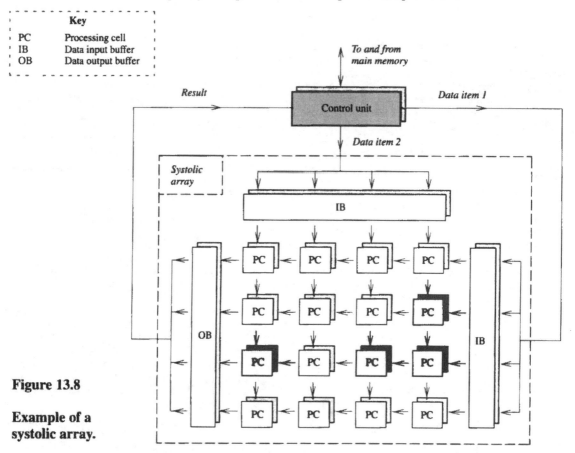

Key

PC	Processing cell
IB	Data input buffer
OB	Data output buffer

Figure 13.8

**Example of a
systolic array.**

The input data items traverse across and down the processing array with the input data
and the intermediate results from previous processing cells being processed and passed from
cell to cell, until the final result appears at the output buffer.

The major features of a systolic array are shown in Table 13.11. At present systolic
systems can either be constructed in one or two dimensions. One–dimensional linear systolic
arrays have been shown to be widely applicable in pipelining applications, while two-
dimensional arrays allow further diversification into rectangular, triangular or hexagonal
structures to make use of higher degrees of parallelism.

Since the processing cells and the interconnections between them are fixed in silicon,
the systolic system must be tailored to execute a particular algorithm. It is also important in
the design of the array that the algorithm to be implemented is arranged to produce a
balanced distribution of work load between the individual elements in the array, while
observing the requirement of locality, i.e. short communication paths. Hence, in designing
a systolic system, it is usual to view the array as a direct mapping of the computations within
a particular problem onto a hardware VLSI structure.

1.	There is a central clock for the system such that the data are moved between the processing elements with each tick of the clock. This synchronizes the processing activities of the whole array.
2.	The processing cells have a small amount of local memory and a basic arithmetic logic unit.
3.	There is no shared memory between the each processor.
4.	Processors are connected via a regular interconnection topology, e.g. a square array, leading to both a modular and regular mode of VLSI construction.
5.	The systolic array uses pipeline processing in that the results from one cell are passes onto another for further processing. In this way the systolic array may be regarded as a special purpose VLSI processor which can maximize processing concurrency by making use of pipeline and parallel processing techniques.
6.	Processors only communicate with a small number of neighbouring processors via data passing. In this way it uses simple and regular data and control flows.
7.	Each input data item is used several times within the array thus achieving high computational throughput with only modest memory bandwidth. This is claimed to resolve the I/O bottleneck usually encountered in multiple processing systems by making multiple use of each data item fetched from the memory.
8.	A systolic system is a scalable architecture in that the size of the array may be indefinitely extended as long as the system synchronization can be maintained.

Table 13.11

Features of a systolic array.

The development of new DSP algorithms tailored to systolic architectures should make it possible to implement sophisticated parallel algorithms to handle processing effectively and economically directly in hardware. The first integrated circuits based on systolic topologies have now been developed, though their restricted architectures will mean that they will probably be used only in high performance image processing applications.

Systolic array

A **wavefront array** is a VLSI arrangement of processing elements connected in a similar fashion to a systolic array but one which uses a data driven, self–timed approach to array processing. This means there is no central clock in the system and the array may be seen as a form of asynchronous network of processing elements which has no central timing requirement, and instead, the system substitutes the requirement of correct timing by correct sequencing. Each processing element operates relatively independently and only synchronizes its activities with its neighbours when it requires to communicate with them.

3. Associative Processor – An associative processor may be characterized as a special form of array processor. This type of system uses content addressing to access stored data items. One of the fundamental differences between an associative processor and the conventional von Neumann type of machine is that the associative processor makes use of an associative memory in place of the contiguous location RAM. It also employ data transformation operations, including arithmetic functions, which can be performed over many sets of data with a single instruction.

Associative processors are comparatively rare. They have tended to be used as very fast database machines, allowing the comparison and searching of input items with data stored in the associative memory.

13.4.4   MULTIPLE INSTRUCTION, MULTIPLE DATA STREAM (MIMD) COMPUTER SYSTEM

The fourth of the Flynn classifications is the MIMD type of architecture. It describes a parallel or **multiple processor** system which uses multiple processing units to execute multiple instructions on multiple data, both independently and concurrently. At the most basic level, multiple processing systems either share the main and secondary memory among a number of processors, known as tightly coupled systems, or have a number of processors loosely connected together onto a shared communication system, known as loosely coupled systems.

Multiple computer architectures that are formed using a shared memory data mechanism are also known as multiprocessors while those using autonomous processing units and a shared communication network are often known as multicomputers or computer networks. Figure 13.9 shows a block layout of a tightly coupled system sharing a common global memory, and a loosely coupled system consisting of independent units sharing a communication network. Many multiple processors have been built using a shared memory structure; however, the most widely accepted form of architecture for multiple processing in the last few years has been the more loosely coupled hypercube structure.

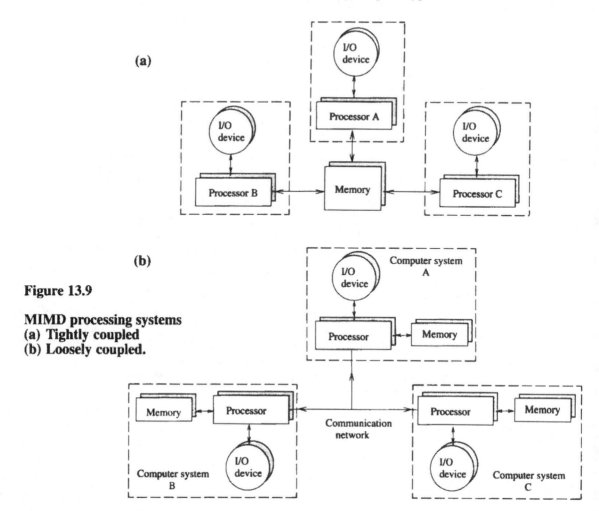

Figure 13.9

MIMD processing systems
(a) Tightly coupled
(b) Loosely coupled.

Company	Current model	Architecture	Microprocessor used at each processing node	Performance per node (MFLOPS)
Intel	Paragon XP/S	MIMD 2–D Mesh	Intel i860 XP	50
Maspar	MP–1	SIMD 2–D mesh	32 custom 4–bit processors per chip	1.2 (per 32 processors)
Meiko	Computing surface	MIMD variable topology	i860 or Sun Sparc	40 with i860
nCube	nCube 2S	MIMD hypercube	64–bit custom scaler processor	2.4
Parsytec	Parsytec GC	MIMD 3–D Mesh with variable topology	Transputer T9000	25
Thinking machines	CM–5	MIMD	Sun Sparc	128 (with max of 4 vector units)

Table 13.12

Massively parallel computers.

A number of multiple processing systems are shown in Table 13.12.

Despite their obvious attractions, multiple processors still have to contend with various difficulties and restrictions on the performance improvement they can expect compared to a single processor. Many factors jointly determine system performance and the modification of some factors affects others.

1. The amount of parallelism inherent in the application problem – The effectiveness of a multiple processing system depends on whether a problem can be identified that lends itself to implementation on such a multiple machine. Problems with inherent sequential operations will not perform well on multiple systems.

2. The design of the programs for the multiple system – Dividing a task to operate in parallel is still a bit of an art; however, better compilers and segmentation tools are being developed. Methods for constructing large programs are also emerging. The complexity, expense and difficulty of writing software is still the major problem in achieving a large scale

MIMD performance factors

speed up with a multiple system. To effectively utilize multiple processing architectures, research is proceeding to develop innovations to make use of any explicit problem visible parallelism within a given task and to develop compilers to discover any implicit parallelism within a problem. This latter task may be regarded as a problem in software. It has become apparent to researchers that although the task of constructing the hardware for a multiple processor is difficult, the task of writing correct efficient software is far more complex. The great lack of suitable programs for multiple systems is so far one of the major drawbacks to the advancement of multiple architectures.

3. The programming language used for the system – There are many distributed processing languages though few which are accepted as standards. The more popular include ADA, Modula–2, Concurrent C and Occam. It can be noted that although these languages are all very different they all contain constructs to define processes and allow communication with messages.

4. The allocation of tasks to processors – The method applied to allocate the tasks within a program to the multiple processors determines how efficiently the multiple system will run.

Glossary 13.1

AMDAHL'S LAW – Amdahl's Law is a rule which says if p is the percentage of a program which is serial, i.e. cannot be broken up into independent tasks, then $1/p$ is the maximum speed up that can be expected with multiple processors.

CONCURRENT – The term concurrent is used in this case to indicate a high level, or global, form of parallelism that denotes the independent operation of a collection of simultaneous computing activities. It is essentially an interactive parallelism that allows the asynchronous operation of a group of processors in a system. Concurrent computing systems are such that parts of them can operate in parallel over the same period of time. The term is sometimes used rather loosely in describing multiprogrammed computer systems where it means the processing of operations rotates between the different programs in the multiprogrammed mix, thereby giving the illusion of simultaneous processing.

DEADLOCK – In a computer system which is multiple tasking, i.e. can handle the processing of a number of separate tasks, deadlock occurs when there is a circular waiting in the system by the individual tasks. This happens when one or more of the tasks are waiting for resources to become available and those resources are held by some other tasks that is in turn blocked until the resources held by the first task are released. Deadlock between two processors is typically called "deadly embrace". A deadlock can be caused by poor software, system faults or errors in a program. There are essentially four conditions that must exist for deadlock to occur in a processing system:
1. processors claim exclusive control of the resources they require;
2. processors hold the resources already allocated to them while waiting for additional resources;
3. resources cannot be removed from processors holding them until the resources are used to completion;
4. a circular chain of processors exist in which each processor holds one or more resources that are requested by the next processor in the chain.
The potential for a deadlock can only occur in systems when multiple paths are required for one information exchange. By utilizing packet switching architectures, for example, each path is obtained in a sequential manner and held until the information transfer is complete, hence deadlocks can never occur. This is one of the reasons why such architectures are proving popular in the design of multiple processing systems.

DUAL PORT MEMORY – A dual port memory is a random access memory which has two separate accessing mechanisms to allow different devices to control access to the device. It is often used in dual processing systems as a common storage area for the two processors.

LOCAL MEMORY – In a multiple processor system where the processors share a common memory for data and program instructions there can be a large amount of contention for access to this memory. A way of circumventing the latency problem caused by many devices using the shared memory is to provide each processor in the system with a small, fast local memory. With each processor having a local memory, program instructions and data can then be moved to them. This relies on the programmer, or the system compiler, explicitly specifying certain instructions to move blocks of data from the shared main memory to the local memory.

LOOSELY COUPLED – The term loosely coupled refers to a multiple processor system in which the individual processing elements are relatively independent units which only communicate occasionally in the processing of any large task. In most implementations a loosely coupled system consists of a network of computers which communicate serially.

MULTIPORT MEMORY – A multiport memory is a special type of random access memory which has many different buses into and out of it. In this way many different processors within a multiple processing system can control access to the multiport memory in order to use it as a message centre, for the processors to leave or collect data. Thus data prepared by one processor can be quickly passed to another processor through the multiport memory. In a multiport configuration the control switching and priority arbitration logic is concentrated at the interface to each of the I/O ports.

MULTIPROCESSOR – A multiprocessor is a computer configuration utilizing two or more central processing units (CPUs) where the programming tasks in the system are divided up between the independent CPUs and then processed simultaneously. The processors in the system have access to a common jointly addressable memory with one unit in the system often being designated the information interchange controller while the others carry out the distinct defined parts of a task.

PARALLEL PROCESSING – Parallel processing is the concurrent, or simultaneous, execution by two or more processors of parts of the same processing task. Parallel processing systems use multiple processing elements which can operate concurrently on different data and instructions streams. They are often referred to as either SIMD (single instruction stream, multiple data stream) or MIMD (multiple instruction stream, multiple data stream) systems under the Flynn taxonomy. There are many types of processing systems that can be referred to as parallel processing: systolic array, array processor, associative processor, pipeline processor, multiprocessor, multifunction processor, data flow processor and demand driven processor.

PASS–BY–REFERENCE – A multiple processor system which uses pass–by–reference techniques to transmit data between processors is such that a pointer to the data location is passed from one processor to another. Pass–by–reference implies that both the transmitter and receiver share a common memory and so is typically used in tightly or closely coupled multiple processing systems.

(Continued)

Glossary 13.1   *(continued)*

Multiple Processing

PASS–BY–VALUE – In a multiple system which uses pass–by–reference techniques to transmit data between processors the value of the data is transmitted from one processor to another and assigned to local variables in the second processor, i.e. they are copied from the sender's local memory area into the receiver's local memory area. In general the receiver processor does not transmit results back to the sending processor. Transmission by value protects the sender processor from side effects introduced by the receiver processor; for example, the receiver processor changing the value of the variable initially transmitted to it from the sender processor. Hence, this method is usually selected when protection criteria are more important than maximum system performance.

SPEED–UP – In measuring the performance of a multiple processor system employing parallel algorithms, one often wants some measure of how fully the parallel algorithms utilize the computational power of the parallel system and how much advantage is gained over sequential algorithms. Hence a relative factor relating to the increase in performance in a parallel system over a uniprocessor system can be defined as the time to solve a problem on a multiple processor using a given parallel algorithm, divided by the time to solve the problem by a sequential algorithm on a sequential machine. This factor is called speed–up. The speed–up that can be achieved by a parallel computer with N identical processors working concurrently on a single problem is at most N times faster than a single processor. However, for most tasks, the speed up will be considerably less due to interprocessor communication, memory contention and synchronization difficulties.

TIGHTLY COUPLED – A tightly coupled multiple processor computer system may be broadly defined as a system of at least two processing units under integrated control which performs the simultaneous processing of two or more portions of the same program. In a tightly coupled system the processing units share a common main memory, I/O devices, programs, tasks and data, and the whole configuration is usually controlled by a single operating system. The term multiprocessor is often used to describe this type of system. Tightly coupled systems which use shared memory systems are limited by memory access issues: access times, communications contention and memory contention while those based on a shared bus are particularly affected by bus–bandwidth limitations.

5.   The grain size of a task executed on each processor – The grain size is an indication of the size of the task handled by each processor.

6.   The possibility of overlapping processing with communication – The major difficulty with multiple systems, comes with the frequent need for one processor to communicate with another processor in order to complete its task. This requires the precise scheduling of the exchanges between the asynchronous processors. The communication between processors is in the form of a transmission of data and usually takes place through the exchange of messages between the tasks. Hence when processors communicate, information flows between them. Clearly the mere occurrence of this communication is informative, which leads to the idea that communication in any form can provide some kind of synchronization signal. The term synchronization implies some control over the ordering of the operations in a processing system. As the number of processors gets larger the interdependency among the tasks becomes more complex and the intertask communication can prove a bottleneck to system performance.

MIMD performance factors

Planned and controlled interaction within a processing system is referred to as process synchronization, a general term for any constraint on the ordering of operations in time. In general it is difficult to separate synchronization from communication since, if one process running on one processor is to synchronize with another process running on another processor, the first must detect an action performed by the second process, which in turn requires a flow of information, or communication, between the processes.

Synchronization and communication within a multiple processing system can be achieved in two ways. Firstly the reading and writing of shared data, typical of a tightly coupled processing system, where the processors share a common data area. Secondly the sending and receiving of messages, typical of a loosely coupled system. It is usual that when message passing is to be used for synchronization purposes, processes send and receive

HYPERCUBE COMPUTERS

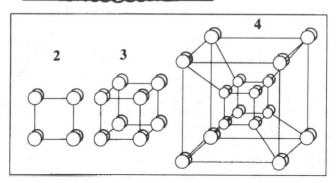

Figure 13.10

Two–, three– and four–dimensional hypercube structures.

It has been known for some time that the hypercube structure has a number of features that make it useful for constructing MIMD computer systems. It is now widely regarded that the hypercube is probably the most straightforward and least expensive way to build a multiple computer capable of operating on hundreds of millions of instructions per second. Examples of the structures of two–, three– and four–dimensional hypercubes are shown in Figure 13.10. It can be seen that the maximum number of nodes in each configuration is equal to 2^N, N is known as the dimensional space in which the hypercube has been designed and built.

The hardware realization of the hypercube consists of a series of processing elements, called nodes, all partially interconnected, i.e. each processor is connected to a fixed number of others, through a number of serial transmission links. The nodes within a hypercube system are connected in a binary N–cube configuration such that each node can communicate directly with its nearest N neighbours.

A hypercube computer architecture calls for individual nodes to operate in an autonomous fashion on a subsection of a larger problem, according to resident process instructions, on data resident in each specific processor's memory. The data can come either through a message passing system from processors resident in other nodes, or from the central hypercube manager. The hypercube architecture therefore achieves concurrency through an ensemble of loosely coupled independent processors executing portions of a larger computational problem simultaneously. Should processors wish to communicate then they can do so through the serial channels. By sending and receiving messages through bidirectional asynchronous communication channels the network of small processing nodes can cooperate on processing large tasks. Each node must also contain routing information in order to transfer messages through the network where the message routing is a function of the node's operating system.

––––––––––––––––––––

The first working hypercube was developed at the California Institute of Technology in the early 1980s. It was called the Cosmic Cube and from the initial development work done on it, many other hypercube structures have quickly been developed. Several are listed below.

1.   Intel iPSC: Shortly after the work on the Cosmic Cube, Intel developed the iPSC commercial hypercube. It consists of a series of microcomputers based on the Intel 80286 or 80386 microprocessor all coordinated by a central cube manager. Further developments have seen the release of the iPSC–VX vector hypercube which also includes a vector processor at each node.

2.   T Series: In the spring of 1986, a supercomputer class machine, the T series from Floating Point Systems Inc. was introduced. This machine is a massively parallel modified vector hypercube structure with a maximum number of 65 536 nodes and a performance of up to 16 MFLOPS per node. Each node in the system contains one T414 transputer, a 64–bit floating point vector processor, 1 MB of dual port random access memory and sixteen serial links all contained on a single circuit board. Eight node boards are connected to each other and to a system board to form a module.

3.   Balance 2000: The Sequent Balance 2000 hypercube contains thirty National Semiconductor NS32032 microprocessors connected to a shared memory by a shared bus with a 80 MB s^{-1} transfer rate. Each processor has an 8 KB cache memory and an 8 KB local memory.

(Continued)

HYPERCUBE COMPUTERS

(continued)

4. N/Cube: The N/Cube system contains a number of specially designed single chip processors. Each processor contains 160 000 transistors and is housed in a 68–pin grid array package. Each node in turn contains the special single processor and six memory chips. Systems of up to ten dimensions (i.e. 1024 nodes) can be built on 16 boards each containing 64 nodes, with the whole ensemble being housed in a single cabinet and connected via a system backplane.

5. Connection Machine: The Connection Machine is a highly parallel processing machine which incorporates 65 536 simple processors. In tandem these processing elements can execute several thousand million instructions per second, attaining this rate over a wide range of applications. The system utilizes a hypercube type structure to route messages between the processing chips. The basic processing unit in the Connection Machine is a specially designed integrated circuit with the processor consisting of 16 small processing units and a device for routing communications among the units. Thirty–two of these processors are then packaged onto a single printed circuit board together with 4 Kbits of memory per processor and 128 such boards incorporated into the complete machine. The boards are arranged in a cube of side 1.5 m.

The flexibility in the way the system can be partitioned and the increased development of software to run on such systems means that the hypercube structure is probably going to be the architecture most often used in future multiple processors.

tokens called synchronization messages instead of reading and writing shared variables. Synchronization is accomplished through the transmission of this type of message because the synchronization message can only be received after it has been sent, which constrains the order of the transmission and reception events.

7. The type of interconnection structure – This is the structure used to allow the processors to communicate. One of the most often used topologies is the Hypercube (Article 13.6).

MIMD performance factors

8. The data access mode – This dictates whether data items are accessed either directly from global memory or first copied to local memories and then accessed from there. For a system intended to handle a wide range of tasks, the use of local memories at each processing node to hold program and data requires a sophisticated switching network to transfer the data and instructions between the processing units. It is then essential to have either a high bandwidth communication network or to determine the proper allocation of data to local memories to eliminate contention within the network.

9. The speed of processors and memories.

10. The need for a reliable operating system – The operating system must be able to cope with a large distributed environment.

All these problems are at present being addressed by many companies and research groups throughout the world.

As a final point on the development of multiple processors it is worth noting that most parallel computers are aimed at large regular, numeric computations which are simple enough to implement with reasonable effort in a concurrent dialect of a conventional language such as FORTRAN. The demand for parallel computers to run a wider class of computations such as those which in practice require programming in higher level languages does not exist yet. Consequently, the technological success of the fifth generation parallel computers aimed at wider and more difficult to program tasks did not resonate in the computer industry. This has meant that the potential market for artificial intelligence (AI) products has not been developed and fifth generation type machines have not been required or desired by industry or the commercial market.

13.5 MULTIPLE PROCESSOR SYSTEMS: EXAMPLES

We will consider two examples of different technologies and implementation techniques used to construct multiple processor systems.

13.5.1　TRANSPUTER SYSTEMS

In the development of commercial multiple processing systems one microprocessor stands out as being unique – this is the Inmos transputer. The transputer devices are a specially constructed family of 16–bit and 32–bit microprocessors manufactured by Inmos which have been specifically designed to handle parallel processing. Table 13.13 shows the different members of the transputer family together with a brief indication of their features.

Table 13.13

Transputer types.

Transputer	Size (bits)	Internal RAM (KB)	Serial Comms Links	Floating Pt Unit	Comment
T212	16	2	2	No	Now obsolete
T222/5	16	4	4	No	
T414	32	2	4	No	Now obsolete
T425	32	4	4	No	Upgraded T414
T400	32	2	2	No	
T800	32	4	4	Yes	Now obsolete
T801	32	4	4	Yes	
T805	32	4	4	Yes	Upgraded T801
T9000	32	16 (cache)	4 + 2	Yes	Released 1994

All members of the transputer family can be programmed in a special high level language called **Occam**. Occam and the transputer have been designed together in an attempt to produce a language and a processor ideally suited to parallel processing applications. Occam is based on communicating sequential processes (CSP) which is an approach to parallelism in which asynchronous processes communicate by sending messages through named point–to–point channels. All communication between processes is synchronous, the process that reaches the communication first is blocked until a complementary process reaches the same operation.

The original T414 transputer, it derives its name from the amalgam of the words transistor and computer, contained a 32–bit processing section, 2 KB of fast random access

Table 13.14

Transputer features.

1.	On–board RAM	6.	Event (external) management
2.	16– or 32–bit CPU	7.	Embedded microcoded priority scheduler to share the processor's time.
3.	T8 and T9 models have on–board floating point hardware		
4.	2 or 4 bidirectional serial communication links	8.	Two levels of priority, 64 and 1 μs time slice for low and high priority respectively.
5	Internal timers for real time processing	9.	Fast context switching

memory, an autonomous I/O processor, a DMA interface and four serial processor–to–processor interfaces on the same chip. The T414 was manufactured in HCMOS, came in an 84–pin package and dissipated 0.5 W. It has been superseded by the T425. Table 13.14 indicates in more detail the features of the transputer.

The T805 transputer is similar to the T414 but contains 4 KB of memory and an on–board floating point unit. A block layout of the T805 transputer is given in Figure 13.11.

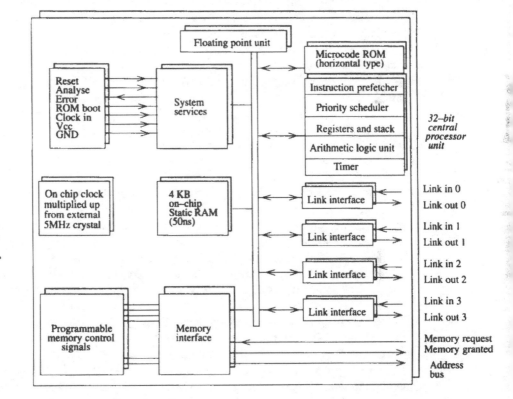

Figure 13.11

T805 transputer (block diagram).

In a broad sense the transputer has a similar architecture to a modern RISM with the exception of the link interfaces. However, it does have a few novel features in the way it operates. In order to exploit programs with a high degree of concurrency the transputer has been designed to use a decentralized model of computation where each transputer can work on an individual task by using its own local memory and then communicate, when it needs to, with other transputers in the network through its communication links. In this way local

Transputer

communication takes place on local data and concurrent processes communicate by passing messages on point to point channels. The four duplex serial communication links, each link can operate at 20 Mbps, are one of the main features which distinguishes the transputer from other microprocessors since they are intended to provide point to point connections between transputers.

By using the links for direct connection to other transputers a network of the devices avoids the physical limitations of parallel bus communications. A multiple processing environment based on transputers then has a number of advantages over conventional microprocessors, such as a simplified board layout and increased communication bandwidth, since many of the links in the system can cooperate concurrently.

Support chips for the transputer family include the C0011 serial link adapter which produces one byte wide input interface and one byte wide output for interface purposes, the C012 serial link adapter provides an interface between a serial link and an 8–bit bidirectional system bus, and the C004 is a crossbar switch which is able to switch 32 serial link inputs to 32 serial link outputs under the control of a 33rd serial link.

The T9000 device, released in 1994, introduced the latest transputer family from Inmos. It is built on 180 mm^2 die and incorporates over two million transistors. It contains a number of on board functional units such as an integer ALU, a floating point unit and a 16 KB cache memory. Internally it runs at 50 MHz from an external 5 MHz clock and has four serial communications links controlled by a virtual channel processor. It also has a 600 Mbps internal instruction and data bus.

The five blocks of logic within the T9000 are all linked by a 32–bit pipeline. A sophisticated instruction grouper which is based around a 32–bit instruction fetch ahead buffer is implemented in hardware and feeds instructions to the pipeline in an order which keeps it full. In theory this lets the device issue eight instructions per clock cycle. The T9000 device is rated at 200 MIPS and 25 MFLOPS.

13.5.2   DATA FLOW COMPUTER

The second of the advanced multiple processors we will consider is the data flow computer. A data flow computer is a parallel processing system consisting of a large number of processing elements. The processing elements perform basic operations such as addition and multiplication, and then pass their results to other processing elements which perform similar operations. As there can be many basic operations which can proceed in parallel within even fairly simple programs, it is then possible to have many processing elements active at the same time, with the results being passed from processing element to processing element. The research into data flow techniques for processing systems has been active for over 20 years and has been motivated mainly by the desire to achieve computational speeds that exceed those possible with conventional processing architectures.

In a data flow machine the program running on the system dictates the flow of data among the individual processing elements in the system. Each processing element then works asynchronously, rather than to a set schedule, calculating the results as soon as it has the data and the instruction to do so and forwarding the results directly to another processing unit. The decentralized, cooperating processing elements within the data flow system are activated on the availability of input data and instruction. The arrival at a processing element of the input data item(s) and the associated instruction, in a packet called a token, triggers the execution of the instruction that consumes the input(s). In this way an instruction is executed by a processing element when it receives all its inputs. This operation may be viewed in terms of a data flow graph in which the instructions pass their results directly to other instructions waiting for the first instruction's results. Hence, to program a data flow processor a graphic representation called a directed graph is used. This consists of a group of nodes connected by arcs.

One of the first advanced research projects in data flow architectures was undertaken in 1976 by John Gurd and Ian Watson at the University of Manchester in England. The Manchester machine has a pipelined ring structure made up from a series of distinct processing sections such that the different tasks of identifying the instructions within a program and executing the instructions can all be carried out one after the other in a pipeline operation. The interest generated by the unique architecture of data flow machines has been such that the Japanese electronics company NEC has produced the first commercially available processing chip to be designed for implementation in a data flow system. The NEC µPD7281 data flow microprocessor has a similar architecture to the Manchester machine but is contained inside a 40–pin integrated circuit. A block diagram of the microprocessor is shown in Figure 13.12.

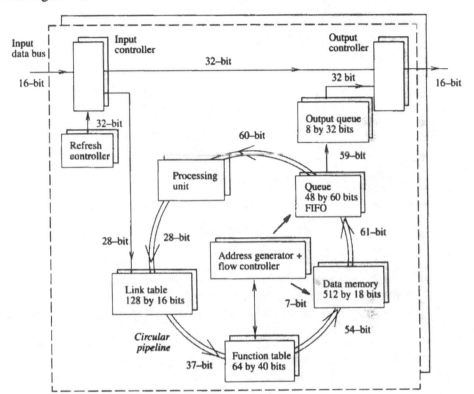

Figure 13.12

Block diagram of the µPD 7281 data flow microprocessor.

It consists of a circular pipeline around which pass tokens of work. As the tokens pass round the pipeline they are changed and augmented with data. When fully constructed each token consists of an instruction plus a number of data items. When the token arrives at the processing unit the instruction is executed and the data items used up. The µPD7281 device is primarily intended for image processing and uses an internal circular pipeline and a powerful instruction set to allow image processing at a nonstop rate of 5 MHz. When a number of 7281s are connected together to form a multiple processing system they can be arranged in a straight pipeline where the data goes in one end is processed by the units and then comes out the other end. Within the pipeline each chip passes its output directly to the input of another chip such that no interface hardware is required.

Data flow computer

13.6 INTEGRATED CIRCUIT: FUTURE DEVELOPMENT

As a final section in the book it is worth looking to the future to see the way integrated circuit technology will develop. It can be noted that by the 1960s the techniques to commercially implement IC technology into computers had been developed and the IC was then a collection of circuits fabricated on and in a single crystal, or substrate, of semiconductor material by etching, doping and diffusion. At the start of the 1970s the integrated circuit became the basis of most new computer designs with for example the production of the Amdahl 470V/6 computer, in 1972, which was the first computer to use medium scale integration technology for the logic circuits of the central processing unit.

An indication of the development of IC technology is given in Table 13.15, where we can see the development with time of the number of transistors incorporated into the Intel range of microprocessors. In the late 1970s the increasing density of gates on the IC chip reached the VLSI level with up to 50 000 gates able to be incorporated onto a small piece of silicon.

Table 13.15

Transistor count for Intel microprocessors.

Year	Intel Micro	Transistor Count
1971	4004	2.3 K
1974	8080	6 K
1978	8086	29 K
1982	80286	134 K
1985	80386	275 K
1990	80486	1.2 M
1993	Pentium	3.1 M

The major factor in increasing the number of transistors which can be used on a VLSI chip is the **minimum feature size,** called the design rule. Minimum feature sizes have decreased from 50 μm (microns) in 1960 to 0.8 μm in 1990. Since the number of components per chip is a function of the square of the feature size, the number of components per chip has dramatically increased.

IC development Improvements in the production of ICs, due to the enhanced capability of process equipment and an improved manufacturing environment have made possible a steady reduction in device feature sizes. The increase in circuit density within ICs has been achieved due to the advances in fabrication processes which have caused reductions in device sizes and **line widths** within an IC. The line widths relate to the distance apart that the components and their connections can be made. With the decrease in the size of the gates within the IC devices has come not only the ability to pack more and more gates onto a single piece of silicon but also to increase the IC's operating speed and to lower its power dissipation.

The technology feature size therefore affects the performance of a microprocessor. Table 13.16 notes the feature sizes and the MIPS performance that could be expected from a microprocessor using these feature sizes. The majority of 32–bit microprocessors use 1μm design rules whereas the previous generation of microprocessors used 2 μm design rules.

Table 13.16

The effect feature size has on the performance of a microprocessor.

Technology feature size (μm)	Performance (MIPS)
1.0	9 – 12
0.7	13 – 17
0.5	27 – 35
0.35	37 – 48

Along with the improvements in circuit density there has also been a continuous increase in wafer sizes:

1. 2 in diameter wafer in 1960
2. 4 in diameter wafer in 1977
3. 5 in diameter wafer in the early 1980s
4. 6 in wafer in the late 1980s
5. 8 in wafer in the early 1990s.

Hence with the increase in the number of chips that can be obtained from a single wafer and advances in minimizing the imperfections in wafers, giving an improved reliability in the devices, the cost of ICs has been reduced.

It is expected that the decreasing feature size used in ICs will continue in the future, hence the complexity of ICs will increase. By increasing the complexity of an integrated circuit not only is there a reduction in the size and cost of the final product, but there is also an increase in the operating reliability of the system. Hence, the recent success in manufacturing integrated circuits that can perform complex logical transformations on the one chip has made it possible to speed up the development of many complex computer systems. It is also to be noted that by creating different combinations of elements on IC chips, and by combining them into the necessary circuitry, extremely diverse computer components can be obtained.

Figure 13.13 indicates a number of categories in which IC technology is projected to develop. With this development will come the opportunity to produce much more powerful computers than the present machines.

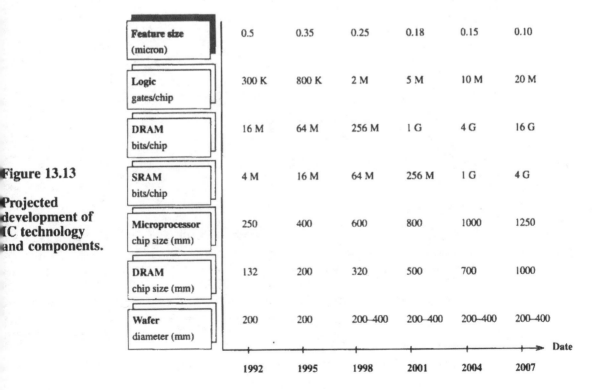

Figure 13.13

Projected development of IC technology and components.

	1992	1995	1998	2001	2004	2007
Feature size (micron)	0.5	0.35	0.25	0.18	0.15	0.10
Logic gates/chip	300 K	800 K	2 M	5 M	10 M	20 M
DRAM bits/chip	16 M	64 M	256 M	1 G	4 G	16 G
SRAM bits/chip	4 M	16 M	64 M	256 M	1 G	4 G
Microprocessor chip size (mm)	250	400	600	800	1000	1250
DRAM chip size (mm)	132	200	320	500	700	1000
Wafer diameter (mm)	200	200	200–400	200–400	200–400	200–400

The first dynamic RAM (DRAM) was the 1103, a 1 Kbit device released by Intel in 197
Since then, due to improving IC technologies, the size of the DRAM has increase
dramatically. In 1986 the first 1 Mbit DRAM chips became available in commerci
quantities. Table 13.17 notes the projected schedule of increases in the memory density c
IC development DRAM devices up to the year 2005. It can be seen from the table that as the feature siz
decreases so the does the density of the DRAM devices. Design rules using less than 1μι
are now possible using X–ray and ultraviolet lithography to produce the patterning of th
ICs. Equally important is the fact that circuit characteristics such as switching time an
power dissipation also improve with decreasing feature size.

The information in Table 13.17 can be plotted in a graph to give Figure 13.14.

Table 13.17

**Projected schedule
of increases in
memory density
per chip.**

Memory density (Mbits)	Feature size (μm)	Type patterning radiation	Commercially available
4	0.8	Visible light	1989
16	0.5	Visible light	1992
64	0.33	Electron beam	1995
256	0.2	X–ray	1998
1024	0.1	X–ray	2005

Figure 13.14

**Projected increase
in density and size
of DRAM.**

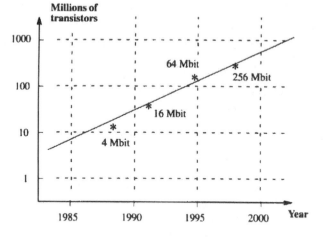

The 16 Mbit DRAMs currently available have been designed on 8 in wafers using 0.5
micron integrated circuit technology. The 64 Mbit devices, however, will require 0.3!
micron line widths, while it is envisaged that sub 0.2 micron line widths will be required t(
produce Gbit DRAMs. To achieve high density DRAMS, improved production techniques
have reduced the minimum size of the transistor elements, but the storage capacitor has to
have a certain surface area in order to possess enough capacitance to be reliable. What has
evolved is a technique for increasing the surface area of the capacitor by creating grooves
or cavities in the surface of the silicon.

The future of IC technology will see the use of wafer scale integration and galliur
arsenide to improve IC performance.

13.7.1   WAFER SCALE INTEGRATION

Wafer scale integration is the use of a complete silicon wafer, up to 8 in diameter, to form a single massive electronic circuit, essentially made up from a large number of interconnected small circuits. This technique was tried unsuccessfully by the Trilogy Corporation in the early 1980s. However, they encountered problems with quality control since any small defect in one part of the wafer could mean the complete circuit would have to be rejected. The technique of manufacturing these devices has been developed to the extent where this problem can be dealt with and complete wafer scale integrated circuits have been produced.

13.7.2   GALLIUM ARSENIDE (GaAs)

Gallium arsenide is a compound of the elements gallium and arsenic which has recently been used as the base material for constructing a wide range of semiconductor devices. Table 13.18 lists some of the properties of GaAs while Figure 13.15 notes some of the application areas.

Table 13.18

Properties of gallium arsenide.

Melting point	1238 °C
Atomic weight	144.6
Atomic density	4.43×10^{22} cm^{-3}
Lattice constant	5.654 A
Density	5.316 g cm^{-3}
Thermal conductivity	0.4555 W cm^{-1} °C^{-1}
Band gap	1.4 eV
Hall mobility	400 cm^2 V^{-1} s^{-1}

Figure 13.15

Applications of GaAs.

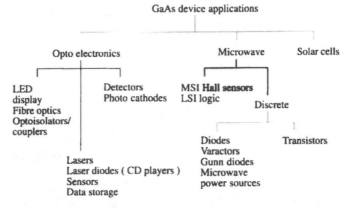

GaAs semiconductor chips have several major advantages over their silicon counterparts.

1. They can work at higher temperatures, with an operating range from –200 to +300 °C, silicon stops at 150 °C.

2. For space applications GaAs has proven to be more resistant to radiation effects than silicon by a factor of a thousand.

3. GaAs electron mobility is five times that of silicon. This gives a fivefold increase in the speed–power performance of GaAs over silicon. Currently for the same power consumption, GaAs is about half an order of magnitude faster than emitter coupled logic (ECL), the fastest silicon family.

The new technology of designing logic devices with gallium arsenide promises to giv
higher speeds to ICs in the future. It should be possible to make GaAs gates with delays o
100 ps and very large scale integrated devices with very low power dissipations.

However, there are disadvantages in using GaAs technology as compared with silico
technology.

(a) GaAs crystals are hard to grow and much more expensive than silicon.

(b) GaAs circuits are difficult to fabricate due to the lack of a native oxide, which is a
essential factor for the fabrication of MOS transistors.

(c) GaAs wafers presently exhibit a large density of dislocations, that is a larger number o
irregularities. Consequently, compared to the equivalent silicon wafer, GaAs chips hav
to be smaller in area, have a smaller transistor count, and have a worse yield.

GaAs wafers are only now reaching 4 in diameter compared to 8 in silicon wafers.

(d) GaAs wafers are mechanically difficult to work with.

Gallium Arsenide

In the future it is expected to see GaAs technology being applied in the area o
microprocessor design to produce extremely fast limited–application devices. It woul
appear that the single chip reduced instruction set computer type of design will be the avenu
where GaAs technology will find an appropriate market. Texas instruments, for example, ha
designed an experimental 150 MHz pipeline RISM which contains 13 000 GaAs gates. I
1990 GaAs integrated circuits with over 100 000 transistors were produced. One of the firs
supercomputers to use GaAs technology is the Convex C3800 where the performance of th
eight–processor machine is claimed to be 2000 MFLOPS for vector operations.

Another avenue open to GaAs exploitation has come in the area of electro–optica
components. GaAs can be used to produce lasers and photodetectors both of which may b
used in the proposed optical computers of the 21st century. GaAs is therefore becoming
popular material for the construction of a wide range of very fast components. However, in
the area of very large scale integration design it is generally accepted that GaAs will neve
totally replace silicon but will be used in selected aerospace, defense and supercomputin
applications.

13.7 KEY TERMS

Amdahl's law	Connection machine
ARM microprocessor	DAP computer
Arithmetic pipelining	Data flow computer
Array processor	Deadlock
Associative processor	Digital signal processor
Balance 2000 hypercube	DSp microprocessors:
BBN butterfly computer	architectural features
Complex instruction set	DSP microprocessors:
computer (CISC)	programming
Complex instruction set	Dual port memory
microprocessor (CISM)	Fixed point arithmetic
Complex instruction set	Floating point arithmetic
Computer architectures	Flynn's taxonomy
Computer taxonomy	Gallium arsenide (GaAs)
Concurrent	*(Continued)*

Harvard architecture	Pipelining
Hypercube	Reduced instruction set
Instruction pipelining	microprocessor (RISM)
Integrated circuit: future	SIMD computer system
development	SISD computer system
Intel iPSC hypercube	Speed up
Line widths	Superpipelined techniques
Local memory	Superscaler techniques
Loosely coupled	Systolic Array
MIMD computer system	Texas instruments DSP
Minimum feature size	microprocessors
MISD computer system	Tightly coupled
MIPS R2000	TMS 32010
Motorola MC 88000/881110	TMS 32020
MPP computer system	TMS 320C25
Multiple processor	TMS 320C30
Multiport memory	TMS 320C40
Multiprocessor	Transputer
N/CUBE hypercube	T series hypercube
Occam	Vector computer
Parallel processing	Von Neumann architecture
Pass–by–reference	Wafer scale integration
Pass–by–value	Wavefront array

13.8 REVIEW QUESTIONS

1. What were the factors which brought about the development of CISMs?
2. Name four microprocessors which can be regarded as having complex instruction sets.
3. Name the five different parts within instruction pipelining.
4. Briefly describe how arithmetic pipelining operates.
5. Why are floating numbers harder to process than integers?
6. What is the disadvantage of using software to handle floating point calculations as opposed to using dedicated hardware?
7. What were the main design decisions that evolved the first RISC computer?
8. List the six elements that are essential to a RISC philosophy.
9. What is the difference between a von Neumann architecture and a Harvard architecture?
10. Why do RISMs require a more complex compiler than a CISM?
11. Why is the development time for a RISM shorter than for a CISM device?
12. List five advantages of digital signal processing compared to analog signal processing.
13. What are the main architectural features in a DSP microprocessor that give it its high numerical performance?
14. What are the advantages of having a wide range of I/O interfaces on a DSP microprocessor?

15. Name Flynn's four classifications of computer systems.
16. What does the term stream refer to in the Flynn classification?
17. What does the term vector refer to when used to describe a vector computer?
18. Name the three most common forms of a SIMD computer system.
19. Describe the operation of a systolic array.
20. What are the differences between tightly coupled and loosely coupled multiple processors?
21. What do the following terms mean when used to refer to multiple computer systems: deadlock, multiport memory and Amdahl's law?
22. List five factors which determine the amount of parallelism which can be exploited in a task.
23. What is a transputer? Note four of the features of a transputer.
24. In a data flow machine what are tokens?
25. Where is the main application area for the NEC μ7281 microprocessor?

13.9 PROJECT QUESTIONS

1. Contrast the differences in superscaler and superpipelined architectures.
2. Contrast a RISM and CISM device in terms of their internal architectures.
3. What advantage does a fixed instruction format have for a RISM?
4. As a project choose one of the RISC processors noted in the text and discover as much about the device as you can.
5. What are the application areas for DSP microprocessors?
6. Why has the programming language C been chosen as the main high level language for DSP microprocessors?
7. DSP processing used to be accomplished using devices called bit–slice processors. What advantages does a DSP microprocessor have over a bit–slice design?
8. List three special instructions that DSP microprocessors have which conventional CISM devices do not have.
9. Is the vector computer a true MISD computer? Explain your answer.
10. What are the main features of a systolic array which distinguishes it from an array computer?
11. Do you think Amdahl's law is still valid?
12. Select one of the computers noted in the section on MIMD computers and through suitable reference describe its main features.
13. What are the advantages that the hypercube form of architecture has over other loosely coupled structures?
14. Describe the requirements for an efficient parallel processing system.
15. What do you think will ultimately limit the speed of a computer?
16. How does the transputer provide flexibility in producing a multiple processor system?
17. What is a crossbar switch?
18. In a multiple transputer system why does the communication bandwidth increase as processors are added?
19. What are the difficulties in using data flow computers?

20. Which new computer technologies appear to be most promising?
21. Describe the computer system you expect to see in the future.
22. How much faster are GaAs logic circuits compared to TTL devices?
23. Investigate the process of wafer scale integration and its potential relevance to future computers.
24. Investigate the potential for integrated optical circuits. Will they ever replace conventional ICs?
25. Discover why Josephson transistors are claimed to be superior to silicon ones.

13.10 FURTHER READING

This chapter has provided an overview of modern microprocessors and computer systems. It has tried to introduce the terms and present a brief description of the reasons for the development of the particular architectures. All contemporary computer magazines and periodicals provide articles on the developments in modern computer systems. A selection of these articles and books is provided below.

Multiple Processing and Advanced Computers (Books)

1. *Multiple Processing: a Systems Overview*, A.J. Anderson, Prentice Hall, 1990.
2. *Computer Architecture Design and Performance*, B. Wilkinson, Prentice Hall, 1991.
3. *Technology of Parallel Processing Volume 1: Parallel Processing Architectures and VLSI Hardware*, A.L DeCegama, Prentice Hall, 1989.
4. *Parallel Computers*, 2nd Edn, R.W. Hockney and C.R. Jesshope, Adam Hilger Ltd, 1983.
5. *Computer Design and Architecture*, L.H. Pollard, Prentice Hall, 1990.
6. *Systems Design with Advanced Microprocessors*, J.R. Freer, Pitman, 1987.
7. *Computer System Architecture*, 3rd Edn, M.M. Mano, Prentice Hall, 1992.

Future Computing (Articles)

1. Computer technology and architecture: an evolving interaction, J.L. Hennessy and N.P. Jouppi, *IEEE Computer*, September 1991, pp. 18–29.
2. The drive to the year 2000, W. Myers, *IEE Micro*, February 1991, pp. 10–74.
3. Essential issues ·in multiprocessor systems, D.D. Gajski and J. Pier, *IEEE Computer*, June 1985, pp. 9–27.
4. Multiprocessor surf up, B. Ryan, *Byte*, June 1991, pp. 199–206.
5. Scaling up:get the message, R.M. Syein, *Byte*, June 1991, pp. 231–240.
6. All systems go, D. Pountain and J. Bryan, *Byte*, August 1992, pp. 112–136.
7. Popular and parallel, M. Robinson, *Byte*, June 1991, pp. 219–228.

RISC Architectures (Books)

1. *Computer System Architecture*, 3rd Edn, M.M. Mano, Prentice Hall, 1992.
2. *A Guide to RISC Microprocessors*, Edited by M. Slater, Academic Press, 1992.

RISC (Articles)

1. Reduced instruction set computers, D.A. Paterson, *Comms of the ACM*, January 1985, pp. 8–20.

2. A comparison of RISC architectures, R.S. Piepho and W.S. Wu, *IEEE Micro*, August 1989, pp. 51–62.
3. Third generation RISC processors, R. Weiss, *EDN*, 30th March 1992, pp. 96–108.
4. RISC enters a new generation, R.L. Sites, *Byte*, August 1992, pp. 141–148.
5. A call to ARM, D. Pountain, *Byte*, November 1992, pp. 293–298.
6. How RISCy is DSP, M.R. Smith, *IEEE Micro*, December 1992, pp. 10–23.

Integrated Circuits (Articles)

1. Farewell to chips, B. Ryan, *Byte*, January 1990, pp. 237–248.
2. Chips for the nineties and beyond, J.J. Barron, *Byte*, November 1990, pp. 342–350.
3. The high octane semiconductor, P. Robinson, *Byte*, January 1990, pp. 251–258.

Appendix

Part Seven

Appendix

The Evolution of the Computer

Chapter Structure

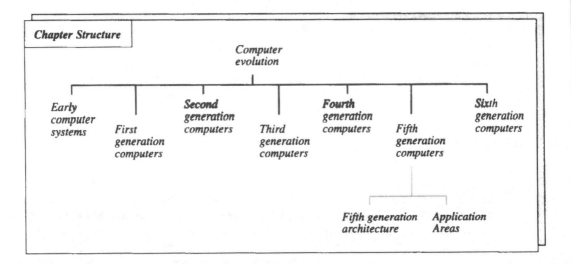

Learning Objectives

In this chapter you will learn about:

1. The history of computers and the major landmarks in their development.
2. The characteristics of the computers developed in the various computer generations.
3. The reasons for each step in computer development.
4. The individuals who pioneered the development of computer hardware.
5. The role of IBM and Intel in computer development.
6. The desire for AI characteristics in the computers of the future.

'Artificial intelligence is what we don't know how to do yet'

Larry Tesler

Appendix

The Evolution of the Computer

Perhaps the most rapid growth of technology that mankind has seen over the past 50 years has occurred in the area of electronics. This can probably best be seen in the evolution of the computer. Present computers did not spring into being as complete working systems, but instead the computer has evolved through many years into many different forms. Although this development has been a continuously changing environment, many commentators have chosen to view this evolution as occurring in various stages. Although these stages are somewhat approximate in their time scale and their extent, and are subjective in the criteria chosen to distinguish the stages, they do allow the different types of computers to be graded and ordered into some kind of taxonomy.

A.1 EARLY COMPUTER SYSTEMS

The term computer was originally used in the late 18th century to refer to a person who calculated various tables which could be used for mathematical calculations. The calculations were usually done by hand (in some cases using logarithms) but were still very tedious and prone to error. Logarithms had been invented in the late 1500s by the Scottish mathematician John Napier (1550–1617) and have the advantage that they enable multiplication and division to be handled by addition and subtraction operations.

The first mechanical **calculator** is credited to the German mathematician Wilhelm Schickard (1592–1635) in 1623. It accomplished multiplication and division by converting numbers to logarithms, adding or subtracting them and then converting back. The first automatic mechanical calculator was developed in 1642 by the French mathematician Blaise Pascal. The device was called the Pascaline and used extremely precise interconnected gears to perform the addition and subtraction calculations; it could not perform multiplication or division. By 1673 the German mathematician Leibnitz had improved Pascal's design to handle multiplication and division; however it was not very reliable. In 1820 Charles Thomas de Colmar in France began selling his Arithmometer a refined version of the Liebnitz machine. He sold approximately 1500 machines over a 30–year period.

The concept of using a machine which could be programmed to generate mathematical tables can be traced back to the Englishman Charles Babbage at the beginning of the 19th century. He proposed a machine called a difference engine which was designed to calculate complex polynomial expressions (Article A.1). The Babbage machine was to use punched cards to store its program, the idea of using cards coming from the Frenchman Jacquard who had used the idea in his loom.

In 1885 that the first successful key–driven calculator was developed by Dorr Eugene Felt (1862–1930). This device is often seen as the forerunner to all modern key–driven calculators. However, up until the 1930s the calculating machines were all mechanical and based on cogs and gears.

CHARLES BABBAGE

The fundamental concept of a machine that could perform general calculations automatically can be traced back to the 19th century to the work of the English mathematician Charles Babbage, born 1791. He initially proposed a machine similar to a basic calculator, designed to handle tedious repetitive calculations, called a **Difference Engine** of which he began to build a prototype in 1823. He initially devised the difference engine to calculate equations of the form: $y = a_1x + a_2x^2 + a_3x^3 + ... + a_6x^6$

where a and x were the entered data and y is the result. Earlier difference engines by Schickard, Pascal and Leibnitz were limited to typically 6 or 8 figures and were of doubtful reliability. Babbage's designs were different in that he proposed a complete design for an automatic calculator – one which would successfully incorporate a mathematical rule in the mechanism. The difference engine was designed to calculate and print mathematical tables automatically, using the method of finite difference. Babbage's motive was to eliminate errors in printing tables upon which anyone requiring more than a few figures of accuracy relied. The results would be impressed on paper strips or soft metal for making printing plates from which the results of the calculations could be replicated. Babbage's difference engine was a system of cogs wheels, cams and levers and would have measured some 7ft long, 8ft high and 3 ft deep. It would have consisted of an estimated 25 000 parts and would have weighed several tons. Less than one seventh of the difference engine was actually built (due it was claimed to the lack of suitable engineering skills of his day) though most of the essential parts were completed. Babbage abandoned the construction of the difference engine in 1833. However in 1853 (after Babbage's death) a difference engine based on Babbage's original design was built by G. Scheutz and E. Scheutz in Sweden.

In 1834, following on from this work, he planned to construct a very complex device called an **analytical engine** which was to both contain a form memory and an arithmetic unit and include the principles of programming. The analytical engine was even bigger than the difference engine and would have stood 15 ft high and 10–20 ft long, depending on the number of variables catered for. The difference engine was hand powered, whereas the analytical engine would have been steam driven. The analytical engine was to be capable of performing the basic arithmetical functions for any mathematical problem and wad to do so at a speed of 60 additions per minute.

The machine was to be composed of five parts:
1. a store in which to hold numbers;
2. an arithmetic unit which Babbage called the Mill since all the operations were to have been carried out automatically through the rotation of cogs and wheels;
3. a control unit for ensuring the machine performed the desired operations in the correct sequence;
4. an input device;
5. an output device to print the numbers.

The analytical engine was not only automatic but also general purpose. It was also supposed to have a central processing cog with a number of smaller drums to store the results. With the exception of an internal stored program, the instruction sequence was stored on external stitched punched cards; the analytical engine had most of the logical features of the modern computer. One of Babbage's associates was Ada Augusta, the countess of Lovelace, who worked with Babbage on devising methods to program the analytical engine. The high level programming language **Ada** was named after her. Only a small part of the analytical engine was ever built, mainly due to lack of funding.

It was not until 1944 that machines which could be called true computers were built. These were developed at Cambridge (England) and Harvard (USA). Both devices put into practice many of the concepts formulated by Babbage, but made use of up–to–date electromechanical techniques which resulted in far more sophisticated machines than that originally planned by Babbage.

Early computer systems

In the 1930s several research groups began to construct electrical calculators based on the use of electromechanical relays. In the mid–1930s **Konrad Zuse** in Germany produced a calculator based on electromechanical relays and Boolean algebra, i.e. the machine used binary rather than decimal arithmetic. Programming was done on paper tape and a multiplication took of the order of 5 s.

Later in the 1930s machines using valves were developed. These could operate 1000 times faster than the relays. In 1936 **Alan Turing**, a British mathematician, published a research paper in which he laid down some fundamental ideas in the area of computing. In the paper he set out the abstract design for a computer which was not just a general purpose device, but one which could in theory solve any problem that could be stated in the form of what we would now call an algorithm. The work of Turing suggested that a number in a computer was nothing more than a string of symbols which could have many interpretations and that data in a computer could be regarded as sequences of symbols. This sequence of symbols could be manipulated by the computer, but would then require interpretation by a user before they could provide information. Turing also noted that a string of symbols could be interpreted as an instruction to inform the machine to perform a certain task. In this way a computer was essentially a symbol–manipulating machine where the interpretation of the symbols used by the machine was left to the programmer or user.

Early computer systems

The Z1 electromechanical calculator was designed by Konrad Zuse in Germany in 1939. Also in 1939, Dr. John V. Atanasoff and Clifford Berry at the Iowa State University in the USA began development of the first electronic digital computer. The device was called the **ABC** (Atanasoff Berry Computer) and was the first computer to use vacuum tubes for the logic circuits, it had 300 vacuum tubes. The first general purpose program controlled computer is generally credited to the Konrad Zuse and Helmut Schreyer who in 1941 built the Z3 electromechanical relay computer. It was a binary computer with a 64–word store.

In January 1943 the **Mark 1** calculating machine which had been designed and built by Howard Aiken at IBM's Endicott Labs in the USA was operating. It utilized 3304 electromagnetic relays, and mechanical counters, was over 15 m long by 2.5 m high and was reckoned to have 800 km of wire. It also had three million electrical connections, contained 0.75 million mechanical parts and weighed about 5 tons. Information was fed into the Mark 1 by punched tape and the results were produced on punched cards. It was called an automatic sequence controlled calculator since it could perform any sequence of operations involving addition, subtraction, multiplication, division, comparison of numbers and reference to stored tables. Addition of two numbers, each containing 23 digits, could be performed in 0.3 s, multiplication in 6 s and division in 12 s. The device stayed in use for about 15 years.

At about the same time the **Colossus** computer designed and built at Bletchley Park in England began working. The Colossus computer, which contained 1500 valves, was used to decode German war codes, primarily the codes produced by the Enigma and more complex Lorentz coding machines. It utilized a fast photoelectric tape reader for input. In all, ten Colossus machines were built. Allan Turing was part of the team which worked on Colossus.

The first fully automatic electric computer known as ENIAC was constructed in 1945 (Article A.2). One of the pioneers of modern computer systems, the mathematician **John von Neumann** was involved in the ENIAC project. In 1945, during the construction of the ENIAC computer, he published his now famous report on the **EDVAC** (electronic discrete variable automatic computer) computer. This suggested a design concept for an internal electronically stored program which could control a computer with a storage capacity of 1000 words of 10 decimal digit capacity. EDVAC was constructed at Pennsylvania in the USA and was working by 1952. EDVAC was far more versatile than the ENIAC computer in that it was not only a stored program machine but also used a punched paper tape input.

Article A.2

ENIAC COMPUTER

The ENIAC (Electronic Numerical Integrator And Computer) computer, which is widely regarded as the first true electronic computer, was started in March 1943. The ENIAC project began with a proposal by Dr. J. Presper Eckert and Dr J.W. Mauchly to the US Army Ordinance Department to construct an electronic digital computer to speed the calculation of artillery firing tables. The proposal was taken up and with government aid the project was finished and operating 30 months later. More than 200 000 man hours went into the building of the machine which was not surprising when considering its sheer physical size, it measured 18 ft high by 80 ft long, weighed 30 tons, contained over half a million solder joints and had about 17 468 thermionic valves; it also used 130 KW of power. The machine was completed at Pennsylvania in the USA in 1946.

Instructions to ENIAC had to be conveyed to the machine through a series of wired plug boards and switches, all of which was a time consuming business.

The machine operated on data and instructions in a binary code, stored together in memory. This had the effect of saving memory space by allowing the memory to be partitioned flexibly between data and instructions and also allowed the mechanisms for fetching both the data and instructions from memory to be the same. More fundamentally it meant that symbols could be treated as data in one context and instructions in another, one of the concepts proposed by Alan Turing. ENIAC could perform 5000 additions per second and could perform a multiplication in 3 ms.

Within the computer a small number of registers were required, to hold such things as the current program instruction and the current data item being processed. The machine operated in a repeated cycle of steps to fetch and execute program instructions in sequence. ENIAC was retired from service in the early 1960s and now rests in the Smithsonian Institute in the USA.

In 1946 von Neumann in collaboration with A.W. Burks and H.H. Goldstein, he wrote a paper that delineated the concepts on which nearly all computers have been built since. The paper, entitled 'Preliminary Discussion of the Logic Design of an Electronic Computing Instrument', advanced the concept of the stored program and introduced the idea of the program counter. In this paper and in subsequent writings von Neumann laid out some of the fundamental concepts inherent in the design of computer systems. These features have been regarded as the prerequisites for a standard von Neumann computer.

Early computer systems

In 1949 the first stored program computer was built at the university of Manchester. It was known as the Manchester MARK 1 computer and was built by Williams and Kilburn to test the Williams cathode ray tube, this tube was used as a memory store. In the same year the EDSAC computer was developed by M.V. Wilkes at the University of Cambridge; it used punched paper tape as an input. Also in 1949 the first random access core memory appeared, allowing comparatively fast, but expensive, main memory.

In the visualization of the development of computers the use of stages also allows a compartmentalizing of the view of technological advances in the area of computer architecture. The various stages in computer evolution are shown in Figure A.1.

A.2 FIRST GENERATION COMPUTERS

The first generation of computers is usually regarded as starting at the beginning of the 1950s; February 1951 saw the introduction of the Ferranti Mark 1 version of the Manchester University computer while a month later the first **UNIVAC** computer was delivered to the US Census Bureau. UNIVAC 1 was a descendant of the ENIAC device and was the world's first commercial computer.

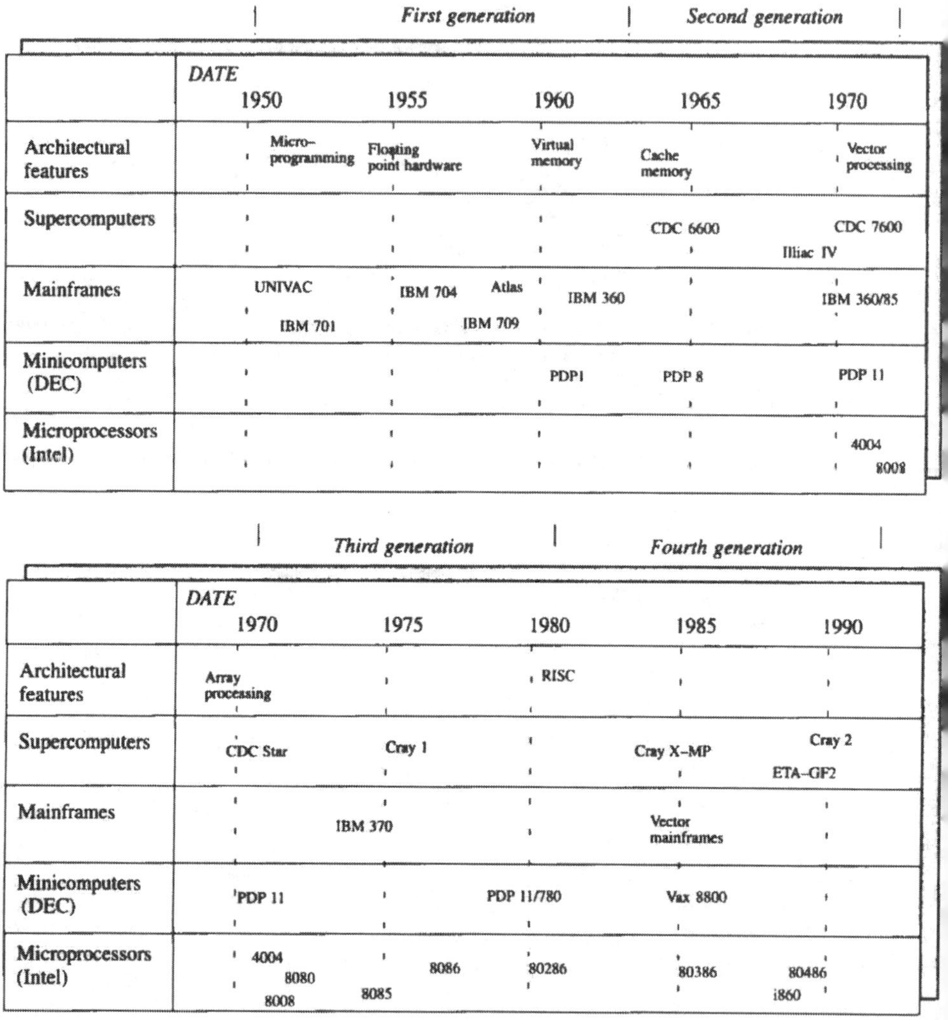

	First generation			Second generation	
DATE	1950	1955	1960	1965	1970
Architectural features	Micro-programming	Floating point hardware	Virtual memory	Cache memory	Vector processing
Supercomputers				CDC 6600	CDC 7600 / Illiac IV
Mainframes	UNIVAC / IBM 701	IBM 704 / IBM 709	Atlas	IBM 360	IBM 360/85
Minicomputers (DEC)			PDP1	PDP 8	PDP 11
Microprocessors (Intel)					4004 / 8008

	Third generation			Fourth generation	
DATE	1970	1975	1980	1985	1990
Architectural features	Array processing		RISC		
Supercomputers	CDC Star	Cray 1		Cray X-MP	Cray 2 / ETA-GF2
Mainframes		IBM 370		Vector mainframes	
Minicomputers (DEC)	PDP 11		PDP 11/780	Vax 8800	
Microprocessors (Intel)	4004 / 8080 / 8008	8085	8086 / 80286	80386	80486 / i860

Figure A.1

Computer evolution.

UNIVAC could execute 10 000 instructions per second and had a 1000 word main memory where each word was 60 bits wide and held 12 5–bit characters, it also used magnetic tape for I/O. UNIVAC's designers used 100 memory modules consisting of mercury delay lines, where each memory module was a shift register that could hold ten words. Within each module the circuitry introduced electrical pulses at one end of the delay line and detected them 404 microseconds later at the other end. After the circuitry detected the pulses, it amplified, cleaned and reinserted them at the beginning of the delay line.The average access time for a word was 202 microseconds. In all, 48 UNIVACs were sold. In 1951 LEO (Lyons Electronics Office) was the first computer in Britain to be used for commercial applications.

First generation computers

The 1950s was the decade when the importance of the computer was first noted on a large scale and by the close of the decade the computer had been established as an important processing device that had evolved from the research labs and universities into the commercial world.

The different generations of computers and their architectural features are given in the table in Table A.1. From Table A.1 it can be seen that most first generation computers used vacuum tube technology, were slow and bulky, required air conditioning to dissipate the huge amount of power they required, and could only accommodate a limited number of input/output devices. Magnetic tape was the predominant input/output medium, with the data access time measured in milliseconds. However, in 1956, the RAMAC 305 hard disk device was introduced. It was the first storage system able to permit files to be randomly stored and accessed. It used 50 magnetic disks and could store 5 million characters with an access time to any record of less than one second. It was the forerunner to modern disk based systems.

Table A.1

Computer technology for the different generations of computers.

Property	First generation	Second generation	Third generation	Fourth generation
Hardware	Vacuum tubes	Transistors	Integrated circuits	Very large scale integrated circuits
Main memory	Magnetic drum	Magnetic core	Bubble memory	Semiconductor memory
Memory capacity (characters)	8 K	64 K	4 M	32 M
External storage	Magnetic tape	Magnetic hard disk	Magnetic Winchester disk	Mass storage optical disks
Processing unit	Valves	Transistors	Micro–processors	Micro–computers, parallel processing.
Speed (millions of instructions/s)	0.01	1	10	1000
Failure rate	Minutes	Days	Weeks	Months
Software	Machine code	High level languages	Advanced languages, structured programming.	Expert systems, object orientated languages.
Operating system	Single user	Batch/ multiprogrammed	Timesharing	Multiple user distributed networks.

First Generation Computers

The early computers also had small main memories and processors matched to the memory's performance. In these machines the data was typically read in or written out via a small memory block called a register located in the arithmetic unit. This however prevented the machine from carrying out any further calculations while input/output (I/O) was taking place and so limited performance. The problem was solved with the introduction of a specialized I/O device called an I/O channel which could transfer data to and from the slow peripherals such as magnetic tape and disk units or handle the output to the printers. The I/O channel had its own instruction set suited to I/O operations. The IBM 709 mainframe computer introduced in 1958 had six I/O channels.

It is to be noted that many of the basic features that are currently associated with modern computers originated in the 1950s. Several examples are noted in Table A.2.

Table A.2

Architectural features and the computers which introduced them.

FEATURE	YEAR	COMPUTER
1. Index registers	1949	University of Manchester computer
2. Parallel arithmetic	1952	The IBM 701
3. Floating–point hardware	1954	IBM 704
4. Interrupts	1954	UNIVAC 1103
5. General purpose register sets	1956	Ferranti Pegasus
6. I/O channel processor	1958	IBM 709
7. Indirect addressing	1958	IBM 709
8. Virtual memory	1959	University of Manchester ATLAS.

The early computers were primarily used to perform scientific calculations, since the power of these processors was thought to be efficiently used only on large computational problems which could not be effectively performed otherwise. However, compared with modern programming tasks, the problems attempted in the mid 1950s had to be small because of the limitations imposed by the small central memory size, slow peripherals and the need to protect program execution from system failure due to the hardware's low mean time between failure. Since the expensive resources were hardware, its efficient use dominated the thinking of the time and shaped the way programs for the computers were written.

First generation computers

The first generation devices were also difficult and time consuming to program as they relied on machine coded programs. Machine language, or machine code, is a form of binary code that is suitable for direct execution on a computer. Assembly language was developed later to aid with this programming. It is a form of symbolic code representing the machine language instructions.

By the close of the 1950s a number of high level languages had been developed. FORTRAN was introduced by IBM in their IBM 700 series machines and in 1958 the European Association for Applied Mathematics and Mechanics and the American Association for Computing, meeting in Zurich, began research into the development of a universal programming language that eventually produced ALGOL 60. Just at the turn of the decade the language COBOL was also introduced. This language had been developed under the leadership of Grace Hopper and was sponsored by the USA government in conjunction with computer manufacturers and users, for commercial data processing.

A.3 SECOND GENERATION COMPUTERS

By the mid 1960s and the second generation of computers, the computer had evolved architecturally to having a single set of general purpose registers combining accumulators, index and base register and subroutine linkage registers, while the control unit was a straightforward microprogrammed read only memory (ROM). The **Atlas** computer for example designed at the University of Manchester was working by 1962 using 80 000 germanium transistors. It was released as a commercial product by Ferranti in 1963. This device introduced many ideas which were incorporated into the computers of the 1960s, such as the following.

1. Multiprogramming based on virtual memory. The virtual memory scheme allowed the programming range of the machine to be much larger than the physical address space of the primary memory.
2. A banked memory store.

IBM COMPUTERS

IBM is the largest computer manufacturer in the world. It manufactures a wide range of systems from personal computers to large supercomputers. IBM can trace its formation back to 1911 when a number of small companies merge to form the Computing Tabulating Recording (CTR) company. In 1924 the company changed its name to the International Business Machines (IBM) Corporation. The company entered the computer market in 1952 when it introduced the IBM 701. In 1953 the IBM 650 general purpose computer was introduced; this computer established IBM as one of the leading computer manufacturers, with over one thousand 650s eventually being sold. The IBM 7030, called the Stretch, was one of the first computers to be built from transistors. It contained 169 100 transistors and was delivered in 1961.

IBM came to world prominence in the computer market when in April 1964 it introduced the IBM **System/360** series of mainframe computers. This series proved to be one of the most influential ranges of systems in the evolution of the computer, with far reaching implications for the structure of computer development. The 360 series was designed to handle both scientific and commercial computing, in effect combining in a single range of computers a series of systems that could satisfy all the needs of IBM customers. It was through this range that IBM captured its dominance of the computer mainframe market. The 360 series was also an attempt to produce a compatible range of computers varying in price and performance but with a single architecture to maximize the compatibility of the peripherals, programs and operating systems among the different models in the series.

The characteristics of the IBM 360 family were
1. similar or identical instruction sets;
2. similar or identical operating systems;
3. increasing speed with increasing model number (see Table A.3);
4. increasing number of I/O ports with increasing model number (see Table A.3);
5. increasing memory size with increasing model number (see Table A.3);
6. increasing cost with increasing model number.

Characteristic	IBM 360 Model				
	30	40	50	65	75
Max memory (KB)	64	256	256	512	512
Processor cycle time (micro-seconds)	1.0	0.625	0.5	0.25	0.2
Maximum number of I/O channels	3	3	4	6	6

Table A.3   **Characteristics of the IBM 360 family of mainframe computers.**

The IBM 360 series were the first major mainframes to use microprogramming. A microprogrammed processor carries out its functions with a program stored in read only memory, that interprets the larger instruction set. This allowed the design to be more flexible and easier to both adapt to changes and also to verify its operation.

The IBM 360/20, the first of the series, had an instruction set which was more complete and complex, compared to any previous design, and included instructions to handle both integer and floating point arithmetic, it had 143 instructions. The IBM 360 series eventually produced a range of models (20, 30, 40, 44, 50, 65, 75, 91 and 95) and as the range of devices developed so did their architectural features. The 360/91, for example, released in 1967, had an instruction look–ahead capability and separate execution units for floating point arithmetic, integer calculation and address generation, each of which was pipelined and could operate in parallel. The IBM 360/91 and 360/95 were the top of the range devices and were aimed at scientific applications. The 360 series eventually contained over 10 different models and influenced the design of the later IBM 370 series of computers.

IBM continued the production of high performance computers with the GF11 in 1985, a modified single instruction multiple data stream (SIMD) parallel computer. It has 566 processing elements, interconnected via a nonblocking Benes network. The network can activate any of 1024 data path permutations among the processors. The switch settings and control decisions are made at compile time. It can sustain more than 1 GFLOPS, i.e. 10^9 floating point operations per second. Each of the 566 processor boards contains 642 chips and a 300–pin connector.

3. Paging with 512 words/page within a 16 KB main memory and 96 KB of secondary memory. The secondary memory utilized magnetic drums and magnetic tapes.

4. An interrupt system which allowed slow I/O devices to work autonomously and only to interrupt calculations in the central processing unit when it was absolutely necessarily.

The second generation machines developed using random access core memories, transistor technology, and multifunctional units with hardware services such as floating point arithmetic. Interrupt facilities, mathematics packs and special purpose I/O equipment were also designed and built to improve the performance of the computers and aid the programming environment.

With the second generation machines came the introduction of the visual display unit to replace the teletype terminal for inputting and outputting data from/to the computer. A simple computer of the 1960s operated at 250 000 instructions per second and had a data access time measured in microseconds.

The major computer family of the mid–1960s was the **mainframe** computer. The most popular series of mainframes was the IBM System/360 family. This series of devices was released on April 7th 1964 and during the rest of the decade IBM sold over 30 000 of the different system models. The 360 series with its wide range of different types and prices had set the standard for computer development since then (Article A.3).

Second generation computers

Within this era large **supercomputers** emerged, such as Control Data Corporation's CDC 6600. This computer contained 400 000 silicon transistors, utilized high density packaging, incorporated sophisticated cooling mechanisms and made use of multiple functional units to address the speed imbalance between the main memory and the processing sections. The CDC 6600 could run at a peak speed of 9 MFLOPS. The later CDC 7600 was an even more powerful computer with a peak speed of 40 MFLOPS.

At the other end of the scale the **minicomputer** was introduced in 1963 with the DEC PDP–8 device. It could perform similar functions to a mainframe but only cost $16,000 and could sit on a lab bench. DEC sold 50,000 units over a 10 year period. This device allowed computer systems to become much more widespread in their use. The 16–bit PDP–11 minicomputer from DEC was introduced in 1970.

With the second generation machines more high level procedural languages evolved. BASIC, for example, was developed in 1964 at Dartmouth College in the USA by J. Kemeny and T Kurtz. Standard subroutine libraries, batch monitors and special purpose I/O facilities also arose as support services for programmers. However, programming in this era still depended on individual experienced programmers or several programmers working to write the software. With decreasing hardware costs and increasingly complex software projects a complexity barrier was created in software development which caused the management of the software/hardware complexity to become the primary practical problem in computer science. Because of this, problems arose when several programmers were brought together to work with systems analysts, engineers and operators in devising large scale information processing systems. The complexity of this task led to the birth of systems/software engineering as a formal science.

A.4 THIRD GENERATION COMPUTERS

The start of the 1970s, and the third generation of computers, saw the increased development of semiconductor random access memory and read only memory and, with the introduction of mass produced magnetic disks, computer systems at last appeared with a comparatively large amount of cheap primary and secondary memory.

The early 1970s also saw the development of the 'computer on a chip' or **microprocessor.** While considering the design of a calculator chip set, Ted Hoff of Intel proposed that instead of using a dedicated set of integrated circuits (ICs) or chips to perform a given function it would be preferable to design a general purpose programmable chip which would perform the calculator functions as well as many other logical operations. In 1971 the 4004 single chip microprocessor was produced by Intel (Article A.4). The second generation of microprocessors appeared in 1972 with the introduction of the Intel 8008 8–bit microprocessor.

With the advances in very large scale integrated (VLSI) technology in the mid 1970s the active devices of the microprocessors themselves became smaller, faster and consumed less power. The speed up of operation of VLSI circuits was due partly to the reduction in gate–switching time and wire length, and partly to the miniaturization of circuits, which allowed cache memories and extended sets of registers to be located physically close to the processor on an IC. The result was that the microprocessor did not have to access the main memory so often for instructions or data but could obtain the data internally, i.e. fewer main memory accesses and a better matching of the processor cycle time to the memory access time. Also with main memory capacities that had become larger, more programs and data could reside simultaneously in the main memory. This led the major USA manufacturers, who had developed the 8–bit microprocessor, to proceed to develop microprocessors with 16–bit data paths. These offered greater performance not only because of their wider word width but also as a result of improved architectures and the use of special techniques such as instruction pipelining, i.e. fetching instructions from memory into a sort of queue to speed up the sequence of operations.

Third generation computers

The top of the range computers, or supercomputers, continued to develop during this generation. The Cray 1 supercomputer was released in 1976 while the CYBER 205 supercomputer, operational by 1979, was capable of processing up to 50 million instructions per second.

The use of multiple processing elements to enhance the performance of a computer system came to a practical reality during the third generation of computers. The first was the large **Illiac IV** array processor which came on–line during this generation. The Illiac IV was a large supercomputer project which was started in 1966 at the University of Illinois under the direction of Professor Daniel Slotnick in collaboration with the Burroughs Corporation. The original specification for the Illiac IV was to have a speed of 1000 million floating point operations per second (MFLOPS) based on the use of 256 64–bit processing elements (PEs) arranged in four reconfigurable arrays consisting of 64 PEs, an eight by eight array, with a control unit for each quadrant.

The four arrays were to be connected together under program control to allow single or multiple processing operations. The system program resided in a Burroughs B6500/B6700 general purpose computer, which supervised program loading, array configuration changes and I/O operations, both internal and external. Each PE had two thousand 64–bit words of primary memory while back up storage for the arrays was provided by a large directly coupled, parallel access disk system. Each processor was reconfigurable, so it could also serve as two 32–bit or eight 8–bit processors. However, due to the cost of the project ($31 million, four times the original estimate) only one quadrant was built, though it was claimed that it could achieve a speed of 200 MIPS and a performance of 50 MFLOPS.

The single Illiac IV was installed at the NASA Ames Research Centre at Moffett Field California in 1973. After a long time testing and correcting hardware faults it became fully operational in 1975 and was operational until August 1981 mainly being used to execute

Article A.4   **INTEL**

One of the major companies in the development of the microprocessor is the Intel Corporation of the USA. INTEL (Integrated Electronics) was founded in 1968 by Bob Noyce and Gordon Moore who had already founded Fairchild in 1957. The company was set up to exploit the semiconductor memory market. However, its real claim to fame is that it was the first company to make a real attempt to use the technology of integrated circuits (ICs) to reduce the requirements of a computer to a single IC. This general purpose programmable chip was later given the title 'microprocessor'.

In 1969 the Busicom Japanese calculator company came to Intel looking for a custom IC manufacture. They wanted a set of approximately 10 custom circuits for the heart of a new low cost desktop printing calculator. Ted Hoff, of the application research department, at Intel thought that it would be better to have a programmable chip set. The idea of a 'central processing unit (CPU) on a chip' had been around since the mid 1960s. During the autumn of 1969 Hoff aided by Stan Mazor, an application engineer at Intel, defined an architecture consisting of a 4–bit CPU, a read only memory (ROM) to store program instructions, a random access memory (RAM) to store data and several I/O ports to interface with external devices such as a keyboard, printer, switches and lights.

However, it was Federico Faggin who took the initial ideas of Hoff and Mazor and converted them into the final chip set, initially called the '4000 family'. It consisted of four 16–pin chips.

1. The 4001 was a 2 Kbit ROM with a 4–bit mask programmable I/O port.
2. The 4002 was a 320–bit RAM with a 4–bit output port.
3. The 4003 was a 10–bit serial in parallel out shift register to be used as an I/O expander.
4. The 4004 was a 4–bit CPU. It had 2250 transistors on a 1/6 in long by 1/8 in wide chip. It had a performance rating of 0.06 million instructions per second (MIPS).

The chip set was installed in the calculators and was working by March 1971. In November 1971 the chip set, the MCS–4, was released.

In April 1972 Intel introduced the 8–bit 8008 microprocessor with a group of supporting chips called the MCS–8. The 8008 was an 18–pin package. The 4004 and the 8008 are both complete central processing units (CPUs) on a chip and have similar characteristics but, because the 4004 was designed for serial binary coded decimal (BCD) arithmetic and the 8008 for 8 bit character handling, their instruction sets were different. The 4004 and 8008 were fabricated in PMOS.

The first run of the 8080 came in December 1973, Intel introduced the product in March 1974. The 8080 microprocessor had 6000 transistors manufactured on a 4.5 micron process and delivered 0.64 MIPS. The 8080 was an enhanced version of the 8008; it could handle 16–bit data types mainly for address calculations capable of addressing 64 K. It also featured BCD arithmetic, enhancement of addressing modes and improvements of interrupt processing. In 1976 Intel introduced the 8085, which was an enhanced 8080 which removed the triple power supply required by the 8080 to a single power supply.

matrix based fluid dynamics applications. The Illiac IV was also the first US machine to use semiconductor memory instead of the older magnetic core memory.

During the 1970s other research projects in the USA considered how multiple computer architectures could be advanced. The following are some of the prototype multiple processors which were developed in the USA during the late 1970s and early 1980s and which have been used as experimental test beds for parallel processing research.

Third generation computers

1. C.mmp at Carnegie Mellon University. The 16 processor C.mmp research project started in 1971 and was operational by 1974. This considered various topics within a tightly coupled multiple processing system where the processors were connected by a cross-point switch.
2. Cm* at Carnegie Mellon University. The later Cm* computer system, operational by 1976, investigated a loosely coupled design using 50 processors connected by a hierarchical switch.

3. The finite element machine at NASA's Langley Research Centre.
4. MIDAS at the Lawrence Livermore Berkeley Laboratory.
5. TRAC at the University of Texas.
6. ZMOB at the University of Maryland.
7. The cosmic cube at CalTech. The cosmic cube developed at the California Institute of Technology investigated hypercube architecture.

A.5 FOURTH GENERATION COMPUTERS

In 1981 Intel released the 32–bit Intel iAPX 432 processing system. Apart from being an introduction into the era of 32–bit microprocessors this device was so radically different from anything that had gone before that most commentators would regard it as constituting the start of what could be called the fourth generation of computers. The iAPX 432 system architecture was designed to be programmed entirely in the high level language Ada. Although it did not prove to be a commercial success and is no longer in production, many of the techniques it made use of were incorporated into later Intel products. Other 32–bit processors introduced during the early 1980s were the Hewlett Packard Focus microprocessor which achieved a density of 450 000 transistors in 1981, the National 32032 microprocessor which was introduced in 1983, the Motorola 68020 in 1984 and the Intel 80386 at the end of 1985.

The increasing ability of semiconductor technology in the 1980s to pack more and more transistors onto a piece of silicon allowed a number of on–chip architectural features such as large register sets, cache memory and sophisticated I/O controllers to be incorporated on–chip. These features in turn made it possible to reduce the number of off–chip accesses a microprocessor had to perform, thereby increasing the overall speed of the device. The present 32–bit devices are extremely complex systems with very large instruction sets making them very flexible and useful for producing general purpose computers. However, their complexity in itself tends to restrict their use in high speed applications where a large amount of number crunching is required, such as in signal processing situations. For this reason special integrated circuits began to be developed in the 1980s – the digital signal processing microprocessors. These special purpose microprocessors are targeted at high speed number crunching applications in signal processing.

During the fourth generation supercomputers evolved to produce ever more impressive peak performance figures. The Cray devices produced during this period such as the Cray X–MP and Y–MP pushed the peak performance figures up towards the 1 GFLOPS mark (10^9 floating point operations per second).

In this generation software also advanced with the standardization by the USA military on the multitasking language Ada, developed to cope with multiple processing applications.

A.6 FIFTH GENERATION COMPUTERS

In the time scale of computer evolution the **fifth generation** of computer development began in 1981 and lasted until about 1992. This era saw a shift in the way computers were supposed to operate. In broad terms a fifth generation computer was conceived as a series of interconnected databases and parallel processing machines accessed by means of an intelligent inference machine which could accept problem statements in a natural language.

Its main function was not information processing in the conventional sense, but drawing inferences from its knowledge bases. The computer would then be able to incorporate a much higher degree of artificial intelligence than contemporary machines, approaching that of a human expert in certain circumstances.

The most widely accepted definition of **artificial intelligence** (AI) is that due to Marvin Minsky 'Artificial intelligence is doing with computers that which if done by humans would be said to require intelligence'. This definition is however limiting, as many computers can perform tasks to out–perform humans but lack that spark of intelligence. Another definition might be that artificial intelligence is that body of knowledge which enables engineers to design machines that perform human like tasks with ever increasing competence or which improve the interaction between men and machines in areas where this integration is largely informational. The potential of artificial intelligence is that it allows the explicit representation of many of the notions currently hidden within more general representations of physical systems and has the potential for increasing the functionality of engineering systems.

Fifth generation computers

The main application area of fifth generation computers was to be the solution of highly complex problems which would normally have required a considerable measure of reasoning, intelligence and expertise when carried out by people. Due to the fact that most of the applications for this type of system were not simply number crunching, as had been typical of the previous computer generations, the hardware and the software was to be entirely different.

Two different approaches to using AI within computers have been identified for the fifth generation machines. The first was to use AI to solve problems that were previously only solvable by humans, i.e. those that appeared to require imagination, intuition or intelligence. To handle this, symbolic processing was required within an AI device for proper recognition and understanding of graphics, images, speech and text. This involves comparison, selection, sorting, pattern matching, logical operations and arithmetic operations. AI could also look at automated reasoning and pattern recognition problems.

The second was to use a specific set of programming techniques known as rule based programming to develop application programs. The programs were designed to solve problems the way a human would seem to solve them. The programs tended to perform computations on symbols rather than on numbers.

The idea of fifth generation computers was introduced in 1981 at an international conference in Tokyo, Japan. The initial development for this form of computer had been undertaken at the Japan Information Processing Development Centre (JIPDEC). In 1984 there were five major research programmes in different countries with the Japanese project co-ordinated by ICOT (Institute for New Generation Computer Technology) probably being the most ambitious. It was a 10 year program which was to culminate in the development by 1992 of a fifth generation computer.

A.6.1   FIFTH GENERATION ARCHITECTURE

Figure A.2 illustrates a block diagram of what a fifth generation computer was to look like. It can be seen that there were supposed to be a number of different software sections within it including an intelligent user interface which would be able to interpret data from a number of different sources and provide both speech, video and text output.

One of the most important parts of a fifth generation machine was to be the software interface between the programmer and the hardware of the device. This software was to be exceptionally sophisticated both in communicating with the operator and in its handling of

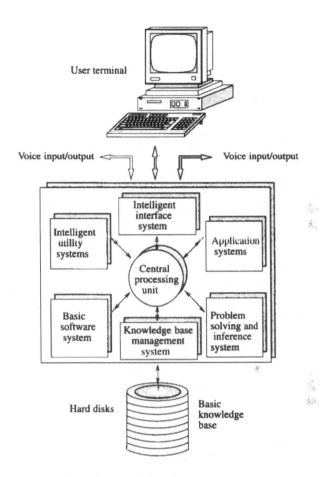

User terminal

Voice input/output ⟵   Voice input/output ⟶

Intelligent interface system

Intelligent utility systems

Application systems

Central processing unit

Basic software system

Knowledge base management system

Problem solving and inference system

Hard disks   Basic knowledge base

Figure A.2

Fifth generation computer system.

the large knowledge base that these devices would require.

The many software sections within such a computer would vary from interface software and knowledge database managers to inference processing software.

1. Intelligent User Interfaces – These enable communication between operator and computer to take place in a way which is simple and natural to the user. Intelligent interfaces are also aimed at improving the conceptual communications between user and computer at a visual level and reduce to a minimum the amount of keyboard control needed to operate the computer and its programs. In this way computers can be used in a much more flexible manner than before. In particular speech recognition is important and the ability to handle the information in speech would require the use of large knowledge bases related to particular fields.

2. Database Managers – The databases that were to support the fifth generation computers would contain knowledge of the following types.

(a) General knowledge base: This would include knowledge of basic words of everyday use, basic sentence patterns and basic scripts, a base of dictionaries of various languages, sentence construction rules and a base related to natural languages, i.e. knowledge of the languages to be used for man to machine communication.

(b) System knowledge base: This base would gather knowledge related to the system itself.

(c) Applied fields knowledge base: This base would gather together knowledge about certain applied fields, i.e. knowledge on the problem areas to be solved.

Fifth generation computers

Such advanced machines were also expected to have the ability to learn from their interaction with people hopefully being able to infer certain information from what a particular operator says.

3. Inference Processing Unit – The central element of the fifth generation machine was to be the inference processing element. This was to interact with the intelligent interface system to communicate with the user, and with the knowledge base mechanisms to draw on the knowledge resources. As well as handling these two tasks it would also address itself to the fundamental task of solving the kinds of problems which would fall within the remit of fifth generation machines, primarily drawing conclusions from evidence. The software for inference processing was to be based on developments in the area of logic programming, exemplified by the programming language PROLOG.

Figure A.3

AI computer architectures.

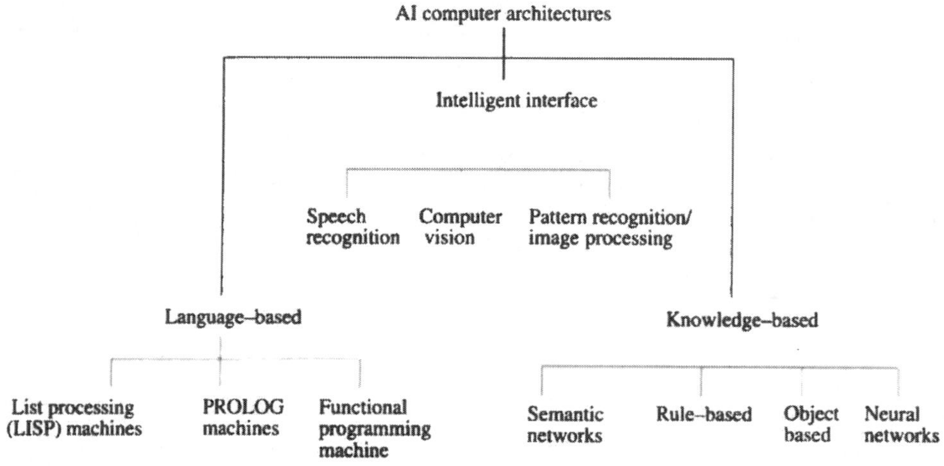

The architecture of the computers designed for AI applications can be classed broadly into the three groups shown in Figure A.3. These are:

(a) Language–based machines: These machines are designed to efficiently execute a computational model of some high level AI language such as LISP. They incorporate special mechanisms in hardware to implement the primitive operations of the languages they supported.

(b) Intelligent Interfaces: The acquisition, representation and intelligent use of information and knowledge was fundamental to the AI processing of the fifth generation computers. An intelligent interface should be able to acquire information from various external sources (visual, vocal and written) despite the fact that input from these sources was often incomplete, imprecise, or even contradictory. While intelligent interfaces are not regarded as true AI machines they offer interfaces onto other computer systems to allow such tasks as speech recognition, pattern recognition computer vision and many other AI based tasks.

Fifth generation computers

(c) Knowledge based machines: These machines are governed by models for the representation and manipulation of knowledge. The knowledge was stored as intentional data (rules) and extensional data, or facts, and manipulated by rule–based systems. The representation of knowledge within such an AI device involved the encoding of information about objects, relations, goals, actions, and processes into data structures and procedures. The encoding facilitated adding to, altering, and manipulating the knowledge.

EXPERT SYSTEM

Expert systems are problem–solving programs that contain knowledge databases and intelligent reasoning mechanism which closely match the knowledge and procedures used by human experts within well defined domains. They are the product of the application of artificial intelligence techniques to specific fields in which the rules followed by a human expert are encoded in a computer program. Such a program embodies sections of organized knowledge about specific areas of human expertise.

The expert system is then able to assimilate some of the expert knowledge of a person such as a doctor, geologist, analytical chemist or social administrator. This knowledge can subsequently be used, usually in an interactive way, by either a highly skilled technician or by someone with only the basics of training, to make the expert's knowledge available to others. Alternatively it may be used to help an expert remember certain rules. In this way the computer provides users with advice on some specialized topic. The expert system has a goal that it should make decisions or make plans as well as, or better than, a human expert. These systems usually also provide human–oriented I/O in the form of audio and video interaction. This can for example allow an expert system for medical diagnosis to operate in a manner analogous to the way a physician and a patient interact in attempting to make a diagnosis of a particular problem.

The first expert system was developed in 1971 at Stanford Research Institute. It was called DENDRAL and mimicked the behaviour of an expert chemist
determining the structure of molecular compounds. It was written by Edward Feigenbaum and Bruce Buchanan who began the work in 1965.

A typical expert system program contains a rule base and a generalized inference engine.

1.   Rule Base – A rule base is a memory store which contains the decision rules of an expert for a particular application.

2.   The Inference Processing Unit – The inference engine is the control structure within the computer that decides which of the candidate rules from the rule base will take part in handling the problem. It takes input data and assumptions, explores the inferences derivable from the rule base then yields conclusions and advice and offers to explain its results by retracing its reasoning for the user. The inference engine typically has to deal with 'fuzzy' or probabilistic data and rules in addition to purely deterministic logic. This processing performed by the engine uses a rule–of–thumb approach, known as heuristics, which is more practical than abstract reasoning for machine applications.

A simple expert system therefore takes the world as it finds it, examines the various systems that exist in it and then infers relationships and regularities about the observations that can be made. This method is essentially empirical in that it represents compiled experience attained over many years. It lacks deductive strength and the models obtained by the empirical observations are normally specialized and are difficult to generalize or to apply to a new situation outside of that covered by the original representation.

This processing of knowledge is required for problem solving, making logical deductions and information retrieval.

Fifth generation computers were required to have an improved cost/performance ratio, to be more applicable to a wider variety of situations and to be more reliable than their predecessors. To aid with the development of these advanced concepts many countries set up large research programs funded on a national and international level, such as ESPRIT in Europe and DARPA in America.

Fifth generation computers

A.6.2   APPLICATION AREAS

There are many possible application areas envisaged for fifth generation machines. Such things as:

1.   Expert Systems: This type of system contains the knowledge and experience of particular human experts (Article A.5).

2. Robots: The new robots which would be introduced due to the fifth generation work were envisaged to have a self learning capability and be able to carry out a variety of different tasks. They would display some form of intelligent behaviour and be able to react to changes in their environment. They would also have a high level communication interface.

3. Integrated Office: The design of an intelligent office workstation was one of the primary goals of the fifth generation initiative. The system was to have a voice activated word processor and an integrated communications network. The office computer would be able to handle all the different aspects of the business of a company such as production, stock control, sales and purchase. In addition the computer would probably contain a knowledge base, to enable a strategic evaluation to be made of business information.

Fifth generation computers

4. Language Translation: The creation of a system to cope with language translation was to be of great value to the international business community, since it would allow the direct translation of articles from different parts of the world almost at the instance they were transmitted.

5. Design Systems: Fifth generation computer aided design systems were to allow the production of a product simply from an operator defining the specifications and product guidelines. These systems would rely on a large and complex design database and knowledge bases of design methodology.

6. Computer Aided Learning: The advent of computers with enhanced intelligence operating on knowledge bases rather than information bases was to bring about major advances in the area of computer aided learning.

The ICOT project in Japan finished in June 1992. Although it was initially intended to produce a 1000 processor user friendly system, after ten years of research Japan failed to produce any large scale parallel machine able to meet the fifth generation specifications. The general conclusion drawn by many of the researchers was that the original aims were too ambitious and it was impossible to create a highly intelligent system in ten years. It can be noted ,however, that the first part of the project, to build parallel computers was successful although the second part, to build intelligent database machines, was not.

A.7 SIXTH GENERATION COMPUTERS

At present many commentators regard us as having entered the sixth generation of computer systems. In the previous section we noted the failure in the previous generation to produce powerful AI based computers. However, AI research has yielded a new avenue of work which may influence the types of computers which will be built during this 6th generation.

In the area of artificial intelligence, the early work in the 1950s was on designing learning machines and adaptive systems which would learn and adapt to their environment. In the 1970s AI then moved on to solving more complex problems on the representation of knowledge. During the late 1980s and early 1990s the investigation has returned to the development of simple knowledge representation systems using brain style computations and a large number of processing cells with a large set of connector values.

In the mid–1980s the large number of very powerful parallel computers that had been developed and the existence of new mathematical formulations allowed the expansion of neural network simulation.

Neural Networks consist of a large number of small simple processing units interconnected by a large number of weighted connection. This is shown in Figure A.4. The network is constructed such that:

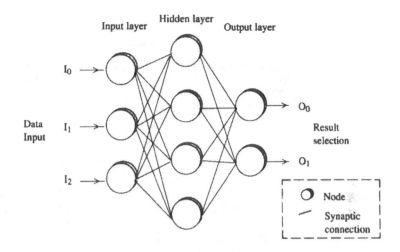

Figure A.4

Example of a Neural network

1. the outputs of some nodes are used as inputs to others,
2. the output of each node in the network is formed from a reaction to its combined inputs;
3. the programming of the network is done by adjusting the extent to which each processor's output is allowed to influence the reaction of the other processor to which is connected.

A neural network is supposed to behave in a similar manner to the neurons (processing nodes) and synapses (interconnections) which make up the brain of a living organism. The neural network is constructed from a large number of processing cells with internal thresholds and a large number of weighted inputs. The output from each neuron is passed to other cells with the stored information being represented in the strengths of interconnections between cells. Such a network can be trained to react to certain input combinations to produce a particular output by altering the weighting on the interconnection pattern, i.e. the synaptic strength of connections can be increased or decreased according to the learning situation. Iterative algorithms are used to change the strengths of the interconnections between the different processing nodes. The topology and strength of the connections influence what information processing functions the network can carry out. In this way learning may be seen as the modification of the pattern of interconnections between neurons and experience leads to the modification of the interconnection pattern. Hence a neural network is not programmed with step by step algorithms but learns through supervised training. It is fed raw data, perhaps through sensors along with feedback on how well or how badly it has done. As it runs through the material again and again it makes myriads of mistakes; but learning from them it finally organizes itself to carry out the desired task on new data.

Sixth generation computers

In the 1990s people's perception of the future of computers has also been turned to massively parallel computing; however, the programming of such machines is very difficult. With neural networks the question as to how to program massively parallel machines may be approached.

A.8 KEY TERMS

ABC computer	Fifth generation
Ada	IBM computers
Analytical engine	Illiac IV computer
Artificial intelligence (AI)	Inference processing unit
Atlas computer	Intel
Babbage (Charles)	Intelligent user interfaces
Calculator	Knowledge base
Colossus computer	Mainframe
Early computer systems	Mark 1 computer
First generation computer	Microprocessor
Second generation computer	Minicomputer
Third generation computer	Neural network
Fourth generation computer	Rule base
Fifth generation computer	Supercomputer
Sixth generation computer	System/360 (IBM)
Database managers	Turing (Alan)
Difference engine	UNIVAC computer
EDVAC computer	Von Neumann (John)
ENIAC computer	Zuse (Konrad)
Expert system	

A.9 REVIEW QUESTIONS

1. What was the contribution to the development of computers made by:
   (a) Charles Babbage
   (b) J. von Neumann.
2. Why did Babbage never develop the analytical engine?
3. List the important features of the ENIAC computer.
4. What was the difference between the Mark 1 computer and the ENIAC?
5. What was the significance of the IBM 360 series of computers on computer development?
6. What was the major technologies used in the construction of the third generation of computer systems?
7. How do you account for the success of IBM and why is there no guarantee that IBM can continue its dominance in the marketplace?
8. What are the chief characteristics which distinguish first to fourth generations computers?
9. Match the description with the computer:

(a) Used in the late 1940s in the USA.	Stretch
(b) A code breaking machine used in World War II.	PDP – 8
(c) A supercomputer .	Colossus
(d) Computer with parallel functional units.	ENIAC
(e) Computer developed at the university of Manchester	CDC 7600
(f) Experimental computer developed by IBM.	Atlas
(g) First minicomputer.	Illiac IV

10. What were the aims and objectives for the designers of the fifth generation computers?
11. List the components of a fifth generation computer.
12. How does the software developed for a first generation machine differ from that proposed for fifth generation devices?
13. Can a computer expert system out–perform a human expert?
14. Which word best describes the role of an expert system: calculate, transform, advise or process?
15. What era in computer evolution are we presently in?

A.10 PROJECT QUESTIONS

1. Explain how changing technology has made the desktop computer a resource that can manipulate information.
2. The Church–Turing thesis contends that mental processes of any sort can be simulated by a computer program. Have any computers passed this test?
3. Describe the process a knowledge engineer would use to write expert system software.
4. What application would decision support software be best suited for?
5. Are chess playing programs examples of AI applications? Justify your response.
6. With the aid of suitable references discover the extent of the integration of robots into the car industry.
7. Note a number of reasons for the apparent failure of the fifth generation project.
8. By obtaining information from current periodicals find out how close we are at present to producing fifth generation type computers.
9. Have expert systems proved popular? Explain your answer.
10. What features can we expect to see in sixth generation computers?
11. Investigate the different models which neural networks are based on?
12. Note a number of tasks which neural networks targeted at?
13. What are some of the benefits of using neural networks to process information?
14. How will fuzzy logic and neural networks be related in the future?
15. Investigate how optical computers and neural networks are being linked.

A.11 FURTHER READING

Most books on computer systems contain a chapter on the development of computers; however, there are also a large number of books detailing the history of computers. A few are listed below together with several articles from popular periodicals and journals.

Computer Development (Books)

1. *A History of Computing in the 20th Century*, Edited by N. Metropolis, J. Howlett and G. Rota, Academic Press, 1985.
2. *A History of Computing Technology*, M.R. Williams, Prentice Hall, USA, 1985.

3. *A History of Personal Workstations*, A. Goldberg (Editor), ACM Press, Addison Wesley, 1988.
4. *The Supercomputer Era*, S. Karin and N. Parker Smith, Harcourt, Brace and Jovanovich Publishers, 1987.
5. *The Making of the Micro*, C. Evans, Victor Gollancz, 1981.
6. *Breakthrough to the Computer Age*, H. Wulforst, Charles Scribers and Sons: New York, 1982.
7. *Engines of the Mind: a History of the Computer*, J. Shurkin, Norton Publishing New York, 1984.
8. *The Computer Comes of Age*, R. Moreau, MIT Press, USA, 1984.

Computer Development (Articles)

1. RISC: back to the future?, C.G. Bell, *Datamation*, 1st June 1986, pp. 96–108.
2. Charles Babbage's engines: the genius of failure, D. Swade, *The Computer Bulletin*, June 1991, pp. 12–14
3. The *IEEE Annals of the History of Computing* is a regularly published journal which contains detailed articles on the evolution of the computer.

Fifth Generation Computers (Articles)

1. The fifth generation project: personal perspectives, E. Shapiro and D.D. Warren, *Communications of the ACM*, March 1993, Vol 36, No 3, pp. 46–101.

Fifth Generation Computers (Books)

1. *Fifth Generation Computers*, P. Bishop, Ellis Horwood Ltd, 1984.
2. *Fifth Generation Computer Systems*, T. Moto Oko, North Holland Publishing 1982.

ACRONYMS

A list is given of the acronyms used within the text is given below.

AC	Alternating current
ACIA	Asynchronous communication interface adapter
AI	Artificial intelligence
ALU	Arithmetic logic unit
AM	Amplitude modulation
ANSI	American National Standards Institute
ARQ	Automatic request for repetition
ASCII	American Standard Code For Information Interchange
ASIC	Applications specific integrated circuit
A/D	Analog to digital converter
CAD	Computer aided design
CAE	Computer aided engineering
CAM	Content addressable memory or Computer aided manufacture
CASE	Computer aided software engineering
CBX	Computerized branch exchange
CCD	Charge coupled device
CCITT	International Telegraph and Telephone Consultative Committee
CD ROM	Compact disk read–only memory
CISC	Complex instruction set computer
CISM	Complex instruction set microprocessor
CMOS	Complimentary metal oxide semiconductor
COAX	Coaxial cable
CODEC	Coder/encoder
CPM	Control program for microcomputers
CPU	Central processing unit
CRC	Cyclic redundancy check
CRT	Cathode ray tube
CSMA	Carrier sense multiple access
DC	Direct current
DCE	Data circuit terminating equipment
DFD	Data flow diagram
DIL	Dual in line
DMA	Direct memory access
DOS	Disk operating system
DRAM	Dynamic random access memory
DSP	Digital signal processing
DTE	Data terminal equipment
D/A	Digital to analog converter
EAROM	Electrically alterable read only memory
ECL	Emitter coupled logic
EEPROM	Electrically erasable programmable read–only memory
EIA	Electronics industries association

EMI	Electromagnetic interference
EPROM	Erasable programmable read–only memory
FDM	Frequency division multiplexing
FDX	Full duplex
FET	Field effect transistor
FFT	Fast Fourier transform
FM	Frequency modulation
FPGA	Field programmable gate array
FPLA	Field programmable logic array
FSK	Frequency shift keying
GaAs	Gallium arsenide
GPIB	General purpose interface bus
HDL	Hardware description language
HDX	Half duplex
HSLN	High speed local network
IC	Integrated circuit
IEEE	Institute of Electrical and Electronic Engineers
ISDN	Integrated services digital network
ISO	International organization for standardization
I/O	Input/output
I^2L	Integrated injection logic
JCB	Job control block
JCL	Job control language
LAN	Local area network
LCD	Liquid crystal display.
LED	Light emitting diode
LIFO	Last in first out
LRU	Least recently used
LSI	Large scale integration
Macro	Macroinstruction
MBM	Magnetic bubble memory
MFLOPS	Million floating point operations per second
MIPS	Million instructions per second
MMS	Memory management system
MMU	Memory management unit
MODEM	Modulator/demodulator
MOS	Metal oxide semiconductor
MS–DOS	Microsoft disk operating system
MSI	Medium scale integration
MTBF	Mean time between failure
MTTR	Mean time to repair
NMI	Non maskable interrupt
NMOS	Negative channel metal oxide semiconductor
OEM	Original equipment manufacturer
Op code	Operation code
OSI	Open system integrated
PAL	Programmable array logic
PBX	Private branch exchange

PC	Personal computer
PCB	Printed circuit board
PGA	Programmable gate array
PIA	Peripheral interface adapter
PLA	Programmable logic array
PLD	Programmable logic device
PROM	Programmable read only memory
PSTN	Packet switched transport network
QAM	Quadrature amplitude modulation
RAM	Random access memory
RISM	Reduced instruction set microprocessor
ROM	Read only memory
RTOS	Real time operating system
RWM	Read write memory
SBC	Single board computer
SCSI	Small computer standard interface
SNR	Signal to noise ratio
SRAM	Static random access memory
SSI	Small scale integration
SVC	Supervisor call
TCB	Task control block
TTL	Transistor transistor logic
UART	Universal asynchronous receiver/transmitter
UNIX	Uniplexed information and computing service
USART	Universal synchronous/asynchronous receiver transmitter
USRT	Universal synchronous receiver/transmitter
VDU	Visual display unit
VLSI	Very large scale integration
WORM	Write once read many

INDEX

There are a large number of acronyms used in describing computer systems. If you wish to locate a reference on a particular topic referred to by its acronym, first use the Acronym section on the previous pages to get the expanded title. This can then be used in the Index section to locate the reference within the text.

N

O